Expanding Nationalisms at World's Fairs

Expanding Nationalisms at World's Fairs: Identity, Diversity, and Exchange, 1851–1915 introduces the subject of international exhibitions to art and design historians and a wider audience as a resource for understanding the broad and varied political meanings of design during a period of rapid industrialization, developing nationalism, imperialism, expanding trade and the emergence of a consumer society. Its chapters, written by both established and emerging scholars, are global in scope, and demonstrate specific networks of communication and exchange among designers, manufacturers, markets and nations on the modern world stage from the second half of the nineteenth century into the beginning of the twentieth.

Within the overarching theme of nationalism and internationalism as revealed at world's fairs, the book's essays engage a more complex understanding of ideas of competition and community in an age of emergent industrial capitalism, and investigate the nuances, contradictions and marginalized voices that lie beneath the surface of unity, progress, and global expansion.

David Raizman is Distinguished University Professor of Art and Art History in the Westphal College of Media Arts & Design at Drexel University in Philadelphia.

Ethan Robey is Assistant Professor of the History of Decorative Arts and Design at Parsons School of Design, and Associate Director of the MA Program in the History of Design and Curatorial Study run by Parsons and Cooper Hewitt, Smithsonian Design Museum.

Routledge Research in Art History

https://www.routledge.com/Routledge-Research-in-Art-History/book-series/RRAH

Routledge Research in Art History is our home for the latest scholarship in the field of art history. The series publishes research monographs and edited collections, covering areas including art history, theory, and visual culture. These high-level books focus on art and artists from around the world and from a multitude of time periods. By making these studies available to the worldwide academic community, the series aims to promote quality art history research.

The *Gamin de Paris* in Nineteenth-Century Visual Culture
Delacroix, Hugo, and the French Social Imaginary
Marilyn R. Brown

Antebellum American Pendant Paintings
New Ways of Looking
Wendy N.E. Ikemoto

Expanding Nationalisms at World's Fairs
Identity, Diversity, and Exchange, 1851–1915
Edited by David Raizman and Ethan Robey

William Hunter and his Eighteenth-Century Cultural Worlds
The Anatomist and the Fine Arts
Helen McCormack

The Agency of Things in Medieval and Early Modern Art
Materials, Power and Manipulation
Edited by Grażyna Jurkowlaniec, Ika Matyjaszkiewicz, Zuzanna Sarnecka

National Identity and Nineteenth-Century Franco-Belgian Sculpture
Jana Wijnsouw

Expanding Nationalisms at World's Fairs

Identity, Diversity, and Exchange, 1851–1915

Edited by David Raizman and Ethan Robey

Routledge
Taylor & Francis Group

LONDON AND NEW YORK

First published 2018 by Routledge

2 Park Square, Milton Park, Abingdon, Oxon, OX14 4RN
605 Third Avenue, New York, NY 10017

Routledge is an imprint of the Taylor & Francis Group, an informa business

First issued in paperback 2020

British Library Cataloguing-in-Publication Data
A catalogue record for this book is available from the British Library

Library of Congress Cataloging-in-Publication Data
A catalog record for this book has been requested

ISBN: 978-1-138-50175-1 (hbk)
ISBN: 978-0-367-78716-5 (pbk)

Typeset in Sabon
by codeMantra

Contents

Illustrations

Contributors

M. Elizabeth Boone is Professor in the History of Art, Design, and Visual Culture at the University of Alberta. She works on nineteenth and early twentieth-century art in the Western Europe and the Americas and is particularly interested in trans-national relations, the role of art in the development of national identity, and the politics of display. She is the author of exhibition catalogues and articles on a number of topics, including the nineteenth-century reception of Vermeer (1992), paintings of Spain by Mary Cassatt (1995), depression-era murals in San Francisco (2002), American variations of Velázquez's *Las meninas* (2003), the use of illustrations by Joseph Pennell to mask international political controversy in turn-of-the century travel literature (2005), and the metaphor of travel in the work of Spanish-born Mexican surrealist Remedios Varo (2005). She published *Vistas de España: American Views of Art and Life in Spain, 1860–1914* (Yale, 2007), and is currently working on a book relating to her chapter in this volume (*Spain at the World's Fairs and American Centennial Celebrations*).

Susan R. Fernsebner is Associate Professor of History at the University of Mary Washington in Fredericksburg, Virginia. She writes and teaches on themes related to material culture, colonialism and capitalism, and also the history of childhood in nineteenth- and twentieth-century China. Her recent publications include "Contextualizing the Visual (and Virtual) Realities of Expo 2010," in *Visualizing China: Image, History, and Memory, 1750-Present* (Lexington Press, 2014) and "Child's Play: Notions of Amusement in Early Republican China," in *The Discovery of the Child: the Problem of the Child in Modern Chinese Literature and Culture*, eds. Andrew Jones and Lanjun Xu (Peking, 2011).

Jørn Guldberg is Associate Professor (emeritus) in the department of design at the University of Southern Denmark (Odense). He has published several books, catalog essays, and articles on a range subjects in art and design history. His current research interests are Scandinavian design and design semiotics. Recent publications in English include "Singular or multiple meanings? A critique of the index/ Anzeichen approach to design semiotics," in Lin-Lin Chen et al. (eds.), *Design and Semantics of form and movement* (Lucerne, 2010), and "Legacy, Heritage, or History? A Study of Artistic Agency in the Art Scene of GDR, 1949–1989 and Beyond," in Mikkel Bolt Rasmussen & Jacob Wamberg eds., *Totalitarian Art and Modernity* (Aarhus, 2010).

Debra Hanson is Assistant Professor of Art History at Virginia Commonwealth University's Middle Eastern campus in Doha, Qatar. Her current research interests include the late-nineteenth-century world's fairs, Thomas Eakins's early paintings of women and girls, and the art and architecture of the US Capitol in Washington. She was designated a Tocqueville Fellow and participated in an initial two-week "Tocqueville Seminar for Transnational American Studies" and three subsequent conferences on this subject. In correlation with this program, Dr. Hanson developed a class on "Gateways to Globalism: The Late Nineteenth Century World's Fairs" that has been taught at VCU and Georgetown's School of Foreign Studies in Doha. She has conducted research on this subject in London, Paris, and Washington, DC, and has delivered conference papers internationally on a variety of world's fair-related topics.

Christian A. Hedrick is an architectural historian and a full-time Visiting Lecturer in the School of Architecture at Northeastern University. He received a Ph.D. in the History, Theory and Criticism of Architecture at MIT (2014), where he was a fellow in the Aga Khan Program for Islamic Architecture. His research examines architecture as a site of cultural exchange over the course of the long nineteenth century. His current work examines the reception and interpretation of Islamic architectural forms within German architectural publications and practice from 1780–1870. Before coming to Northeastern, he held a Digital Humanities Research Associate position at the Aga Khan Documentation Center at MIT where he led a new research initiative focused on the pedagogy of architectural history. He is a founding member of the Global Architecture History Teaching Collaborative and was recently a co-recipient of a grant for a project entitled "Islamic Architecture: A Global History." He is also preparing his book manuscript based on his dissertation: *Modernism with Style: Form, Meaning and the Origins of Modern Architecture in Berlin, 1780–1870*.

Rebecca Houze is Associate Professor of Art History at Northern Illinois University specializing in the art and design of Central Europe in the nineteenth and twentieth centuries. She was co-editor of *The Design History Reader* (Berg, 2010) and author of *Textiles, Fashion, and Design Reform in Austria-Hungary Before the First World War: Principles of Dress* (Ashgate, 2014). Her articles on world's fairs have been published in *Studies in the Decorative Arts*, *Centropa*, and the *Journal of Design History*. Professor Houze teaches courses on world's fairs, fashion, modern architecture and graphic design. Her current research focuses on the relationship between architectural and ethnographic display at international exhibitions of the late nineteenth century and the subsequent popularization of the open-air museum.

Anca I. Lasc is Assistant Professor in the History of Art and Design Department at Pratt Institute, New York City. Her work focuses on the invention and commercialization of the modern French interior, the development of the profession of interior designer in the nineteenth century, and the art of commercial window dressing around the turn of the twentieth century in France and America. She has published essays in the *Journal of Design History* and *Interiors: Design, Architecture, Culture*. *Designing the French Interior: The Modern Home and Mass Media*, a volume that Dr. Lasc co-edited with Georgina Downey and Mark Taylor (Bloomsbury), was published in 2015. Another edited volume, *Visualizing the Nineteenth-Century Home: Modern Art and the Decorative Impulse* (Routledge), was published in 2016.

Daniela N. Prina is a postdoctoral research fellow of the Fund for Scientific Research-FNRS at the University of Liège. Her research interests span nineteenth- and twentieth-century architecture, urbanism and design in Belgium and Italy, with a particular focus on the training of architects and engineers and on the transmission of architectural culture in art academies and applied arts museums. The outcomes of her studies are regularly presented at national and international conferences, and are widely published in international journals, including the *Journal of Design History* and the *Journal of the History of Collections*. She was a contributing author to *Made in Italy: Rethinking a Century of Italian Design* (Bloomsbury, 2013) and *Van Academie tot Universiteit. 350 jaar architectuur in Antwerpen* (University Press Antwerp, 2013). She also was a contributing author for the recently published *Bloomsbury Encyclopedia of Design* (2015).

Bart Pushaw is a PhD student in the Department of Art History and Archaeology at the University of Maryland. His research focuses on the intersections of race, gender, and identity in Baltic and Nordic art in the context of nineteenth-and early twentieth-century global modernism. He has recently published articles on the Finnish Symbolist painter Beda Stjernschantz as well as women artists in fin-de-siècle Estonia. He is currently working on another project concerning images of the black body in the art of Ants Laikmaa and Akseli Gallen-Kallela.

David Raizman is Distinguished University Professor of Art and Art History in the Westphal College of Media Arts & Design at Drexel University in Philadelphia. He is the author of *History of Modern Design* (2ed, Laurence King and Pearson Publishing, 2010) and, with Carma Gorman as co-editor, of *Objects, Audiences, and Literatures: Alternative Narratives in the History of Design*, (Cambridge Scholars Pub., 2007). His work on world's fairs includes the article "Giuseppe Ferrari's Carved Cabinet for the 1876 Centennial Exhibition: Presentation Furniture in the Cultural Context of World's Fairs," *West 86th Street* 20, no. 1 (Spring-Summer, 2013): 62–91. He directed the 2015 NEH Summer Institute "Teaching the History of Modern Design: The Canon & Beyond," and was a subject editor for the *Bloomsbury Encyclopedia of Design* (2015).

Ethan Robey is Assistant Professor of the History of Decorative Arts and Design at Parsons School of Design, and Associate Director of the MA Program in the History of Design and Curatorial Study run by Parsons and Cooper Hewitt, Smithsonian Design Museum. He has published widely on fairs and expositions, including "Souvenirs of the Invisible: Display of Energy at Twentieth-Century World's Fairs," in *The Journal of Decorative and Propaganda Arts* 27 (2015) and the essay "Kings, Peasants, Dragons and Flowers: Varieties of National Symbolism in World's Fair Objects," in *Inventing the Modern World: Decorative Arts at World's Fairs, 1851–1939* (Skira Rizzoli, 2012). He has also written chapters on nineteenth-century trades fairs in *The American Bourgeoisie* (Palgrave Macmillan, 2010), and on the 1876 Centennial exhibition in *Philadelphia's Cultural Landscape* (Temple, 2000), as well as a section of the New York Public Library's educational app, *The World of Tomorrow: Exploring the 1939–1940 World's Fair Collection* (2011).

Hannah L. Sigur studies internationalism and cross-cultural exchange in material culture, with special focus on national identity and ideology in the architecture of international expositions 1851–1915. She is a lecturer at Santa Clara University and University of California, Davis and writes online for Khan Academy and SmartHistory AP Art History. Her 2008 book, *The Influence of Japanese Art on Design* (Gibbs Smith) examines Japonisme, Arts and Crafts, Art Nouveau, and early Contemporary art and design. An essay on the 1893 Chicago World's Fair appears in Volume III of *What Happened? An Encyclopedia of Events that Changed America Forever*, (ABC-Clio, 2010). *Christian Science Monitor* listed her co-authored *A Master Guide to the Art of Floral Design*, (Timber Press, 2002, 2010), in "Year's Best Garden Books." She is currently preparing an article on early British industrial designer Christopher Dresser's journey to Japan. She holds a Ph.D. from the Institute of Fine Arts, New York University.

Introduction

Communities real and imagined: world's fairs and political meanings

David Raizman and Ethan Robey

Mapping diversity

Nineteenth- and early twentieth-century world's fairs provide a rich material and literary record of the dynamic intersection of the arts, education, commerce, tourism, technology, and national politics and international relations of the era. Between 1851 and 1915, more than twenty major international exhibitions and fairs, displaying a wide range of raw materials and manufactured products, were held on every continent of the world, representing as many as 60 participating nations and territories, and attracting millions of spectators—more than 50 million people visited the 1900 Exhibition Universelle in Paris, for example.[1] In the twentieth century, the list of host nations grew to include Brazil (1922), Haiti (1949), Israel (1953, 1956), Canada (1967), Japan (1970, 1975, and 2005), and many other countries. In our new millennium, host nations include China (Shanghai, 2010), South Korea (Yeosu, 2012), Kazakhstan (Astana, 2017), and the United Arab Emirates (planned for 2020 in Dubai).[2] This recent trajectory of world's fair venues provides a barometer of post-Cold War economic development.

Showcases for raw materials, means of production, and finished goods from all over the world, the fairs were significant vehicles of international communication among nations, manufacturers, and the millions of visitors and potential consumers. In the past half century, great and small world's fairs have been subject to several bookcases' worth of study, in many disciplines, including art history, cultural anthropology, economic history, architectural history, political science, post-colonial studies, gender studies, history of technology, and the list goes on.[3] The present volume, however, explores constructions of nationalism at nineteenth- and early twentieth-century world's fairs in terms of a global design history. The essays consider complex representations of collective identity among interest groups outside of the main loci of power, marginalized in terms of economic or political structures, and/or geography—representations promulgated in visual terms, by displays of material goods. Exploring the world of commodities, this book describes networks of exchange, economic and symbolic, that make it difficult to see design as isolated from a fertile and interconnected international context. It brings together essays by both established and younger scholars who address the intersection of design, politics, and commerce among nations and manufacturers at the world's fairs between 1851 and 1915, a period of unprecedented economic expansion, free enterprise, social change, nationalism, imperialism, technological innovation, and the emergence of a consumer society.

Stemming from, as well as expanding upon, a College Art Association annual conference session on the politics of manufactures at world's fairs in the second half of the nineteenth century, the essays take up the challenge of the developing discipline of global design history, to consider the world of goods as manifestations of connections between societies, and work against models of center and periphery.[4]

Each chapter explores multiple and interrelated dimensions of nation-building, competition, commercial expansion, taste, and technology as they are revealed in the forms, selection, and display of manufactured goods. As a unit the essays deny the assumptions of a Eurocentric viewpoint, giving voice to alternative nationalisms and networks of international exchange in order to reveal a wealth of diverse strategies and intentions on the part of participating countries and manufacturers.[5]

Historiography: modernism and pluralism

From the first formulations of the discipline, world's fairs had some place in the history of design, but largely as a foil to a particular narrative of design evolution rather than as a subject of investigation on their own terms. Nikolaus Pevsner expressed what became a common take on the 1851 Great Exhibition of the Works of Industry of All Nations (usually known at the Crystal Palace Exhibition) in his *Pioneers of Modern Design* (first published 1936 and still in print).[6] Pevsner dismisses the British products displayed as "ludicrously over-ornate," a goad to design reformers such as A. W. N. Pugin, Owen Jones, and Henry Cole to forge principles of "good taste" amid the cacophony of competing fashions—a kernel that would later grow into a universalizing modern style. Pevsner's characterization of the Great Exhibition became part of an overarching narrative that cast the majority of its displays as the antithesis of a rationalist, production-centered paradigm for creating a history of design.

Nevertheless, Pevsner acknowledged the positive popular response to the Crystal Palace exhibition through the media of the time, an excitement and attention that only grew with the series of subsequent international exhibitions in Paris, Philadelphia, Vienna, and elsewhere during the latter half of the nineteenth century. He also included an excerpt from Prince Albert's celebratory and progressive address to the Society of Arts in Britain as the exhibition began. Yet throughout *Pioneers*, Pevsner promulgates the thesis that industrialization and competition were to blame for the poor quality and the lack of stylistic unity among the goods on display.[7] His focus was upon a linear narrative leading inexorably to the "modern movement," the culmination of reform efforts to purge architecture and design of historicism and decoration. From this perspective, the world's fairs did not fare particularly well.

Pevsner's history came from a Modernist moment that saw functionalist design principles as a self-evident social benefit and had little patience for the ornamental tastes of a previous generation. By the mid-twentieth century, historians began to reevaluate both the Crystal Palace and Modernist constructions of design history. Later authors criticized Pevsner's narrow selectivity in constructing a Modernist narrative as well as his architecturally based paradigm for the study of industrial design in relation to consumer products. In the 1960s, Pevsner's student Reyner Banham questioned the relevance of Modernist architectural theory to the study of consumer goods produced by mechanization, taking seriously the aesthetics of a "throwaway" culture and the realities of a consumer economy. For Banham, the rationalism Pevsner had tried in vain to see in the goods displayed at the Crystal Palace was not fundamental to an understanding of design, but a historically bound rhetoric of the Modernist era.[8] His head-on appreciation of mass-produced excess subsequently informed Penny Sparke's short but provocative *Design and Culture in the Twentieth Century* (first published 1986), which incorporated a popular and consumer-based approach to the narrative of design history as it applied to product housings and styling in mass-produced manufactured goods.[9] John Heskett's *Industrial Design* (1980) includes an acknowledgment of Banham as well, and is also concerned with the realities of the process of mechanized industrial production in a consumer society as they inform design history.[10] Although neither Heskett nor Sparke revisit the later nineteenth-century world's fairs as sites for a more pluralistic historical understanding of design, authors from fields such as anthropology and economic history were turning to world's fairs as a rich vein of study in the 1980s, unpacking these events as expressions of imperialism, politico-social theory, and the development of consumerism. Robert Rydell, Paul Greenhalgh, Tony Bennett, Timothy Mitchell, and others investigated the roles of fairs in affirming racial classifications and validating the colonialist project.[11] Greenhalgh, for instance, in *Ephemeral Vistas: the Expositions Universelles, Great Exhibitions, and World's Fairs, 1851–1939*, saw the world's fairs as opportunities for a carefully orchestrated spectacle of national propaganda for the host as well as participating nations, pointing to the attention lavished upon colonial exhibits and the rationalization of imperialism against the backdrop of emerging industrial capitalism as a civilizing force amid the quest for new markets, abundant natural resources, and labor. Tony Bennett coined the phrase "exhibitionary complex" to refer to the process of masking hegemonic ambitions under the guise of aesthetic appreciation, and Mitchell unpacks the Orientalist perspective that casts non-Western cultures as exotic and appealing on one hand but also primitive and inferior in comparison with a Eurocentric, progressive, and industrialized civilization. The direct influence of the first world's fairs upon the emergence of a broadly based consumer and advertising culture is traced in Thomas Richards's history of British consumerism, *The Commodity Culture of Victorian Britain: Advertising and Spectacle 1851–1914* (1990).[12] Scholarship from disciplines *outside* of art and architectural history helped to give the fairs alternative meaning for design historians. And yet, *Pioneers* remains in print and the Modernist narrative of the world's fairs as a pluralistic diversion from the pure stream of design flowing toward functionalism continues to exert an influence in the history of design.

A nineteenth-century world view

It is perhaps all too easy to overlook the most remarkable as well as obvious feature of the first world's fairs: their international scope. Against the tide of much conventional wisdom of the time, the mid-nineteenth-century organizers of international fairs of the arts and industry recognized the advantages that accrued from the combination of free enterprise, free trade, and competition: the very features that constitute our modern economic system, in which reside both the promise, as well as the uncertainties and fluctuations, of that system in social and political terms. This keystone of capitalist economic policy and national and international political will was a novel idea at a time when tariffs and guild structures protected manufacturers and governments *against* imports, effectively restricting competition, and allaying fear about the flow of inexpensive and poorly made manufactured goods. And the time was right: manufacturers, especially those in Great Britain, generally seemed eager to embrace the challenges of competition for the possible benefits of its rewards. For instance, a committee representing cotton-printers in Manchester, England, wrote in 1849:

> It is very necessary that all parties should know what the French and all nations are doing, and should compare their manufactures with our own. The comparison would show what our manufacturers could do, and by generating increased knowledge and appreciation in our consumers would induce the production of a much higher class of work.[13]

Prince Albert (1819–1861), the German-born husband of Queen Victoria and guiding force behind the 1851 Crystal Palace exhibition, was a champion of this confident brand of internationalism. In an 1849 speech promoting the idea of a world's fair, he expressed a hope that technological advances might lead to international unity, re-marking, "[t]he distances which separated the different nations and parts of the globe are gradually vanishing before the achievements of modern invention."[14] The Prince's globalism, given material form in the staggering abundance and variety of goods on display at the Crystal Palace, was linked to the spread of knowledge, while competition linked technological with social progress; what economists and politicians today might call the benefits of an "innovation economy."

Eurocentrism, militarism, and imperialism

Forging and reinforcing a unified and liberal ethos amid a dizzying diversity of language and custom required a clear, confident ideological framework.[15] Thus, the logistical complexity of planning and staging such vast enterprises was taken by its organizers as validation of their progressive world view, an ideology allegorized and made visible in institutions and monuments inspired by the Great Exhibition, including the memorial in London erected in memory of Prince Albert (built 1872–1876) and the South Kensington (later Victoria and Albert) Museum (established in 1852). In the Albert memorial (Figure I.1), the natural world, represented by figural groups identifying the four continents, was understood as a rich storehouse of raw materials waiting for its potential to be realized. The peoples, races, flora and fauna of the world were all classified and sorted into neat and comprehensible categories, with European civilization naturally assumed as the yardstick. Human enterprise in agriculture, manufacturing, commerce, and engineering are also figured in sculptural groups, presented as the drivers of progress and civilization itself. Carved in monumental relief on the podium supporting the seated figure of Albert are the poets, craftsmen, composers, painters, architects, sculptors and other knowledge-makers from all times and places whose work formed the era's assumed universal canonical culture.

No reference is made in the memorial to wars or military campaigns; rather than traditional symbols of power or authority, the effigy of the late Prince Consort holds a giant catalogue of his greatest achievement: the 1851 Great Exhibition. The absence of conflict seems taken for granted in the Albert Memorial but the timing of the first world's fair in 1851, just three years after the political and social upheavals of 1848 in Europe, suggests that the threat of armed struggle, whether between states or social classes, loomed behind the metaphoric conflicts enacted in the exhibition hall as competition. Indeed, throughout the third quarter of the nineteenth century, military

Figure I.1 Prince Albert Memorial, Kensington Gardens, London (Photograph date: ca. 1876-ca. 1885).

Source: Photograph courtesy of Cornell University Library, Andrew Dickson White Architectural Photographs Collection. https://digital.library.cornell.edu/catalog/ss:3874583.

metaphors were frequently employed in descriptions of the competition among objects displayed at world's fairs even as international participation was taken as a realization of Albert's vision of global unity. Militarism was key to many expressions of nineteenth-century nationalism; at the same time participation in a world's fair was a symbolic endorsement of moving beyond armed conflict toward a politics wherein design, production, and consumption would determine the relative merits of nations.[16] Displays of weaponry amid other industrial products in the exhibition halls were thus fraught with contradiction, representing national solidarity and sovereignty, as well as internationalist industrial innovation.[17] Such idealism about a post-militaristic nationalism remained ingrained in the ethos of international exhibitions, extending to all the various national groups that sought to define themselves though exhibitions, from the British to the Chinese. For example, Susan Fernsebner notes in this book that Chinese officials promoting the 1910 Nanyang Exposition argued that international contests were no longer just battles of armies, but the competition of *shengji* (livelihood).[18]

Behind this reassuring and unified image, however, lay considerable tensions and struggles among competing interests, both within and among European nations, and between them and the rest of the world. Notably, economic competition among participant nations extended outward to supplies of raw materials and workers, and inexorably to colonialism, underpinned by the same military hardware so admired in the exhibition halls. Eruptions of armed conflicts between nations and within imperial territories—including the Crimean War (1855), the Franco-Prussian War (1870–71), the Boer Wars (1880–81, 1899–1902), the Russo-Japanese War (1904–05), and the Balkan Wars (1912–13) to name just a very few—stain the whole era treated in this book, and demonstrate the fragility of the ideal of peaceful competition among nations, which culminated in World War I in August 1914.[19]

Inasmuch as the world's fairs can be read as in terms of international economic competition, it is not difficult to link Greenhalgh's thesis in *Ephemeral Vistas* (1988) with contemporary scholarship in history and the social sciences that emphasized the alignment of a competitive and expanding capitalist economic climate with the emergence of modern nationalism, notably in Ernest Gellner's *Nations and Nationalism* (1983) and Benedict Anderson's *Imagined Communities* (1983).[20]

And yet, curiously, while the significance of vernacular print culture in nation-building was acknowledged by Anderson and others, this important strain of historical scholarship tended to marginalize or fail to invest the parallel "language" of material culture with political meaning, ignoring the world's fairs and those displayed goods that were essential to the achievement of national and colonial aspirations.

In some ways, more recent design historical scholarship has begun to address this oversight, advancing more nuanced approaches to framing the issues surrounding the representation of Western culture's "other." For example, examining several viewpoints expressed about the display of goods from India at the Crystal Palace, Lara Kriegel noted the combination of domestication as well as *integration* of the Indian sub-continent at the Great Exhibition, a mixed reaction rather than a single overriding imperialist narrative.[21] Moving beyond the Crystal Palace and the ideologies of host nations and economic leaders, the present volume extends these kinds of questions to a wider geographical framework, revealing more about the local voices that were at work in contesting and shaping national identities on a world stage.

Owen Jones and multi-cultural awareness

Exhibitions of objects were central to the ideology of nineteenth-century imperialism. While the motivations for establishing overseas colonies were varied, condescension and brutality by colonial administrations in Asia, Africa, South America, and elsewhere were often accompanied by active forms of engagement with design in the form of study, publication, collecting, and exhibiting the work of artisans from subjugated cultures.[22]

As an example of this kind of complex cultural relationship, one might look at Owen Jones's exquisitely illustrated and world-embracing *Grammar of Ornament* (1856), the culmination of extensive travels (he had studied and published the decoration of the Alhambra Palace in Granada in 1842).[23] Through a careful process of selection, the apparent diversity of decorative motifs and compositions, divorced from their own cultural, technical, or functional context, was reduced to a series of thirty-two "principles" of design, brought together from remote cultures and historical periods. Thus, as in the Albert Memorial, an expanding and diverse world of goods was unified by a grand, unified, and universal narrative—everything was different, and yet everything was the same.

Part of the rationale for Jones's project was educational: along with the improvement of "taste" as defined by design reformers, Jones's principles offered designers and manufacturers an opportunity to improve English production by profiting from access to exemplary models of appealing and well-designed goods from all over the world.

Jones was a member of the organizing committee for the Great Exhibition and designed the decorations of the vast glass and iron palace according to ornamental principles laid out in his *Grammar*. As Debra Hanson points out in the current volume, Jones's interior was designed to harmonize and organize the multitudinous displays, again mediating difference with totalizing principles. In adopting ornamental motifs largely from "Oriental" sources, Jones reinforced associations among his compatriots between the world's fair and the lavish hoardings of Arabian princes in the Western imagination.[24] Jones's admiration for the products of non-Western cultures seems to acknowledge difference and yet also *deny* it under a dominant, rationalizing European gaze. Such domination was also carried out politically as Britain asserted increasing control over the flow of raw materials from the Indian sub-continent and elsewhere, backed by military intervention and fulfilling the world's fairs' dominant narrative of advancing "civilization."

While the concept of "otherness" and its political ramifications in imperialism inform the political interpretation of world's fairs and their displays, it does not exhaust the combination of wonder, envy, fear, curiosity, and reflection that was aroused by the presence of so much diversity concentrated in a single place for a period of six months, attracting increasing millions of visitors. The fairs produced a wide variety of responses, competing and contested, rather than a single, all-embracing narrative thread, whether mediated by progressivism, imperialism, or populism.

World's fairs and "imagined communities"

In treating a wide range of nations as well as international exhibitions, the essays contained in *Expanding Nationalisms at World's Fairs: Identity, Diversity, and Exchange, 1851–1915* engage the themes of national identity and international relations, moving beyond a Eurocentric point of view to examine the political dimensions of the fairs as networks of exchange rather than as independent univocal displays. This global approach examines the more familiar themes of competition, taste, and the materials and production technologies that were often touted by fair organizers and critics alike; but interprets these topics in light of national interests and reciprocity as practiced and promoted during the period examined in this book.

It is not surprising that the global turn in design history has been stimulated through interdisciplinary scholarship. Anderson's concept of nations as "imagined communities" is useful in creating a sense of continuity among the chapters in the present volume; it signals the necessity of constructing national identities, selecting from among groups and sub-groups those products and displays that served particular local, national, and international interests. The debate and intersection among competing narratives, what Patricia Mainardi called, in reference to the Second French Empire, a "polyglot of diverse interest groups,"[25] is what concerns our authors and editors, along with an expanded understanding of international awareness and reciprocity: a combination of resistance and envy that characterizes the experience of difference and reveals the social and political dimensions of design through the pavilions and displays at the fairs. The interpretive richness and complexity of the fairs emerges through the study of their artifacts.

The fairs and contemporary global culture

As noted above, "peaceful competition" was a phrase frequently employed by world's fair observers in the mid- and later nineteenth century. It affirmed the hope that free trade was a benevolent enterprise, permitting economic growth through competition and engendering national pride and solidarity. Further, it reminded readers and visitors and observers then (and now) that the material benefits brought by trade are based upon an acknowledgment of shared economic interests among governments and between governments and their constituencies. Thus, however clouded by past, present, and future tensions, the connection between competition and peace is both explicit and implicit in the understanding of world's fairs, much as it is in the Olympic Games today.

There are further analogies with today's Olympic Games: the attending ceremonies and celebrations, the awards, the distinctive branding of athletes' clothing, the masking of tensions and contradictions through patriotic display, the public works and buildings, the tourism, all attended the world's fairs during the period under investigation in this volume of essays. And rather than reflecting national pride, the world's fairs contributed to *building* national identity as well; writers, manufacturers, and local and foreign visitors, experienced the combination of envy, pride, and hope that forged national interests and helped to reinforce patriotism and a sense of the shared values that comprise a national culture. World's fairs were part of the structure of the nation-state; that is, of the modern world.

The chapters

Our volume spans the era between the Crystal Palace Exhibition and the Panama-Pacific International Exhibition, a few months after the outbreak of World War I in Europe, when the foundation of "peaceful competition" crumbled and when the destructive power of nationalism shook to the core the liberal ideologies that had shaped the era.[26] Each chapter of *Expanding Nationalisms at World's Fairs* investigates the political meanings of world's fair displays in this period, representing an expanded field both materially and geographically, and introducing global perspectives of representation, networking, and exchange. Its purview includes both the major fairs in world capitals and lesser-known national exhibitions in Europe, Asia, and South America, such as the 1880 Jubilee Exhibition in Brussels, the 1888 Nordic Exhibition in Copenhagen for Denmark, the Antwerp exhibition of 1885/1886, or the 1910 Centenary exhibitions in Argentina, Chile, and Uruguay. The histories of these overlooked fairs supply a wealth of new information as well as a back-story for different nations' participation in the major events held in London, Paris, Vienna, St. Louis, and Chicago.

Jørn Guldberg, in "The Nordic Spectacle of 1888," looks to the 1888 Nordic Exhibition of Industry, Agriculture and Art in Copenhagen as part of a series of efforts to define Danish identity in the wake of territorial losses and the increasing concentration of political power. Focusing upon the architecture of the exhibition halls and the clever display of packaged agricultural products (Figure I.2), Guldberg examines how the fair affirmed Danish agricultural and industrial reforms and played a role in shaping the national political identity in a time of struggle between a more liberal manufacturing class and a conservative, land-owning elite.

Figure I.2 Rosenborg Mineral Spring Water Establishment Display, 1888 Copenhagen Exhibition
of Art, Industry, and Agriculture.
Source: Photo courtesy Jørn Guldberg.

Arrangements of ordinary packaged consumer goods, even among more grandiose
and monumental exhibits, were an effective way to symbolize a nation's economic
potential. In her chapter, "When the Local is the Global," Susan Fernsebner mentions
how manufacturers used orderly stacks of bars of branded soap to accentuate the
display of Chinese chemical industries at the 1910 Nanyang Exposition in Nanjing.

In "The 1910 Centenary Exhibition in Argentina, Chile, and Uruguay," M. Elizabeth
Boone turns her attention to the fine arts, and to its role in diplomatic relationships
between the United States and those South American nations. Boone considers how
the United States was perceived, through its artwork, in the Latin American context,
and how this collection was received against a backdrop of contemporary anxieties
about industrialization and definition of a national character in Argentina, Chile, and
Uruguay.

Comparison and competition were key to the political purposes of world's fairs in
this period. Displays of goods not only embodied national traditions and styles, but
were also held up against products of rival countries and ranked in terms of tangible
and intangible qualities, such as skill and taste. In "Converging Views," Daniela Prina
considers the complex relationship between Italy and Belgium in the later nineteenth
century, exemplified through Italy's participation in the 1880 Jubilee Exhibition in

Brussels and the 1885 world's fair in Antwerp. Prina's investigation contributes to our understanding of Italian identity in the wake of unification and the nation's efforts to reconcile strong regional craft traditions with an image of modern industrial progress and state-sponsored design education. While the chapter primarily focuses on "official" views of world's fairs and the issues of reform and government-sponsored design education that are familiar in the better-known historiography of the Great Exhibition, Prina places these issues in their precise and complex political context and charts new ground in extending this discourse beyond Britain and France.

As with the other authors of this book, Prina unravels the webs of intersecting interests guiding the development of the Italian system of design schools and participation in international exhibitions. A key concern among the authors collected here is specifying the individuals who promote and guide the assembly of a national display and the varied audiences to whom these exhibitions speak.

In the case of China in the early twentieth century, for example, international exhibitions were forums for assertions of identity against European colonizers, but also for the reification of elite groups' claims to define and control the cultural narrative. Susan Fernsebner's chapter discusses the new Chinese mercantile elite, including Chen Qi (1878–1925) who was increasingly influential in Chinese participation in world's fairs, from the Saint Louis exhibition in 1904, to China's own 1910 Nanjing Exposition, as well as China's participation in the 1915 Panama-Pacific International Exhibition in San Francisco. Fernsebner examines how the national image was shaped by Chen and his colleagues and by the displays' varied reception, among non-elite Chinese, expatriate merchants and Westerners.

Similarly, Hannah Sigur in "A Neoclassical Translation" considers the role of Kuru Masamichi (1855–1914) the architect of Japan's Hôôden pavilion at the 1893 World's Columbian Exposition in Chicago, and the academic institutions in Japan that influenced him. Sigur bases her analysis of the building and the selection of paintings on display inside in contemporary debates among Japanese academics about the political significance of national artistic heritage. Rather than simply exhibiting "traditional" styles, Sigur argues, the Japanese invented canny pastiches, aimed at constructing a narrative of Japan as a heritor of the Western Classical tradition and thus in cultural parity with Western Europe.

In addition to being the sites of such national definition, world's fairs were, by design, international, and provided opportunities for communication among participating designers and displayed the interdependence of actors in an expanding global marketplace. The Prussian architect and designer Carl von Diebitsch (1819–1869), for example, who designed an Orientalizing pavilion for the 1867 Paris Exposition Universelle, received significant commissions from both German and Egyptian clients, while the Hungarian architect Pál Horti (1865–1907) designed furniture in the United States following his participation in the 1904 St. Louis World's Fair. Such career trajectories suggest that designers' professional aspirations and commercial opportunities provided by the world's fairs to work abroad produced hybrid results that challenge readers to rethink the very notion of a unified "national" style.

Christian Hedrick's chapter, "From London to Paris (via Cairo)," uses the career of Carl von Diebitsch to reach broader understanding of the Islamic architecture and decoration that captured European taste in the mid-nineteenth century. Hedrick finds that the architect's interest in modern metallurgy, combined with finely cast Islamic arabesques, was directed toward attracting the interest and patronage of a progressive Egyptian elite rather than appealing to a Western, Orientalist taste for the exotic. Thus von Diebitsch can be understood as advancing a form of hybrid modernity, only superficially clothed in Islamic ornamental revivalism.

In "A Revelation of Grace and Pride," Rebecca Houze examines the malleability of a national character as assembled in the trappings and goods displayed at several world's fair pavilions at the turn of the twentieth century. The colorful wooden objects and textiles representing Hungary in the 1900 Paris Exposition Universelle bespoke a country close to its ethnic roots and folk traditions. By contrast, two years later at an exposition in Turin, the nation of Hungary chose instead to display itself as a cosmopolitan nation represented by sophisticated and progressive decorative arts styles. Two years after that, at the St. Louis exhibition in 1904, Hungary's displays emphasized vernacular styles and ornamental wooden architecture. In analyzing these exhibitions, both in pavilions designed or decorated by architect Pál Horti, Houze explores the varied nature and complexity of Hungarian nationalism in the decorative arts. At times expressing national pride in internationalism and at times in ethno-traditionalism, Hungary's displays serve as a case study in the role of international dialogue happening through the exhibition. Himself influenced by American crafts, Pál Horti was, toward the end of his life, producing decorative arts for several North and South American firms.

For many nations, economic and cultural identities involved complicated, even contradictory, relationships of industrial progressivism and traditionalism. The chapters here also provide some insight into the contested nature of modernity itself. Celebrating industrial progress, participation at the fairs entailed constructing a carefully, often nuanced past. In many cases, progress is more easily construed as continuity than a radical break. National unity, even political stability, and hence economic advancement often depend on keeping historicism close at hand. While the rejection of the past became a cornerstone of Modernist theory, the notion of engaging and invigorating history was integral to the construction of national identity in the world's fairs and the economic and political interests of participating nations leading up to World War I. In many of the exhibitions discussed in this book, organizers debated and sought to reconcile an imagined past with a promising, if uncertain future. Such is the dynamic explored in Anca I. Lasc's chapter "Paris, 1900," which examines the role of period rooms at that exhibition in the elaboration of French national identity. By "re-branding" period styles to conform to the recent rather than distant past, period rooms communicated a combination of coherence as well as continuity, consonant with asserting national confidence in the aftermath of defeat in the Franco-Prussian War.

Lasc's chapter also touches upon the popular aspects of world's fairs, in particular the role of immersive experience as "info-tainment" with great attention paid to detailed recreations to engage audiences, a key aspect of the world's fair experience. Similarly, Daniela Prina, in her contribution, examines the significance to the promotion of Italian artisanship of a detailed reconstruction of a medieval village at the 1884 Esposizione Generale Italiana in Turin, and Susan Fernsebner describes how an immersive attraction at the 1915 Panama-Pacific International Exhibition unified various Chinese interest groups in their opposition to it.

National unity can also involve a careful circumscription of boundaries between a normative group and its Other. In the first chapter in the book, "East Meets West," Debra Hanson describes how elements of Islamic cultures, real and simulated, were recombined in later nineteenth-century European fairs to create an image of the exotic, conforming to preexisting literary conventions. The exhibitionary structure allowed signifiers of Middle Eastern and North African cultures to be commoditized, an easily consumable image of the East, a familiar exotic, physically close, but culturally distant. Hanson's chapter further considers the meanings of these displays to their organizers. Where Western audiences saw articles of clothing, such as the Tunisian fez, as markers of otherness, to Tunisians, it represented the hegemony of the Ottoman Turks over North Africa. In a similar vein, Hanson discusses a Bedouin tent displayed at the Crystal Palace. A symbol to Westerners of primitive handicraft, it was, for Tunisians, a reminder of the tensions between traditionalism and modernization, between nomadic tribes and Ottoman control of space in the form of land registration regulations.

Late nineteenth- and early twentieth-century nationalism often relied on a conflation of nation and race, anchoring both with physical as well as artistic characteristics that constituted an "authentic" expression steeped in an imagined primeval past. In the chapter "Our country has never been as popular as it is now!" Bart Pushaw examines the Finnish pavilion at the 1900 Paris world's fair: Finland's bid to distinguish itself as a coherent nation against its Russian rulers. Pushaw demonstrates that the Romantic Nationalist imagery that pervaded the pavilion came at the expense of Finnish minority communities, such as the indigenous Sámi peoples or the Russian-speaking communities within the rugged Karelian region, the heartland of an imagined pure Finnish folk life.

We hope that readers will share the editors and authors' enthusiasm for the subject at hand and will recognize in the world's fair displays of manufactures and fine art the political dimensions of design and culture as they are revealed in the forging of national identity, in the investigation of competition and free trade, and in the complex and expanding interaction and interdependence among participants.

Notes

1 Robert Rydell, *The Books of the Fairs: Materials about World's Fairs, 1834–1916, in the Smithsonian Institution Libraries* (Chicago: American Library Association, 1992), 1; Richard D. Mandell, *Paris 1900: The Great World's Fair* (Toronto, ON: University of Toronto Press, 1967), ix.

2 See www.bie-paris.org/site/en/expos/about-expos/expo-categories/world-expos and www.bie-paris.org/site/en/expos/about-expos/expo-categories/international-specialized-expos, accessed June 14, 2015.

3 For a bibliography of secondary sources, see Alexander C.T. Geppert, Jean Coffey, and Tammy Lau, *International Exhibitions, Expositions Universelles and World's Fairs, 1851–2005: A Bibliography*, 3rd ed. (November 2006), http://fresnostate.edu/library/subjectresources/specialcollections/worldfairs/ExpoBibliography3ed.pdf, accessed June 14, 2015, which currently runs to 94 pages.

4 Glenn Adamson, Giorgio Riello, and Sarah Teasley, *Global Design History* (London: Routledge, 2011); see also Daniel Huppatz, "Globalizing Design History and Global Design History," *Journal of Design History* 28, no. 2 (May 2015): 182–202.

5 Recent scholarship on world's fairs has begun to deliberately look past the larger expositions and Euro-American host countries. See, for example, Marta Filipová, *Cultures*

of *International Exhibitions, 1840–1940: Great Exhibitions in the Margins* (Farnam and Burlington: Ashgate, 2015).

6 Nikolaus Pevsner, *Pioneers of Modern Design: From William Morris to Walter Gropius* (Harmondsworth: Penguin Books, 1960; first published 1936), 41–44.

7 Ibid., 43–46.

8 On Banham see Nigel Whiteley, *Reyner Banham, Historian of the Immediate Future* (Cambridge, MA: MIT Press, 2002), 34–35. A selection of Banham's criticism is found in Penny Sparke, ed., *Design by Choice/Reyner Banham* (New York: Rizzoli, 1981).

9 Penny Sparke, *An Introduction to Design and Culture in the Twentieth Century* (New York: Harper and Row, 1986), 50–55.

10 John Heskett, *Industrial Design* (New York: Oxford University Press, 1980), 7–9.

11 See Robert W. Rydell, *All the World's a Fair: Visions of Empire at American International Expositions, 1876–1916* (Chicago: University of Chicago Press, 1984); Paul Greenhalgh, *Ephemeral Vistas: The Expositions Universelles, Great Exhibitions, and World's Fairs, 1851–1939* (Manchester: Manchester, University Press, 1988); Tony Bennett, "The Exhibitionary Complex," *New Formations*, no. 4 (spring 1988): 73–102; and Timothy Mitchell, "The World as Exhibition," *Comparative Studies in Society and History* 31, no. 2 (April 1989): 217–236.

12 Thomas Richards, *The Commodity Culture of Victorian England: Advertising and Spectacle 1851–1914* (Stanford, CA: Stanford University Press, 1990), 17–72.

13 Henry Cole, "On the International Results of the Exhibition of 1851," in *Lectures on the Results of the Great Exhibition of 1851*, Second Series (London: David Bogue, 1853), 423.

14 Prince Albert, Speech at Mansion House, March 21, 1849, quoted in "The Great Exhibition and its Results," *Exhibition Supplement to The Illustrated London News* 19, no. 523 (11 October 1851): 457.

15 On systems of classification see Steve Edwards, "The Accumulation of Knowledge or, William Whewell's Eye," in Louise Purbrick, ed., *The Great Exhibition of 1851: New Interdisciplinary Essays* (Manchester and New York: Manchester University Press, 2001), 26–52, esp. 35–37. The classification system was not always observed in the displays themselves: see David Raizman, "Giuseppe Ferrari's Carved Cabinet for the 1876 Centennial Exhibition: Presentation Furniture in the Cultural Context of World's Fairs," *West 86th: A Journal of Decorative Arts, Design History, and Material Culture* 20, no. 1 (Spring-Summer 2013): 70–72.

16 See Raizman, "Giuseppe Ferrari's Carved Cabinet," 73 and n. 43. Benedict Anderson discusses the militaristic nature of "official nationalism," a mid-nineteenth-century appropriation of popular cultural nationalism by dynastic empires to shore up their authority; see Benedict Anderson, *Imagined Communities: Reflections on the Origin and Spread of Nationalism*, rev. ed. (London and New York: Verso, 1983, 2006), 86–98.

17 Perhaps the most vivid example is the exhibit of Krupp's artillery at the 1867 Paris Exposition Universelle. The cannons were roundly praised as an industrial marvel in French and international reports; three years later, Krupp guns were bombarding Paris during the Franco-Prussian War.

18 Susan R. Fernsebner, "When the Local is the Global: Case Studies in Early Twentieth-Century Chinese Exposition Projects," p. 179 in the present volume.

19 Faith in economic growth and competition was also shaken by periodic financial panics in the United States and Europe, culminating in a downturn that lasted much of last quarter of the nineteenth century. See p. 56 of the present volume.

20 Anderson *op. cit.*, Ernest Gellner, *Nations and Nationalism* (Ithaca, NY: Cornell University Press, 1983).

21 Lara Kriegel, "Narrating the Subcontinent in 1851: India at the Crystal Palace," in Louise Purbrick, ed., *The Great Exhibition of 1851: New Interdisciplinary Essays* (Manchester and New York: Manchester University Press, 2001), 146–178.

22 On imperialism and the decorative arts, see Michael Snodin and John Styles, *Design & the Decorative Arts: Victorian Britain 1837–1901* (London: V & A Publications, 2004), 319–322. An intriguing recent study on the relationship of design forms to imperialist regimes can be found in Debora L. Silverman, "Art Nouveau, Art of Darkness: African Lineages of Belgian Modernism," *West 86th: A Journal of Decorative Arts, Design*

History, and Material Culture 18, no. 2 (Fall-Winter 2011), 139–181; 19, no. 2 (Fall-Winter 2012), 175–195; 20, no. 1 (Spring-Summer 2013), 3–61.

23 Jules Goury, *Plans, Elevations, Sections, and Details of the Alhambra* (London: Owen Jones, 1842–1845); Owen Jones, *The Grammar of Ornament, Illustrated by Examples from Various Styles of Ornament* (London: Day and Son, 1856).

24 Debra Hanson, "East Meets West: Re-Presenting the Arab-Islamic World at the Nineteenth-Century World's Fairs," p. 17 in the present volume.

25 Patricia Mainardi, *Art and Politics of the Second Empire: The Universal Expositions of 1855 and 1867* (New Haven, CT: Yale University Press, 1987), 1.

26 Paul Greenhalgh, "A Strange Death..." in Paul Greenhalgh, ed., *Art Nouveau: 1890–1914* (London: Victoria and Albert Museum, 2000), 429–436.

1 East meets West

Re-presenting the Arab-Islamic world at the nineteenth-century world's fairs

Debra Hanson

The world's fairs held in the second half of the nineteenth century were a distinctly modern phenomenon. International in scope, they were the largest and most comprehensive intercultural events staged to that date. As such, they initiated a new awareness within the Western worldview of the transnational connections between nations, peoples, and cultures. This awareness included the Arab-Islamic world, most often constructed in accord with Western interests and the emerging global order they championed. In this regard, the fairs reflected and re-enacted the power dynamic that determined the course of East-West political, military, economic, and cultural interactions of this period, but also provided—within tightly controlled parameters—a forum for displaying the emerging nationalisms of countries outside the Western sphere.

The initial section of this chapter examines the manifestations of these dynamics at the first world's fair, the 1851 Great Exhibition in London, and its pivotal role in establishing a visual, material, and experiential model of "the East" that served to negate physical distance while confirming cultural difference.[1] Through a variety of representational and display strategies that blurred the boundaries of fact and fiction and past and present, the objects, peoples, and cultures of these regions were shaped into an exotic product available for Western consumption. Later expositions, such as those held in Paris (1855, 1867, 1878, 1889, and 1900) and Chicago (1893), adhered to many precedents set in London as they continued to advertise the themes of progress, peace, and prosperity introduced there. With regard to representations of the Arab-Islamic world, however, later fairs considerably modified the 1851 model in accord with the expositions' physical expansion, wider popular appeal, and concern for financial profits. The concluding portion of the chapter examines these modifications, their cultural impacts, and long-term legacies.

While the late-nineteenth-century expositions were pivotal in the production, display, and marketing of "the East" as a consumable commodity, the attitudes they advanced were not unique to this era. As Edward Said has shown, European attitudes toward the region had long been rooted in "a style of thought based upon... [a] distinction made between 'the Orient' and...'the Occident.'"[2] He defines this knowledge-producing discourse, Orientalism, as underpinning the "systematic discipline by which European culture was able to manage...the Orient politically, sociologically, militarily, ideologically, scientifically, and imaginatively during the post-Enlightenment period."[3] As a discourse grounded in assumptions of Western superiority, Orientalism paralleled and supported nineteenth-century colonial expansion in the cultural realm. Timothy Mitchell has demonstrated how the world's fairs were a part—indeed, the most publically visible part—of this Orientalist network.[4]

As such, one of their functions was to transform the abstract rhetoric of nationalism, and its promotion of Euro-American superiority, into a visceral reality that could be directly experienced by fairgoers. The spectacles of otherness and exoticism that the fairs provided—at first, simply the novel proximity of foreign artifacts and peoples and later, their incorporation into "villages," theatrical performances, and other interactive, immersive scenarios—were effective tools for creating this reality. When positioned in relation to foreign peoples and cultures—such as those of Tunisia, to be examined in this chapter—Western nationalisms and the vision of collective progress they created were brought into sharper focus. Dominant populations could thus be defined, and imperialist policies justified, in relation to the "colonized object-worlds" on view.[5] The material culture of the fairs—the design of their buildings and grounds, the objects and peoples they displayed, the texts they generated, and the viewing experiences they structured for visitors—thus reflected the political strategies of the host nations who stood at the center of an expanding global order shaped by the prerogatives of imperialism, capitalism, industrialization, and free trade.

At the expositions, expanding nationalisms were communicated through the wide range of objects on display; fairgoers' understanding of those objects needed to be shaped and molded accordingly. Prescriptive guidebook commentary played a part in this process, as did the spatial relations constructed within the expositions. At the Great Exhibition of 1851—officially titled "The Great Exhibition of the Works of Industry of All Nations"—all foreign and domestic displays were housed in a single monumental building of glass and iron dubbed the Crystal Palace by *Punch*, the popular London periodical, due to its reflective glass surfaces (Figure 1.1).[6] Neatly dividing the cruciform structure into two sections, its central transept has been described as functioning "like the equator," an observation typified in a contemporary description of the building's cartography as "a geographical arrangement which places the showy productions of tropical regions nearest to the transept, and removes to the extremities the less gaudy but more useful industry of colder climates."[7] This symbolic mapping signified a foreign exhibitor's position relative to the host nation, encouraging visitors to compare the relative merits of each through the products they displayed. While Britain occupied the western half of the Palace with her colonies and dependencies, those entities—such as Malta, Ceylon, and most prominently, India—were assigned spaces closest to the central transept, in proximity to the other "Oriental" countries— Turkey, Tunisia, China, Egypt, and Persia—located just to the east of the transept. Should any viewer misinterpret this arrangement, the text accompanying a lithograph of the transept (Figure 1.1) clarified its meanings:

> ...a thoughtful observer, standing in the Transept of the Crystal Palace... could embrace the industrial position of the East... In the gorgeous luxury of oriental habits he could trace the germs of the Mahommedan creed so grateful to the senses... he could see the gradually fading image of the Past, whilst in the dazzling display of the contributions of our own land ... he could read all the glorious promises which [this nation is] holding out to the world, of a resplendent Future.[8]

While positioning Britain as modern and progressive, "the East" is cast as its regressive and oppositional "other." In rationalizing its material culture as a "fading image of the Past," the writer implies that the resources of these countries—raw materials and manufactures alike—could be better utilized by Western interests, and so offers

Figure 1.1 "The Opening Ceremony," lithograph.

Source: Reproduced from *Dickinsons' Comprehensive Pictures of the Great Exhibition of 1851 / / from the originals painted for H.R.H. Prince Albert, by Messrs. Nash, Haghe and Roberts, R.A.; published under the express sanction of His Royal Highness Prince Albert, president of the Royal Commission, to whom the work is, by permission, dedicated* (London: Dickinson Brothers, Her Majesty's Publishers, 1854 [i.e. 1852]. Photograph Courtesy of the Smithsonian Libraries, Washington, DC.

cultural as well as economic justification for Britain's imperial ambitions. Within the Crystal Palace, the physical location of national displays enabled the narration as well as the mapping of power dynamics.

Inside the Crystal Palace: Owen Jones and the Arabian Nights

The Crystal Palace that housed the 1851 Exhibition was, first and foremost, a practical solution to the problem of inexpensively constructing a large-scale temporary structure in a short amount of time. Designed by Joseph Paxton, it was built of prefabricated iron and glass components assembled on site in London's Hyde Park in less than six months.[9] While the building's "form follows function" ethos and foregrounding of industrial materials might at first appear antithetical to any re-presentation of the Arab-Islamic world, this proved not to be the case, as shown by the interior scheme devised by Owen Jones, Superintendent of Works for the 1851 Exhibition. Jones, a key figure in the British design reform movement, devised an interior showcasing the universal design principles that he championed: fitness, proportion, and harmony, to be realized through the use of polychromy, flat pattern, and geometric form, best represented by Islamic structures such as the Alhambra, on which he was an acknowledged expert.[10]

Seeking to elevate public taste and "bring the building and its contents into perfect harmony," Jones appropriated visual motifs from a variety of Oriental sources.[11] Color was a key element in achieving the harmony he sought, with primary shades of red, blue, and yellow accenting slender iron colonettes, balcony railings, and other structural details (Figure 1.1). A flat, repeating geometric pattern (as seen in the balcony railing above the iron girders) also helped to unify a monumental space filled with a diverse array of objects and people, as did the textile banners, wall coverings, and curtaining devices Jones utilized.[12] These also served a practical function by dividing the vast spaces of the Crystal Palace into smaller, more comprehensible increments reminiscent of Eastern bazaars and souqs.[13] Initially criticized in some circles, the building's interior received wide praise once the Exhibition opened and its functionality and appeal were proven; in the six-month duration of the Exhibition in Hyde Park, many came to regard the Crystal Palace as the greatest attraction of the event it housed.

Although the "objects of every form and colour imaginable...far as the eye could reach...and...sixty thousand sons and daughters of Adam, passing and repassing, ceaselessly"[14] noted by one visitor surely mediated the legibility of Jones's design, its innovative use of color and pattern, coupled with the sweeping vistas, intensity of light, and other sensory stimuli, no doubt created an exotic atmosphere that contrasted sharply with the everyday lives of most attendees. Charlotte Brontë's description of Exhibition as "such a bazaar or fair as Eastern genii might have created...with such a blaze and contrast of colours and marvellous powers of effect" alludes to the connection many visitors made between the Crystal Palace and *The Arabian Nights*.[15] Like the building's interior, this literary work was a composite drawn from a variety of Oriental sources, locations, and periods that were adapted to suit the tastes of Western audiences.[16] In conjunction with popular travel narratives, Orientalist painting, and other period sources, *The Arabian Nights'* stories of romance, adventure, and magic reiterated prevailing views of the East as an "imaginative geography" of the ahistorical and the anti-modern, and were widely understood as "depicting a true picture of Arab life and culture, past and present..." [17] A satirical letter from one "Mrs. Fitzpuss of Baker Street" published in *Punch* underscores the extent to which this work of fiction served as a touchstone for viewing the Crystal Palace and its contents:

> Since the 1st of May [when the Great Exhibition opened] I've driven directly...to the Palace of that great Jin [genie], PAXTON, in Hyde Park, where for hours I've done nothing but think myself a great Princess of the Arabian Nights...[18]

As did Jones's interior plan, the *Punch* writer intertwines references to modern industrial design (Paxton) with Eastern tropes and motifs (genii). In later world's fairs and other cultural production of the period, these fictional tales would persist as a framework for viewing, constructing, and disseminating re-presentations of the Arab-Islamic world in the West, and for articulating the transformative benefits of industrialization, mass production and other forms of capitalist innovation in a period of rapid social and technological change.[19]

Tunisia at Mid-Century

In 1850, the Middle East, Arabian Peninsula, and parts of North Africa and the Balkan area were aligned with the Ottoman Empire, then uneasily allied with Britain and France. Ottoman trade, financial, and military policies, as well as other issues of international concern, were linked to the mid-century domestic reforms collective known as the *tanzimat*, which were intended to modernize infrastructure, administrative and educational systems, and cultural practices. Eager to demonstrate the success of these restructuring efforts and project a new, modern identity, Turkey and its territories—most notably Egypt and Tunisia—regularly participated in the nineteenth-century world's fairs held in the West. Emulating these models, an Ottoman General Exposition was held in Istanbul in 1863, with a similar event planned (but never held) for 1894.[20]

For Tunisia, which exercised de facto political autonomy but maintained close social and religious ties to the Ottomans, the project of asserting a national identity within a larger political structure *and* fashioning a global presence at the fairs was particularly complicated. Although recognized as "one of the most cohesive societies in the Muslim Mediterranean" early in the century, its internal unity and independent status were increasingly threatened following France's 1830 occupation of Algeria, which upset the North African balance of foreign and regional powers.[21] The rapid expansion of European diplomatic and commercial interests within Tunisia further destabilized the country, as did the economic policies that financed the international ambitions of its "enlightened despot" Ahmed Bey (r. 1837–1855).[22] Beset by pressures at home and abroad, he supported Tunisian participation in the Great Exhibition of 1851 to the point of personally overseeing the selection of goods sent to London. While projecting an emerging sense of Tunisian nationalism, the display mounted under his supervision also revealed a number of disparate—and in some cases conflicting—ideas regarding the country's cultural heritage and identity.

Tunisians, tents, and textiles

Whereas the Crystal Palace interior synthesized a number of generalized Eastern references, visitors were afforded a more immediate and particularized experience of the exotic in the displays of individual Arab-Islamic countries. The Tunisian section, which received particular approbation and was awarded a medal "on account of the whole collection," featured around 500 objects.[23] Of these, approximately 200 were "manufactures" (see n. 6); the rest were items such as spices, foodstuffs, minerals, and other raw materials. A large portion of the display area—which was already fragmented into a series of tenuously connected spaces—was configured to resemble a bazaar or souq, which the *Official Catalogue* described as "a series of small Eastern shops, the counters and stalls being fitted, and the articles being grouped in such a manner as to convey this impression."[24] (Figures 1.2 and 1.3) As it reiterated the overall spatial divisions within the Crystal Palace, this "great and glorious bazaar" featured

> ...articles of apparel...constitute one of ... [its] most singular and attractive features... Of these, every portion of dress is represented. The celebrated Fez caps ... are shown... Cloaks, joubbas, mantles, shawls, jackets, &c., of various materials, but all indicative of the peculiar characteristics of Oriental taste and design...[25]

Although displayed with leather goods, jewelry, metalware, and other items, textiles and clothing were accorded the most space and received the most attention from visitors.[26] Their interest was no doubt abetted by the presence of individuals in native dress, such as the man and boy in one view of the display in *Dickinson's Comprehensive Pictures of the Great Exhibition of 1851* (Figure 1.2) who each wear native garb: jacket (montane), baggy trousers (sirouel), and fez. At this time, the fez

Figure 1.2 "Tunis I: The Arab Tent," lithograph.
Source: Reproduced from *Dickinsons' Comprehensive Pictures of the Great Exhibition of 1851 / / from the originals painted for H.R.H. Prince Albert, by Messrs. Nash, Haghe and Roberts, R.A.; published under the express sanction of His Royal Highness Prince Albert, president of the Royal Commission, to whom the work is, by permission, dedicated* (London: Dickinson Brothers, Her Majesty's Publishers, 1854 [i.e. 1852]. Photograph Courtesy of the Smithsonian Libraries, Washington, DC.

was a mandated head covering for all males within the Ottoman territories, where it was understood as a "homogenizing status marker" and symbol of modern identity, as opposed to Western views of this headgear as "Orientalizing."[27] Viewed through a Western lens, an object linked to Tunisian self-presentation became something quite different: a marker of difference and exoticism in the sartorial realm.

While the man and boy studying the Arab Tent appear to be visitors, another view of the display in *Dickinson's* (Figure 1.3) pictures the attendants sent by the Bey to staff the Tunisian display: Sy Hamda Elmkadden and his assistant, Saido Belais, seen here conversing with visitors. While Belais is attired in native dress, Elmkadden wears a frock coat, tailored trousers, and carries a walking cane; only a fez signifies his Tunisian/Ottoman identity. As in the case of Ahmed Bey and other Oriental rulers known to alternate between native and Western dress, Elmkadden's apparel suggests the mix of cross-cultural elements impacting the identities of both individuals and nations in this time period; indeed, a related cultural dynamic can be observed in many parts of the Arab-Islamic world today.[28]

While these figures lent a note of authenticity to the bazaar, their proximity also put visitors in direct contact with "the other," and so foreshadowed, albeit modestly, the spectacular re-presentations of Oriental persons featured at later world's fairs. However, their presence was also questioned, since the sale of exhibited goods was banned, at least in theory, by event organizers who considered the practice detrimental to the intended reform-minded educational focus of the Exhibition.

Figure 1.3 "Tunis II," lithograph.

Source: Reproduced from *Dickinsons' Comprehensive Pictures of the Great Exhibition of 1851 // from the originals painted for H.R.H. Prince Albert, by Messrs. Nash, Haghe and Roberts, R.A.; published under the express sanction of His Royal Highness Prince Albert, president of the Royal Commission, to whom the work is, by permission, dedicated* (London: Dickinson Brothers, Her Majesty's Publishers, 1854 [i.e. 1852]. Photograph Courtesy of the Smithsonian Libraries, Washington, DC.

Although the benefits of free trade were widely promoted, the Tunisians were criticized for displaying sales prices and bargaining with customers, as in a real bazaar. While widely sought, "authenticity" was also subject to limitation and control.[29]

Like the wares they promoted, the attendants were one component of an "object-world" that was, in essence, a simulation of Tunisian culture intended for Western consumption. At the same time, their names were listed in the text accompanying the image:

> The collection was sent to England under the care of his highness's Commissioner, Signor Hamda Elmkadden...accompanied by an interpreter ...and also by an Arab attendant, called Saido Belais, who, from his picturesque costume, loud talk in Arabic, and energetic gestures, attracted no little attention... His portrait was taken, and he appears in the foreground of the accompanying view, squatting, as was his custom, on one of the counters.[30]

While accorded an individual identity, the description of Belais's demeanor nonetheless signals his "otherness" in the realm of etiquette and, by extrapolation, civilizational progress. Just as his exotic garb contrasts with the top hats and frock coats of his British visitors, Belais's "squatting" position opposes their stable, upright posture, and so distinguishes between subjects and objects of power. In an era fascinated with the study of ethnography, imagery detailing body types, poses, facial expressions, and related data was among the elements used to construct the hierarchal systems of racial identification and comparison that supported the Western nationalisms in evidence at the fairs. Understood in this way, Belais's "foreign" deportment acquires deeper meanings in the realm of evolutionary theory, its social applications, and notions of Euro-American progress in relation to the rest of the world.

While the lithographs from *Dickinson's Comprehensive Pictures* depict the cross-cultural exchange taking place within the Tunisian section (and by extrapolation, the Exhibition as a whole) as polite, dignified, and instructive, an illustration from a humorous view-book of the Exhibition provides a very different perspective on the same venue, demoting the "great and glorious bazaar" to the status of Tunisian "Rag Fair" (Figure 1.4). Lunging unsteadily, a Tunisian attendant attempts to bridge the spatial gap that separates him from his potential customers, who peer through their lorgnettes to inspect the foreign "specimen" more closely. While resembling that of Saido Belais in Figure 1.3, the attendant's dress and demeanor are now exaggerated to the point of caricature, calling to mind the stock character of the "Arab trickster/bandit" derived from literary sources.[31]

The slender, upright figures of the proper British ladies—also portrayed in the guise of an easily read national stereotype—serve as effective foils for his physical bulk, foreign dress, and ungainly movements, while their facial expressions and gestures signal their discomfort in his presence. In this image, the rhetoric of expanding and conflicting nationalisms is visualized for a popular audience and linked to the spectacles of "otherness"

Figure 1.4 "A Peep Into Tunis, or a Walk thro' Rag Fair," lithograph.

Source: Reproduced from Thomas Onwhyn, "What I Saw at the World's Fair, or, Notes on the Great Exhibition by Mr. Comic-Eye" (London: Prints in Volume, 1851). Photograph © The Victoria and Albert Museum, London.

enacted at the Exposition. Within the Tunisian "rag fair," physical proximity underscores cultural difference as the cartoonist lampoons the feasibility of a readily consumable "world-in-miniature" as well as the reductive meanings and outcomes it implies.

Unsurprisingly, Owen Jones and other design reformers expressed a very different view of the same objects. Consider, for example, a silk sash purchased by Jones and now in the collection of the Victoria and Albert Museum (Figure 1.5). While exemplifying the universal design principles that he championed, the color, geometric form, and repeating pattern of the sash also signify a national tradition of excellence in textile production. Produced in an urban workshop, the intricate abstracted patterns of the sash animate its flat surface and, when in use, the body of its wearer. Like the permutations of national dress previously noted, this object suggests the global influences informing Tunisian goods and identity during this period, since in all likelihood it was made from imported raw silk that was then processed and dyed by native workers.[32]

Figure 1.5 Woven silk sash, Tunisia, c.1850.
Source: Photograph © The Victoria and Albert Museum, London.

Throughout the Exhibition, visitors were encouraged to judge material objects in comparative terms. Although one impetus for staging the 1851 Exhibition was the introduction of new design models intended to improve British industrial output while elevating public taste, problems arose when it became obvious to perceptive visitors that many of the host country's manufactures—other than machinery, a category in which Britain reigned supreme—were in fact inferior to those produced elsewhere.[33] In an attempt to resolve this rupture in the narrative of British progress, scientist William Whewell and other like-minded critics shifted the terms of comparison from the aesthetic to the socio-political realm. While acknowledging the beauty of goods from "the gorgeous East," he directed attention to the conditions of their production, and how "...the machine with its millions of fingers works for millions of purchasers, while in remote countries where magnificence and savagery stand side by side, tens of thousands work for one. There Art labours for the rich alone."[34] Under these circumstances, he argues, art benefits only the privileged few, whereas if wedded to Western scientific and technological advances, it can be made to serve the common good. Framing his democratizing rationale in terms of self and "other," Whewell's comments illustrate yet again how politics, the rhetoric of nationalism, and material objects were intertwined at the Exhibition.

The Arab tent

Described as "among the most interesting items at the Exhibition," a large Arab tent made of goat and camel's hair, its exterior covered with fur pelts and its interior "decorated with the skins of lions, leopards, antelopes and wild goats" served as the centerpiece of the Tunisian display (Figure 1.2). Inside, domestic objects including "rude husbandry tools, Arab musical instruments, candles, straw hats, foodstuffs, and other household treasures..." were on view.[35] Whereas many of the garments and textiles the Tunisians displayed were luxury items "of great elegance, embroidered with gold" that evoked the fictional world of *The Arabian Nights*, the Arab Tent functioned as a more prosaic but no less important marker of national identity. As such, it surely conveyed a range of meanings to different audiences.

Demonstrating "the ancient and simple method by which the Bedouin Arab is protected from the weather in the desert," the tent could be viewed as an ethnographic specimen indicative of the history, material culture, and social development of the ethnic group that produced it.[36] While some might understand the tent as an object perfectly adapted to its function—providing shelter in a desert environment—others surely saw in its natural materials, unrefined surfaces, and simple construction yet another "fading image of the past." Divorced from its original contexts of use and production—as seen in Figure 1.2, the tent was incongruously surrounded by retail displays "fitted up so as to resemble one of the Bazaars of Tunis"—the Arab Tent became, in essence, a re-presented, hybridized, and mediated object. Displayed in this manner, it confirms Walter Benjamin's claim that "the world exhibitions...create a framework in which commodities' intrinsic value is eclipsed...and open up a phantasmagoria that people enter to be amused."[37] While the prevalence of open guidebooks in the lithograph of the tent suggests instruction more than amusement, the panoptic viewing position of the figures on the mezzanine above suggests a controlling gaze that reenacts the host country's position of authority while recalling Said's assertions regarding Western "management" of the East. This power dynamic is also visualized in spatial/architectural terms; dwarfed by the structure that houses it, the Arab Tent's human scale and organic, malleable appearance—"when packed up, such a tent occupies but little space, and is easily carried on the back of a camel"—contrast dramatically with the monumental size and geometric regularity of the Crystal Palace's iron and glass components. At the same time, both were ephemeral structures designed for maximum portability and functionality.[38]

For many Tunisians, the tent no doubt invoked complex tribal histories and affiliations, the often contentious relationship between Arab and Ottoman-affiliated elements within their society, and a nomadic lifestyle at odds with the modernizing dictates of the state. In Tunisia and other areas within the Ottoman suzerain, Bedouin opposition to land registration and other *tanzimat*-related laws challenged governmental authority, and was also contested by factions within the ruling and urban classes. In the West, the romanticizing of this anti-authoritarian, independent stance contributed to the Bedouins' characterization as "untamed denizens of the desert," picturesque bandits, and other types that re-presented the Arab-Islamic world as a locus of spectacle, performance, and "primitive manhood untouched by the... influences of civilization."[39] In this way, an object that conveyed complex, shifting, and entangled meanings within mid-century Tunisian politics and culture could also serve to uphold the discourses of Oriental exoticism advanced in the West.

Although the array of goods sent to London was selected and managed by Tunisians, the degree of agency they were able to exercise in matters of self-presentation was mediated by the viewing frameworks and narratives constructed within the Crystal Palace. In this foreign environment, the objects and people on display were re-presented, hybridized, and transformed. Assimilated into Britain's imperial vision of the world, they generated new, exoticized meanings that often departed from those understood in their country of origin. But if the "exhibitionary order" sought to control this process, the East-West/regressive-progressive binary it advanced was, as we have seen, simultaneously problematized and upheld by the encounters of people and objects within its precincts. As historian Jeffrey Auerbach has observed, "If Britain used the Exhibition to represent the world, so too did its colonies and other countries use the Exhibition to represent themselves."[40] This was certainly true of Tunisia, a country that had a great deal at stake in matters of self-presentation as it attempted to establish an identity resonant with Western progressivism while advancing its own nationalistic agenda. Its efforts in this arena show how the process of globalization, and the increasingly fluid and relational constructions of national and individual identities it abetted, were accelerated by the significant role the international expositions assumed in the cultural life of the late nineteenth century.

After 1851: structures and spaces

The reign of Ahmed Bey, which ended with his death in 1855, initiated widespread reform within Tunisia while raising the country's international profile. At the same time, his policy of "collaborationist modernization" was beset by political and economic troubles that increased under the rule of his successors. Tunisia's mounting debts led to its bankruptcy; soon thereafter a Foreign Debt Commission took control of its finances, and in 1881 the country became a French protectorate. Although his strides toward modernization had been inspired by those earlier enacted by Muhammed 'Ali in Egypt, the neighboring country followed a similar pattern, coming under British rule in 1882. After 1851, Tunisia's gradual transition from independent governance to French colony paralleled changes in its manner of re-presentation at the expositions, whose appearance and focus was also in the process of change.

In this period, increasing numbers of visitors prompted larger-scale events requiring multiple buildings, new outdoor spaces, services, and amusements. In this way, the expositions soon became cities-within-cities that transformed the urban environments of the municipalities hosting them. Theoretically, education and design reform remained priorities, but the parameters of the fairs expanded into the realm of popular entertainment; by 1889, the artisan workshops, native villages, "Street in Cairo," and other human constructions of the exotic were the most popular and profitable attractions at the Paris Exposition Universelle. This revised scope and scale enabled wider dissemination of the cultural re-presentations the fairs constructed.

The Crystal Palace had been partitioned into individual display areas, with its Western half reserved for Britain and her colonies and dependencies (see n. 7). Later fairs continued to cluster countries by region and colonial status, but with more space available, East and West, "other" and "self," were more definitively separated. At the 1867 Paris Exposition, national pavilions were introduced, and the monumental Main Building was encircled by groups of these smaller, eclectic foreign buildings; the overall scheme resembling a planetary system with the host nation at its center. Located with the Egyptian and Ottoman pavilions in the "picturesque confusion" of the park surrounding the Main Building, the focus of the Tunisian section was a scaled-down version of a section of the Bey's summer palace (Figure 1.6).

Like the Tunisian structures that would be featured in the 1878, 1889, and 1900 Paris Expositions, the 1867 "Bardo of the Bey" designed by French architect Alfred Chapon was, as architectural historian Zeynep Çelik has noted, an architectural collage collapsing "features from different time periods and regions into a single structure." In this respect, it resembles other Arab-Islamic exposition structures

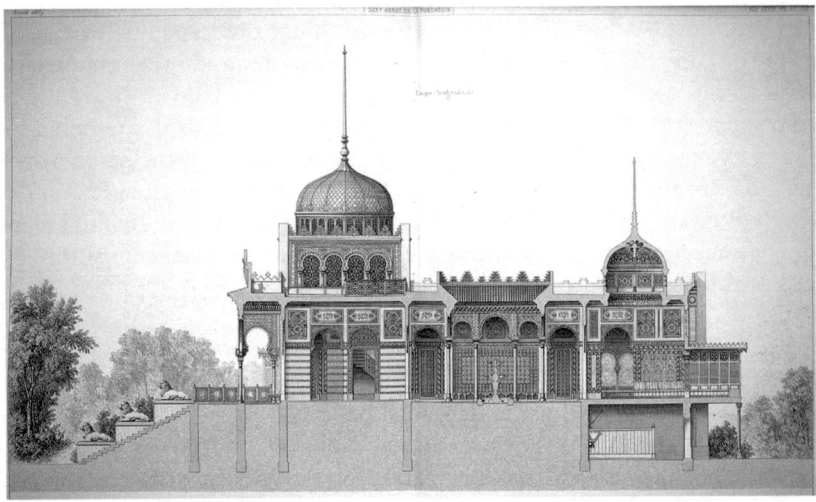

Figure 1.6 Alfred Chapon, "Bardo," Pavilion of the Bey of Tunis, Exposition Universelle 1867, Paris, sectional print.

Source: Reproduced from *Revue generale de l'architecture et des travaux publics*, vol. 27 (Paris: Paris Ducher 1869), pl. 35–36. Photograph: Collection of the Royal Institute of British Architects, London.

planned by Western architects,[41] as well as the paintings of J.L. Gérôme and other popular Orientalist artists of the period, who similarly privileged "authenticating" exotic detail—in the Bardo, horseshoe arches, rooftop crenellation, polychrome tiling and mosaics, a painted and gilded architrave—over historical accuracy, and individual parts over a cohesive whole.[42] The interior of this "specimen of the architectural art of Tunis in its most elevated type" featured "[rooms] where the Bey sits on a splendid ottoman surrounded by his court...the most costly ornamentation of the East has been used, with wall panels of carved plaster, polychrome marbles, carpets, and rich embroideries throughout" that extended the Orientalist trope, creating a setting and ambiance evocative of the "imaginary geography" of *The Arabian Nights*.[43] The basement of the palace housed a museum of antiquities, barber shop, café, and rest areas with divans and *narghiles*, while outside the "industries, trades, and arts" of Tunisia—working artisans, a "long row of graceful and attractive shops fitted with rare eastern productions," stables, horses, camels, and other exotic objects and activities—were on view.[44] While nearby tented areas referenced the Arab Tent shown in London and the Bedouin traditions it symbolized, overall the 1867 pavilion conveyed a far more extensive, spectacular, and complex picture of an increasingly contested Tunisian national identity. But like the Arab tent, the Palace of the Bey could be understood in multiple ways.

With regard to its conception and execution, the building, as directed by Chapon, might be viewed as an example of French encroachment in the cultural realm, and of the "informal imperialism"—domination based on cultural, trade, and economic policy rather than military incursion—regularly exercised over the Arab-Islamic world by European powers.[45] In Saidian terms, the building demonstrates Western "management" of the East, with contemporary viewers finding in its rich surfaces, materials, and interior decoration yet another "fading image of the past" as opposed to Occidental progress and modernization. With a large contingent of artisans and other workers nearby, the palace's splendor might be regarded as evidence of the gulf between Eastern rulers and their subjects, and as confirming prior assertions that "in the gorgeous East...Art labors for the rich alone" (see above, page 24). Viewed from this socio-political perspective, the entire Tunisian display could be understood as "underlining the rhetoric of progress by serving as its counterpart," and as clarifying French national identity even as that of the country it re-presented was increasingly destabilized.[46]

Alternative readings might, however, consider the Palace as a site of cultural negotiation and of hybridity, and in doing so align it with Euro-American architectural revival styles of the period. In adapting components of past styles to contemporary tastes and needs, this mode of building suggested one way of engaging with the present while preserving cultural, religious, and social traditions, an issue—then as now—of great concern throughout the Arab-Islamic world.[47] Thus, it might also be understood as a model for the modern reformulation of Islamic cultures, a process that—despite world's fair re-presentations and other assertions to the contrary—was well underway in this time period.[48] Whether viewed as a performance of national identity in accord with European expectations and the vocabularies of Orientalism, an effective use of these networks to forge an Islamic modernism incorporating elements of East and West, or some combination thereof, all these possibilities underscore the changes occurring in Tunisia and the larger world in the post-1851 era.

After 1851: the human component

While the presence of Tunisian attendants at the 1851 London exhibition furnished, among other elements, authenticating detail, direct experiences of the "exotic other," multi-cultural instruction, ethnographic data, and visible proof of assumed racial hierarchies and Western superiority, they were, as we have seen, few in number and limited in their activities. As national pavilions with extensive outdoor spaces were added in later fairs, the human component became an ever more important part of the exotic product—i.e., the foreign culture on display—made available for consumption at the fairs. Presented as living *tableaux vivants* within the native villages that were initially conceived as educational displays, the artisans working in the Tunisian sections of the late-century Paris expositions demonstrated indigenous crafts such as woodworking, metalwork, jewelry, basket, and shoe making, weaving, embroidery, and other forms of textile production.[49] Situated near or within the now-ubiquitous bazaars and souqs introduced at the Crystal Palace, they were closely identified with the goods they crafted, which were uniformly ethnic, handmade, and pre-industrial in their function and mode of production. Whether viewed as a "fading image of the past," an attempt to preserve native traditions under threat from mass-produced European imports, or a visualization of Tunisia's increasingly precarious political position in relation to French interests, these popular displays demonstrated the increasingly performative roles assigned to indigenous peoples at the fairs. As the century waned and the once-exotic became more familiar, the public desire for (and profitability of) theatrical spectacle intensified, reaching a climax in the "Street in Cairo" staged at the 1893 World's Columbian Exposition in Chicago.

Both the Cairo Street and an extensive Tunisian-Algerian village complex were located in the center of the Midway Plaisance, a mile-long concourse stretching westward from its oppositional "other": a White City composed of large-scale, symmetrical structures built in the Beaux-Arts style based on Classical prototypes.[50] With its grand buildings punctuated by waterways, monumental sculptures, and at night, electrical lighting, this was "the magic city" described as a genie's creation, but one in which "the powers of human (i.e., Euro-American) genius transcend the beauty and opulence of Arabic dreams...and are fabricated into living ideals of grandeur more magnificent than any that ever Rajah or Caliph beheld in vision or fact."[51] Like their British and French counterparts, planners and publicists of the Chicago fair proved adept at defining America's expanding nationalisms in relation to the Arab-Islamic "object-world" on display in their city. While guidebook and other prescriptive commentary continued to play an important role, the placement and appearance of exhibits and their contents—people, objects, and people-as-objects—presented this comparison most effectively.

In contrast to the White City, the Midway—the first instance in which an ethnic/amusement area was spatially segregated from the main body of the fair—was described as "a jumble of foreignness...gorgeous with color, pulsating with excitement... and peculiar to the last degree."[52] Clustered in its center, the location of the Arab-Islamic exhibits midway between the European concessions at the Midway's entrance and the African and Native American areas at its terminus was meant to instruct visitors regarding their place on the "sliding scale of humanity" indicating civilizational progress or lack thereof. Featuring an array of ethnic villages, bazaars, coffeehouses, restaurants, and theatres (including the Egyptian Theatre with its famous belly dancers) with streets and enclaves populated by Arab horse and swordsmen, donkey boys, camel drivers, shopkeepers, artisans, musicians, and other "authentic" inhabitants of all ages and genders, the Midway exhibits were organized and maintained by concessionaires rather than foreign governments. The chaotic "jumble" of the area extended to its architectural components, which, in the case of the Cairo Street, were scaled-down replicas of actual Cairene structures that often incorporated fragments of previously demolished buildings.[53]

With over twenty-seven million visitors in its six-month duration, the Chicago exposition was instrumental in propelling the re-presentations of the Arab-Islamic world it adopted and expanded from past fairs into the mainstream of American society. As in Europe, their reverberations in the political realm may be traced in the troubled history of twentieth-century East-West relations, while commercial interests—makers of cigarettes and other tobacco products, confectioners, and the Hollywood film industry, among many others—continued to perpetuate Orientalist-inflected stereotypes in the marketing of their products.[54] Standing at the vanguard of a globalized, capitalized, and increasingly multicultural world, the late nineteenth-century expositions instigated as well as reflected many of the disruptions, challenges, and possibilities of modern life while re-presenting and problematizing the past, present, and future of the Arab-Islamic peoples and cultures positioned within it.

Notes

Major portions of the research for this chapter and the acquisition of the images it features were supported by Virginia Commonwealth University Qatar Faculty Research Grants awarded in 2011 and 2015. The author also extends her thanks to curator Miriam Rosser-Owen, who helped to facilitate much of the initial research undertaken at the Victoria and Albert Museum, London; to colleagues Radha Dalal and Matthias Detterman in Doha, who provided valuable support and feedback; and to editors Ethan Robey and David Raizman for initiating the *Expanding Nationalisms at World's Fairs* project and their support throughout the writing and editing process.

1 In this time period, the terms "the East," "the Orient" and "the Mohammedan lands," referred to North Africa, the Middle East, Turkey, India, and Persia, or some combination thereof, rather than China and/or Japan. As a general term, "the East" was also associated with the religion—Islam—and ethnic group—Arab—dominant in many parts of these regions, and so conflated religious and cultural traditions, national borders, and ethnic groups. While the title and sections of this chapter also refer broadly to representations of the Arab-Islamic world, it seeks to link this general terminology to specific case studies that illustrate many of the problems and issues at stake in these words and the ideas they represented in this era.

2 Edward Said, *Orientalism* (New York: Vintage Books, 1978), 2.

3 Said, *Orientalism*, 3. While utilizing selected aspects of Said's work in relation to the World's Fairs, the author also acknowledges the alternative views voiced in works such as Joshua Muravchik, "Enough Said: The False Scholarship of Edward Said," *World Affairs* 175, no. 6 (March 2013): 9–21 and Robert Irwin, *For the Lust of Knowing: The Orientalists and their Enemies* (London: Penguin Books, 2006).

4 Timothy Mitchell, "Orientalism and the Exhibitionary Order," in *The Visual Culture Reader* (London: Routledge, 2002), 499.

5 Mitchell, "Orientalism and the Exhibitionary Order," 498–499, states that "the effect of such spectacles [the panorama and diorama popular earlier in the century] was to set up the world as a picture…[and later, at the Expositions] encounters with natives and their artifacts arranged to provide the direct experience of a colonized object-world."

6 At mid-century, the term 'industry' still referred primarily to human labor, while 'manufactures' were goods produced in this manner; neither term was yet associated

exclusively with industrialization or mass production. On the Exhibition floor plan, see James Buzard, "Conflicting Cartographies: Globalism, Nationalism, and the Floor Plan of the Crystal Palace," in *Victorian Prism* (Charlottesville: University of Virginia Press, 2007), 40–52.

7   "The Great Exhibition," *The Times*, February 3 1851, 5.

8   *Dickinson's Comprehensive Pictures of the Great Exhibition of 1851* (London: Dickinson and Sons, 1852), 4.

9   Most of its components were prefabricated; the 2,000 workers on site in London's Hyde Park bolted and welded the iron frame and its attachments, attached 300,000 panes of glass, and assisted in finishing work. *The Building Erected in Hyde Park for the Great Exhibition of 1851* (London: Woodfall and Sons, 1852), i–xi.

10   Jones's reputation derived primarily from his on-site study of key monuments in Spain, Egypt, and Turkey, the 1842 publication of his *Plans, Elevations, Sections, and Details of the Alhambra*, and his lectures on these topics.

11   Owen Jones, "The Interior Decoration of the Crystal Palace," in *Bulletin of the American Art Union*, no. 2 (May 1851): 28–29.

12   In this respect, Jones's scheme reflects the use of fabrics in Oriental interiors as well as the textile-like "wrapped" surface decoration in mosaic, stucco, and other materials featured in buildings such as the Alhambra.

13   While an "authentic" small-scale bazaar was the centerpiece of the Tunisian area at the 1851 Exhibition, the Crystal Palace as a whole referenced this retail model, as did the department stores developing in this era, and later, shopping malls. On this topic see Peter Coleman, *Shopping Environments Evolution, Planning and Design* (London: Routledge Architectural Press, 2006), 23–24.

14   Samuel Warren, *The Lily and the Bee: An Apologue of the Crystal Palace* (London: Blackwood, 1851), 9.

15   Quoted in Clement Shorter, *The Brontes: Life and Letters*, vol. 2 (London: Haddon Co., 1908), 216.

16   In 1851, the most recent "complete" translation was Edward Lane's 1839–1841 edition. On the larger network of interest in Cairo and "Egyptomania" that Lane promulgated in this period, see Derek Gregory, "Performing Cairo: Orientalism and the City of the Arabian Nights," in *Making Cairo Medieval*, eds. Nezar AlSayyad, Irene A. Bierman, and Nasser Rabbat (New York: Lexington Books, 2005), 69–93. In addition to books, the tales were widely serialized in popular periodicals and also performed onstage, making them known to a wide variety of audiences.

17   Husain Haddawy, preface to *The Arabian Nights* (New York: Alfred Knopf, 1990), xxv.

18   "How We Hunted the Prince," *Punch*, no. 20 (June 1851): 222.

19   In this time period, the new power of electricity was often personified as a genie released from the magic lamp featured in the tale of "Aladdin and the Magic Lamp" appended to *The Arabian Nights*. The genie analogy was also used to explain the new and "magical" material abundance derived from industrialization and mass production. On this topic see Susan Nance, "Capitalism and the Arabian Nights," *How the Arabian Nights Inspired the American Dream* (Chapel Hill: University of North Carolina Press, 2009), 19–50.

20   On the Ottoman Exposition, see Zeynep Çelik, "Exposition Fever Carried East," in *Displaying the Orient: Architecture of Islam at the Nineteenth Century World's Fairs* (Berkeley: University of California Press, 1992), 139–151.

21   Carl Brown, *The Tunisia of Ahmed Bey, 1837–1855* (Princeton, NJ: Princeton University Press, 1974), 27. Within this chapter, the term "national identity" is used to communicate the sense of uniquely Tunisian identity, as represented by its distinctive cultural traditions, rather than an advocacy of absolute political independence.

22   He was also known as Mushir Ahmed Basha Bey; 'bey' is the Turkish term for "chieftan." Tunisia was also referred to variously as a "regency" or "beylik." In addition to his other modernizing initiatives, Ahmed Bey's 1846 state visit to France was the first by a Muslim ruler in a non-military capacity; in the same decade he also outlawed slavery and slave markets in his country. Brown, *The Tunisia of Ahmed Bey*, 317.

23   "Tunis II," *Dickinson's Comprehensive Pictures*, vol. II, 9.

24   *Official Descriptive and Illustrated Catalogue of the Great Exhibition of the Works of Industry of All Nations, Part V, Foreign States* (London: Spicer Bros., 1851), 1412.

25　Ibid.

26　"A splendid case of clothes and dresses, embroidered in gold, deserves special notice…" *Reminiscences of the Crystal Palace* (London: Routledge & Co, 1852), 173.

27　*Tanzimat*-related reforms included new sartorial rules "intended to eliminate clothing distinctions and make the state the sole arbiter of identity." Donald Quateret, "Clothing Laws, State, and Society in the Ottoman Empire," *International Journal of Middle Eastern Studies* 29 (1997): 403.

28　Brown, *The Tunisia of Ahmed Bey*, 308. When the Ottoman sultan and Egyptian governor visited the 1867 Paris Exposition, much disappointment was reported in the French press over the fact that, except for the fez, both wore Western clothing. On historical and contemporary modes of Western and traditional dress in the Gulf region, and the alternation between them, see Miriam Cooke, "Performing National Identity," *Tribal Modern: Branding New Nations in the Arab Gulf* (Berkeley: University of California Press, 2014), 123–137.

29　On free trade and pricing debates at the Great London Exhibition, see Jeffrey Auerbach, *The Great Exhibition of 1851: A Nation on Display* (New Haven, CT: Yale University Press, 1999), 62–64, 118–121.

30　*Dickinson's Comprehensive Pictures*, vol. II, 9.

31　Stories of Ali Zaybaq the con man and Juha the trickster derive primarily from Arab, Indian, and Persian folk tales, many of which were incorporated into later versions of *The Arabian Nights*. On the stereotype of the Arab trickster see Nance, *How the Arabian Nights Inspired the American Dream*, 122. The trickster stereotype could also encompass a "trickster /bandit/ shopkeeper" stock character related to the theme of commercialism noted in this chapter.

32　www.vam.ac.uk/content/articles/t/tunisian-textiles-at-the-great-exhibition/ accessed May 1, 2016. On Tunisian textiles, see also Spring and Hudson, *North African Textiles* (London: British Museum Press, 1995), 33–37, and Jacques Revault, *Designs and Patterns in North African Textiles* (New York: Dover Press, 1973), 10–13. By the mid-nineteenth century, a number of craft industries in Tunisia, including textiles, were under threat from European imports, so Ahmed Bey may have also welcomed the London Exhibition as an opportunity to market native crafts to a wider international market.

33　See Owen Jones, "Gleanings from the Great Exhibition of 1851," *Journal of Design and Manufactures* 5, no. 28 (1851): 89–91. Upon viewing the Exhibition, design reformer Willam Dyce observed that "many other nations show better faith and practice in design [than Britain]…does the progress of civilization destroy principles of taste?" "Universal Infidelity in Principles of Design," *Journal of Design and Manufactures* 5, no. 28 (1851): 158–161.

34　Quoted in Patrick Young, *Globalization and the Great Exhibition* (New York: Palgrave Macmillan, 2009), 108–109. Echoing Whewell's stance, a writer for the *Catalogue of the Turkish Section of the Great Exhibition* (London: McKewan & Co., 1851) stated that "the time has come when a display of barbaric magnificence is no longer accounted the test of a wealthy power, but the statistics of imports and exports suited to the wants of the millions are alone taken as an indication of the prosperity or weakness of a nation."

35　*Reminiscences of the Crystal Palace*, 171–173.

36　On Bedouin history and culture see Dawn Chatty and William Young, "Bedouin," *Encyclopedia of World Cultures* (New York: Macmillan Reference, 1996) and, Dawn Chatty, *From Camel to Truck: The Bedouin in the Modern World* (New York: Vantage Press, 1986).

37　*Dickinson's Comprehensive Pictures*," Tunis II," Vol. 1, Part II, 9. Walter Benjamin, *Reflections: Essays, Aphorisms, Autobiographical Writings*, ed. Peter Jernetz and trans. Edmund Jephcott (New York: Harcourt Brace Jovanovich, 1978), 152.

38　*Dickinson's Comprehensive Pictures*, vol. II, 11. Following the close of the 1851 Exhibition, the Crystal Palace was dismantled and moved to the London suburb of Sydenham, where it reopened in 1854 and remained until destroyed by fire in 1936.

39　Martin Glassner, "The Bedouin of the Southern Sinai," *Geographic Review* 64, no. 1 (1974): 31. Susan Nance, *How the Arabian Nights Inspired the American Dream*, 122–123. On the range of Bedouin-derived stereotypes in the American entertainment industry, see Nance, 112–125. The 1893 Chicago exposition—which featured Bedouin swordsmen and horsemen performing in mock battles, equestrian shows, caravan spectaculars, and other extravaganzas rivaling Buffalo Bill's Wild West show just outside the fairgrounds—is but one example of the popularity of these stock characterizations.

40 Jeffrey Auerbach, "Introduction," in *Britain, the Empire and the World at the Great Exhibition of 1851,* ed. Jeffrey Auerbach and Peter Hoffenberg (Aldershot: Ashgate, 2008), xviii.

41 See also Christian Hedrick, "From London to Paris (via Cairo): The World Expositions and the Making of a Modern Architect, 1862–1867," Chapter 2 in this volume.

42 Çelik, *Displaying the Orient,* 56. Çelik's argument derives in part from her reading of Linda Nochlin, "The Imaginary Orient," in *The Politics of Vision: Essays on Nineteenth Century Art and Society* (New York: Harper and Row, 1983). For an example of the "authenticating detail" in Gérôme's work, see the 1879 painting of "The Snake Charmer" (Sterling and Francine Clark Art Institute, Williams, MA).

43 William P. Blake, ed., *Reports of the U.S. Commissioners to the 1867 Paris Universal Exposition* (Washington, DC: U. S. Government Printing Office, 1870), 18–19. Also noted was a hidden staircase leading to the private harem quarters of the Bey. According to several sources, Bey Muhammed al-Sadiq (r. 1859–1882) was briefly in residence at the Palace "during his visit to Paris." Çelik, *Displaying the Orient,* 62.

44 Blake, *Reports of the U.S. Commissioners,* 19.

45 On the topic of informal imperialism see Auerbach, *Britain, the Empire and the World,* xiv–xv.

46 Tony Bennett, *The Birth of the Museum* (London: Routledge, 1995), 78–79.

47 On this topic in the late nineteenth century, see Ahmet Ersoy, "Osman Hamdi Bey and the Historophile Mood: Orientalist Vision and the Romantic Sense of the Past in Late Ottoman Culture," in *The Poetics and Politics of Place-Ottoman Istanbul and British Orientalism* (Istanbul: Pera Muzesi, 2011), 144–155. On the continued relevance of this topic today, see the Qatar National Vision 2030, which lists the balancing of "modernization and the preservation of tradition" as the first of the state's five major challenges.

48 See, for example, Paula Sanders, "Islam for the Modern World: Medieval Cairo between Egyptian Reforms and British Critics," in *Creating Medieval Cairo* (Cairo: American University in Cairo Press, 2008), 59–88; Zeynep Çelik, "Bouvard's Boulevards: Beaux-Arts Planning in Istanbul," *Journal of the Society of Architectural Historians* 43, no. 4 (December 1984): 341–355.

49 Çelik, *Displaying the Orient,* 22.

50 On the Beaux Arts style see David Brain, "Discipline and Style: The École des Beaux Arts and the Social Production of American Architecture," *Theory and Society* 198, no. 6 (November 1989): 807–868. Rather than marble, the steel-frame buildings of the White City were surfaced with staff, a mixture of plaster and hemp that reduced costs and accelerated completion of the structures.

51 James W. Buel, "Gateway to the Magic City," in *The Magic City: A Massive Portfolio of Original Photographic Views of the Great World's Fair* (St. Louis, MO and Philadelphia, PA: Historical Publishing Company, 1894), n.p.

52 Julian Ralph, *Harper's Chicago and the World's Fairs* (New York: Harper and Bros., 1983), n.p.

53 On the buildings replicated and the use of architectural fragments in their construction, see Irene Bierman, "Disciplining the Eye: Perceiving Medieval Cairo," in *Making Cairo Medieval* (New York: Lexington Books, 2005), 22–24; István Ormos, "The Cairo Street at the World's Columbian Exposition, Chicago, 1893," in *L'Orientalisme architectural entre imaginaires et saviors* (Paris: Picard, 2009), 1–16. electronic version: inha.revues.org/4915 accessed March 2, 2016.

54 Holly Edwards et al., "Catalogue of the Exhibition" and "Orientalism in America, 1870–1930," in *Noble Dreams, Wicked Pleasures: Orientalism in America, 1870–1930* (Princeton, NJ: Princeton University Press, 2000), 42; 200–206.

2 From London to Paris (via Cairo)

The world expositions and the making of a modern architect, 1862–1867

Christian A. Hedrick

We know of no other artist of our time who is able to detect and record the noble forms, charm and fine play of lines in Arabic ornament. And who is able to apply them independently with a steady hand because of his long stay in the south. The vase was awarded a medal for its invention and design.[1]

International Exhibition Award Committee (1862)

At the London Exposition of 1862, forty-two-year-old Prussian architect Carl von Diebitsch entered an ornate zinc vase he had recently designed and cast in the category "Iron and Metallurgy."[2] While no images of the vase are extant, a contemporary description provides an insight into its creator's intent. There is also a photograph showing Diebitsch with a similar, although much smaller, metal vase. Accounts of the original vase, which rested upon a pedestal with columns, describe it as "colossal,"[3] standing "not less than 15 feet high,"[4] and that the work took him six weeks to produce. Described in the official report as "polychromatic," the surface was composed of a variety of different colors from a copper matte to gold leaf. The choice of zinc for the vase was significant because it was considered a highly resistant and durable metal, ideal for intricate casting.[5] It also remained relatively uncommon in the British manufacturing industry due to the difficulty the English had in successfully purifying it at the time.[6] By the 1830s Prussia had already compiled a surplus of zinc, which made it inexpensive and readily available.[7] For decades Prussian architects had been taking advantage of the availability and low cost of the material, which architect Karl Friedrich Schinkel (1781–1841) had once advocated as being forty percent cheaper— in some cases—than other materials used for architectural details such as stone or copper.[8]

Diebitsch's award-winning vase depicted allegorical representations of the sciences, arts, industry, and agriculture as well as portraits of Queen Victoria and of the Prussian King Wilhelm I enthroned, with his son, the prince, and his brothers.[9] Also represented with the king were members of the Ministry and products for display at the exhibition, referencing the king's active support of the arts and industrial commerce and trade. Surrounding the relief of Queen Victoria were representatives of the different categories of objects on display from the physical and natural sciences. Combined allegories of architecture, sculpture, and painting, as well as agriculture and industry, with special reference to machine engineering also appeared on the vase. Connecting many of these motifs, and central to Diebitsch's design philosophy, was what one judge in the awards committee referred to as "the fine play of lines in

Arabic ornament realized with a sure hand in an original way."[10] Amid these "arabesques" covering the surface were additional reliefs containing references to German cultural history and a portrait frieze of famous Germans, including Luther, Frederick II, Goethe, Schiller, and Herder.[11] Not only did the subject of this frieze reveal the likely origin of the vase, but the appearance of such content can hardly be ignored in the years preceding the unification of German territories.[12] Indeed, the very association that enabled Diebitsch to attend the exposition was the so-called *Zollverein* ("German Customs Union"), which was established in the 1830s and created a powerful economic union across many German-speaking lands that enabled businesses and entrepreneurs to compete on an increasingly international level. This economic foundation paved the way to the political unification of 1871. Thus, it was through the agency of the Zollverein that Diebitsch was able to display his vase, reinforcing the idea that his work was integral to the organization's broader strategy of industrialization and modernization.[13] The relationship Diebitsch had with the Zollverein may also explain some of the subject matter of the vase, which acknowledged the association in terms of its patriotic function and its collective bargaining power, without whose support Diebitsch would not have had the opportunity to submit his work. The preoccupation of the Zollverein with technological advancement is seen in the organization's commendation of the vase. Whereas the official list in the British source, *Medals and Honourable Mentions*, states the rather generic affirmation that Diebitsch was given the award "For excellence of the zinc vases exhibited," [14] the official report of the Zollverein gives attention to the *material* aspects of the vase, commending Diebitsch's "invention" (*Erfindung*) and "design" (*Zeichnung*), but also noting the "technical execution of the casting," indicating an official concern that its exhibitions present the state as technologically progressive.[15]

Considering the context of Diebitsch's medal-winning vase at the London 1862 exhibition, the use of an Islamic style must have seemed somewhat unusual due to the overwhelming presence of Renaissance and Gothic revival styles at the time. Diebitsch's chosen materials and methods of production, in combination with the iconography on the vase, indicate a specific direction in the architect's design theory. He believed that the kinds of repeated abstract patterns he found so common in Islamic art and architecture easily lent themselves to casting.[16] Furthermore, he claimed in an 1852 lecture that "without any considerable expense [carpenters/builders] could make the forms as rich as they wanted."[17] And since Diebitsch's designs for ornament tended toward geometric abstraction he believed that this broadened their public appeal and made his designs appropriate for the modern age.[18] This idea that Islamic design is somehow predisposed to modern industrial manufacturing processes was echoed almost twenty years later by Matthew Digby Wyatt in his "Orientalism in European Industry," where he observed that "everyone remained all but blind to the value of the East as a source of inspiration for industrial designers."[19]

Diebitsch's unexpected client

It is not surprising that exhibits embodying industrial modernization were also of particular interest to non-Europeans, who desired to take advantage of the latest ideas for their own countries in an effort to keep up with, if not surpass, Europe's industrial hegemony. Among the variety of foreign celebrities and prominent political figures who attended the 1862 London Exhibition was the Viceroy of Egypt, Muhammad Sa'id Pasha, fourth son of Muhammad 'Ali the so-called "Father of Modern Egypt."

Sa'id was in London with his banker Henry Oppenheim, of the well-established Cologne banking family, brokering an historic state loan to help fund, among other things, the Suez Canal project after the Egyptian treasury had all but dried up by the end of 1861.[20] It was during an excursion to the exhibition with the royal entourage that Oppenheim first met Diebitsch. Following their meeting, Oppenheim commissioned Diebitsch to design the interior of his new villa in Cairo, which included a considerable amount of iron and plaster work. Oppenheim hired Diebitsch because he had demonstrated a thorough knowledge of Islamic design, which was gaining in popularity in Cairo at the time.[21] Diebitsch also convinced Oppenheim that he could offer these designs at an architectural scale, in modern materials, at lower costs and in less time than that of traditional building methods utilized in Cairo.[22] Diebitsch's friendship with Oppenheim led to a series of commissions in Cairo and ultimately to the architect's relocation there soon after the exhibition in the winter of 1862. This surge in business would have otherwise been impossible were it not for the unique commercial opportunities created by the London Exhibition. For the next six years Diebitsch's life was substantially different; he ultimately relocated his practice and family to Cairo where he was able to work and manage a moderately successful architectural and design practice.

Cairo in transition

> My country is no longer in Africa; we are now part of Europe. It is therefore natural for us to abandon our former ways and to adopt a new system adapted to our social conditions.
>
> Khedive Isma'il (1879)

Cairo was a city in flux when Diebitsch arrived in the winter of 1862, and was on the verge of one of the greatest modernization efforts of the nineteenth century. Indeed, some argue that the pace of transition was simply too fast, especially gathering momentum after 1863.[23] Well in advance of British colonization in 1882, the Ottoman viceroys of Egypt strove to modernize. The efforts began with Muhammad 'Ali (1769–1849), who introduced industrial manufacturing to the country. This new industrial presence involved the creation of Egypt's own textile manufacturing capabilities, marked by the opening of the first wool factory in 1808. Also, the first iron foundry and printing press in Egypt were founded during this period—in 1820 and 1822 respectively.[24] Ultimately, however, it is with Khedive Isma'il Pasha (1830–1895) that this yearning for a truly "modern" Egypt became an obsession. Under Isma'il modernization took on a particularly Western character: he strove for the modern conditions he witnessed in Europe, not the model advanced by the Ottomans. Indeed, he wanted Egypt to be seen as part of Europe, not as a tributary state of the Ottoman Empire. Throughout the 1860s modernization took several forms and manifested itself in a variety of ways, from political and religious reforms to the built environment.[25]

Diebitsch in Cairo (1862–1867)

In Cairo, Diebitsch finally had the opportunity to explore the ideas he had been working on since his return from his *Studienreise* in 1848. There is little doubt that the main reason Henry Oppenheim hired Diebitsch was because of the promises the architect had made at the exhibition regarding the cost benefits and efficiency of building in a modern Islamic style. However, his transition to Egypt was not an easy one. The assistant with whom he travelled to Cairo, Carl Ohnesorge, kept a journal that provides information regarding Diebitsch's life and work in Cairo.[26] Ohnesorge lists Diebitsch's first three commissions in Cairo as a free-standing mausoleum for Suleyman Pasha, a "large hall of cast iron" for the Prime Minister Sherif Pasha, and various projects for the villa of "Financier" Henry Oppenheim in the "modernized Moorish style" (*modernisiert maurischem Stile*).[27] Ohnesorge specifically refers to the style they are working in as "modernized," which suggests that neither he nor Diebitsch saw their work as simply reviving an old style. This is precisely why Diebitsch's clients were individuals who were, for the most part, involved in the great push to modernize Egypt.[28] These patrons identified with Diebitch's goal to produce a truly modern architecture by rendering abstracted variations of formal Islamic elements in modern materials, a combination that began to define the character of Egyptian modernization in architecture and design in the 1860s.

As noted above, Diebitsch's first commission, and likely also "the initiator of his Egyptian business dealings,"[29] was an interior design commission from Henry Oppenheim in the winter of 1862/1863. According to Ohnesorge, the project consisted of "the renovation and expansion of a large villa... including all of the furniture, draperies, mirrors and chandeliers, vases, fountains and more."[30] As it was Diebitsch's first commission in a foreign country and in a city more than 2,500 miles from his home, his theories on modern production and transportation methods of architecture were immediately put to the test. Architecturally, the project consisted of installing prefabricated ironwork and plaster details on the villa's interior. All of the plaster moulds were prefabricated in Diebitsch's studio on the Hafenplatz in Berlin with the assistance of skilled workers and painters, and the iron was cast at the Lauchhammer foundry outside of Dresden.[31] Aside from a series of distractions and minor setbacks involving the arrival of broken materials—including cast iron parts—and a demanding client, the project was an overall success and established Diebitsch's reputation in the Cairo building industry.

A second, more significant, commission came in the form of a freestanding structure. In 1862 Diebitsch was approached by the Prime Minister Sharif Pasha (who had already placed an order for his own villa) to design and build a mausoleum for his father-in-law, the famous general Sulayman Pasha 'al-Faransawi' ('the Frenchman') (1788–1860). Sulayman Pasha was a central figure in the reorganization of the Egyptian army and subsequently became a national hero, playing a fundamental role in Egypt's modernization process.[32] Ohnesorge records that the commission was for "an octagonal building with a veranda-like portico of iron, crowned with a dome of zinc, richly ornamented on the inside and out and decorated in various colors and gold."[33] The building, which still stands today, was originally set within the garden of Sulayman Pasha's estate adjacent to the Nile (Figures 2.1–2.3). The interior of the cella was originally lined with red and blue tiles with vegetal motifs and the metal ceiling was enameled.[34] Everything—except the stereobate, the stone cella and its contents, comprising a catafalque and the wood supports under the dome—was cast in iron or zinc in the Ilsenburg foundry in Saxony and, like all of his Cairo projects, was shipped to the site for installation.[35] Thus, Diebitsch successfully rendered his ideas, initially proposed in his zinc vase prototype at the London Exposition of 1862, into a fully realized freestanding structure. Equally crucial to his success was the fact that he was only able to achieve this complex feat by taking advantage of a series of emerging international networks of exchange including new modes of transport, such as steamships and railways, and the nascent world financial markets. As Jürgen Osterhammel has demonstrated, the decades after 1860 witnessed an unprecedented growth of these international networks, which precipitated not only substantial intercontinental migrations, but also the expansion of colonial empires.[36]

The role and importance of these networks in terms of Diebitsch's intellectual formation and entrepreneurial ambitions cannot be ignored. From his early experiences at the Berlin Bauakademie he was exposed to different interpretations of the architectural past, primarily due to his formation under his mentor Wilhelm Stier. Stier, unlike his colleagues at the Bauakademie who showed little interest in Islamic architecture, reinterpreted Islamic forms in creative and highly sophisticated ways through his competition designs. After his education with Stier,

Figure 2.1 Carl von Diebitsch, *Mausoleum for Sulayman Pasha*, Cairo, Egypt. c.1863.
Source: Photo by Author (2008).

Figure 2.2 Carl von Diebitsch, *Mausoleum for Sulayman Pasha*, Cairo, Egypt. c.1863. Zinc
        dome.
Source: Photo by Author (2008).

Figure 2.3 Carl von Diebitsch, *Mausoleum for Sulayman Pasha*, Cairo, Egypt. c.1863. Detail
        of iron fretwork at entry.
Source: Photo by Author (2008).

Diebitch continued to evolve during his non-traditional *Studienreise* through
Sicily, Spain, and North Africa in the 1840s, which had only become accessible
over the previous two decades through French colonization and the establishment
of transportation networks. Upon his return to Berlin he pursued his dream of
practicing in an architectural idiom that he believed was modern and Islamic.
Diebitsch's design proposals demonstrate an awareness of both European and Islamic architectural traditions through their formal composition by drawing upon
shared forms such as the octagon, conceived and produced in the latest materials
and most recent technologies. Equally important to Diebitsch's formation are the
individuals he met along the way, including Oppenheim, who led him to a variety

of clients in Cairo. The most important of these clients was Khedive Isma'il himself whose own vision was reflected in Diebitsch's work.

The 1867 *Exposition Universelle*: Diebitsch's dilemma

By the time of the 1867 Exposition Universelle in Paris, Diebitsch had been gainfully employed in Cairo for five years by a variety of clients—all of whom were active participants in Egypt's fast-paced modernization led by then Viceroy Isma'il.[37] Isma'il prioritized his involvement with the exposition, including his own attendance, in an effort to make Egypt's positive association with Europe more apparent.[38] With the Suez Canal preparing to open within two years and the incredible modernization of Cairo's infrastructure and urban fabric, the Khedive had a great deal to show off. The exposition was Isma'il's first official trip as Khedive and Egypt's pavilions were the result of a committee headed by his Minister of Foreign Affairs, Nubar Pasha (for whom Diebitsch also designed a villa). Egypt's display, which contained "enormous examples from both ancient and modern Egypt," included several pavilions.[39] There was a Pharaonic temple (a copy of the temple of Philae chosen to be represented by the French archaeologist Auguste Mariette who was also one of the organizers for the Egyptian section) with a museum of antiquities. The second pavilion was a richly decorated structure in the "Arab style" (*selamlik*) representing Egypt's Medieval Arab history, as well as an "okel," or covered market, featuring the work of artisans and Egypt's contemporary industry.[40] The third pavilion was dedicated to the Suez Canal and featured an elaborate model of the canal, then nearing completion, maps and various displays of information dedicated to explaining the canal's development and progress. Egypt had been invited by France to participate in the exposition and was in charge of its own design. Thus, the representation of Egypt in 1867 was in its own leader's hands and is reflected not only in the choice and style of the buildings (designed by a French architect hired by the Egyptians), but in its overall layout, which was organized along a linear street. This is in contrast to later representations of Egypt at world's fairs, which tended to reflect less "Egyptian" agency.[41]

Expressly for the Paris 1867 Exposition—and on his own initiative—Diebitsch designed and fabricated the so-called "Moorish Kiosk" (Figures 2.4 and 2.5). Given the architect's prior success at the 1862 Fair, this structure was most likely intended to showcase his work on a larger architectural scale and to further his original goal of demonstrating the universal applicability of a modern Islamic style that was a tasteful and an economical answer appropriate for the needs of a growing middle class. His kiosk, submitted under the class "Materials and methods of civil engineering, public works and architecture," was set within the Prussian section of "Le Quart Allemand," an area that was intended to feature small buildings that would exemplify a "nation's" architecture (Figure 2.6), part of an expansion of national pavilions anticipating the "Avenue of Nations" at the Paris 1878 Universal Exhibition. Since the Zollverein was no longer the dominant face of Germany at the fairs due to Prussia's continued consolidation of political power on the eve of political unification, presenters were categorized within regional political entities.[42] Significantly, Diebitsch's kiosk was one of three large structures chosen by the Prussian Exhibition committee to represent Prussia at the exhibition.

Figure 2.4 Carl von Diebitsch, *Moorish Kiosk*, Exterior view, 1867.
Source: *Illustrirte Zeitung* (27 July 1867), 68.

Figure 2.5 Carl von Diebitsch, *Moorish Kiosk,* Interior, 1867.
Source: *Illustrirte Zeitung* (27 July 1867), 69.

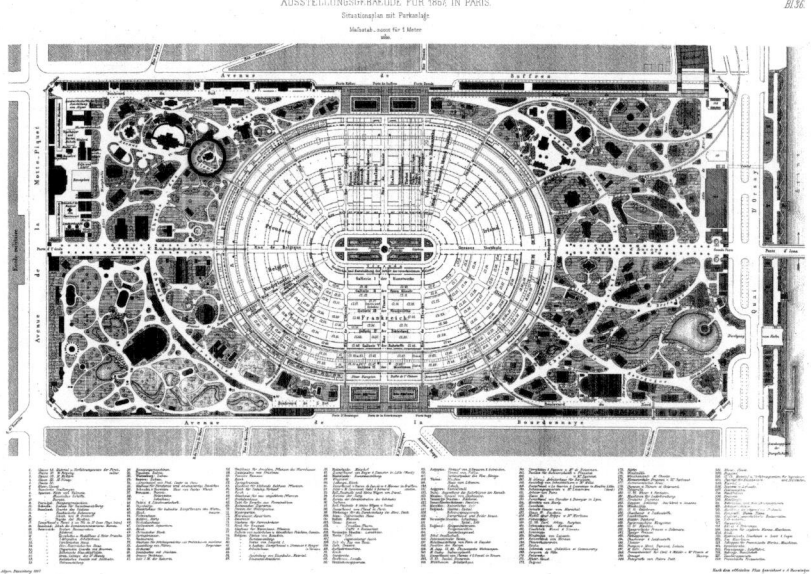

Figure 2.6 Plan of the 1867 Universal Exhibition. The kiosk is in the upper left section (circled).
Source: Heinrich and Emil Ritter von Förster, *Allgemeine Bauzeitung mit Abbildungen* (Vienna: Expedition der allgemeinen Bauzeitung, 1867), Plan 36.

The "Moorish Kiosk" was constructed mainly of cast iron with a variety of additional materials, and capped by a gilded copper dome. It is described as a "Kiosque du Bosphore"[43] by the contemporary observer Hippolyte Gautier:

> We are surprised to find in this area for northern exposures, an elegant oriental kiosk whose walls are covered with arabesque revetments, surmounted by a gilded copper dome, it is indeed a kiosk for the Bosphorus, but made in Berlin, which has passed through Paris before returning to the soil where it should definitely rise.[44]

The Kiosk was set within the "Jardin et kiosque prussiens" as described on the plan of the exposition (Figure 2.7) and its appearance was unique within the Prussian program: the style chosen for Prussia's exhibit inside the main building was a vivid polychromatic neo-Renaissance.

Similarly, the Roman imperial style—complete with triumphal arch—was chosen to showcase the Prussian machine gallery (Figure 2.8). Isabella Fehle has suggested that the kiosk be interpreted not in a "narrow ethnic sense" that one might have derived from the nearby Etruscan-styled schoolhouse, with its overt reference to the development of a renewed education system in Prussia,[45] but instead should be seen in contrast to the other more "temporary" buildings of the exposition as its own expression of current ideas in Prussian architecture. Seen in this way, without the distorting lens of historicism, the kiosk displayed an exemplary use of modern materials, durability and portability, and showcased an elaborate and original use of Prussian iron casting that was also domestically produced. Furthermore, the choice of iron over zinc

AVENUE DE SUFFREN.

| Porte Dupleix. | Agriculture (France). | Laiterie. | Hangars. | Agriculture (Suisse). | Écuries russes. | Porte Kléber. | Porte Suffren. |

.PLAN DU QUART ALLEMAND.

AVENUE DE LAMOTTE-PIQUET.

Pavillon portugais.
Pavillon espagnol.

Boulangerie viennoise.

Beaux-arts (Suisse).

Isbah.
Maison de Gust. Vasa.

Salle de réunion.

Fromage de G^d Foudre. Roquefort.

Maison autrich.

Grande braserie viennoise.

Chalet tyrolien.

Chalet norvégien.

Poteries d'H. Drasché.

Bois des forêts autrichiennes.

PLAN du QUART ALLEMAND. [dans] LE PARC.

Grand buffet omnibus

Locomotives routières.

Pavillon du Wurtenberg.

Jardin et kiosque prussiens.

Hangars et Fermes.

Agriculture.

Docks de campement.

Porte de l'École militaire.

Beaux-Arts (Bavière).

Statue de Guillaume Ier.

PALAIS.

GRANDE AVENUE D'EUROPE.

RUE DE BELGIQUE.

Figure 2.7 "Le Secteur prussien."
Source: Hippolyte Gautier, *Les curiosités de l'Exposition universelle de 1867: suivi d'un indicateur pratique des moyens de transport, des prix d'entrée, etc. avec six plans* (Paris: Ch. Delagrave et Cie., lib.-éditeurs, 1867), n.p.

Figure 2.8 "Prusse et Allemagne du Nord – Galerie des Matières Premières."
Source: *L'Exposition Universelle de 1867 Illustrée*, 10e livraison (29 May 1867), 148.

for the kiosk signified Diebitsch's own evolution, away from an artistic minded manufacturing strategy of (as seen by his 1862 submission), to a more practical, industrial, modern strategy of production. Indeed, it was the display of these newer, industrial materials, in combination with an original take on historical forms, and the end product's portability that contributed to the kiosk's *modernity*. However, as I will demonstrate, the kiosk's placement in the Prussian section, as opposed to the Egyptian one, was at once the ultimate symbol of architectural modernism for Diebitsch, as well as, ironically, the beginning of the architect's decline.

Despite Diebitsch's favorable standing with Khedive Isma'il, who insisted that he exhibit his kiosk in the Egyptian section, Diebitsch instead chose Prussia for his venue.[46] This choice is a clear indicator that Diebitsch did not see his work as only belonging to the specific context of the "Orient." Rather, he seems to have believed that this style and his work were truly international. As a result, he featured his pavilion in the Prussian section in order to reinforce the belief that this style is not limited by geography or religion, but that it is an architecture, which—fashioned in this modern way—is applicable everywhere; that is, an international architecture appropriate for a global context. But this decision had quite unexpected consequences for the architect. The move angered his most important client, the Khedive, who was at the height of his reign and wanted to showcase Diebitsch's elegant and modern kiosk as a product of his patronage on this important international stage. In fact, this tension was played out in the press, which compared Diebitsch's "delicate" (neo-Islamic) architecture (representing Prussia) to the "coarse" (neo-Islamic) architecture they saw emerging from the "Islamic countries" (never mind that the Egyptian section was largely organized by Auguste Mariette and designed by the Frenchman Jacques Drévet).[47]

This new site in the Prussian section, however, was not without its own problems. Set within a garden environment, the kiosk was removed from the main industrial sections of the Expo featuring metals and materials (Figure 2.9). The remote location meant that this "Moorish" kiosk effectively became decontextualized from its association with Prussia and its modern building industries. Furthermore, the Moorish style had become less exceptional due to a sharp rise in its popularity. This can be linked, to some degree, to the extensive and uncritical reproduction of scientifically measured drawings published in scholarly texts by individuals such as Owen Jones, Friedrich Hessemer, Jules Goury, and Girault de Prangey in the 1830s and 1840s. Indeed, what had been approached for years as an area of scholarly research was being quickly appropriated and distorted by popular culture's fascination with Islamic style. This is evidenced by the growing commodification of Islamic forms for a variety of purposes, from consumer goods to interior design, in the mid-1860s, and clearly evident at the 1867 Exposition. Even France's principal building, the Pavillon Impérial au Champ-de-Mars, was designed to look like an Ottoman tent with retractable posts, yet was capped with an eye-catching bulbous dome covered with chevron patterns and other details echoing a Mamluk precedent (Figure 2.10). The increased number of Islamic-styled pavilions and kiosks, and more prevalent casual use of the style in general, increased dramatically at the exposition, making Diebitsch's contribution—and his *modern* proposal—less distinctive and therefore less of an exemplar of modern design "reform."

Figure 2.9 "Exposition Prussienne – Pavilion Oriental."
Source: *L'Exposition Universelle de 1867 Illustrée*, 10e livraison (29 May 1867), 160.

Figure 2.10 "Pavillon Impérial au Champ-de-Mars."
Source: *Grand album de l'Exposition Universelle 1867:150 dessins par les premiers artistes de la France et de l'étranger* (Paris: Michel Lévy Frères, 1868), 13.

The modernity and internationalism of Diebitsch's Islamic architecture and design have been obscured to some degree in contemporary scholarship on Orientalism. Zeynep Çelik, for instance, ignores the 1862 exhibition altogether in her study *Displaying the Orient: Architecture of Islam at Nineteenth-Century World's Fairs* (1992) preferring to "focus on those [exhibitions] where the architectural representation of Islam was significant."[48] Nonetheless, much of what produced the bizarre architectural expressions of the later expositions was formulated many years earlier in more complex ways. And while the critique of Said's *Orientalism* (1978) has certainly led to apposite and productive work and theories, at least in its early stages, it has been superseded by more recent scholarship that has examined and exposed the vast complexities of these "colonial" and or "imperial" relationships as thoroughly entangled.[49]

For our immediate purposes it is only necessary to return to the Exposition Universelle of 1867, highlighting the fact that the Ottoman Empire, French Tunisia and Morocco, Iran, and Egypt all had substantial displays at the fair, rendered in various Islamic styles. In addition to these were a variety of individual pavilions and kiosks with Islamic elements serving purposes from cafés to tobacco sales. As a result of the ubiquity, proliferation, and variety of these Islamic styles at the fair, Diebitsch's kiosk, and subsequently his all too brief success in Egypt, was suddenly obscured by what had become a somewhat crass and popular style trend in Europe. Indeed, how could a visitor to Diebitsch's pavilion—intended to exemplify modern Prussian industry by showcasing the latest ideas of efficiency through modern materials and methods of production—tell the difference between his aim and that of, for example, Alfred Chapon's elaborate palace pavilion, a copy of the fifteenth-century Bardo Palace, for the Bey of Tunis, (see Figure 1.6) whose decorative scheme was described as "endlessly luxurious and reminds one of the monuments of Spanish Arabian art of the fourteenth century," or the Turkish café for that matter?[50] Indeed, there were differences in their construction, assembly, and overall impression. For Diebitsch, the kiosk represented the portability, flexibility, and efficiency of modern building production, whereas the Bardo, with its luxurious and well-known "traditional" materials such as "[b]rilliantly colored tiles sent from Tunisia" and "beautiful carpets and embroideries; gold-painted elements," represented a "static," "traditional" orient.[51] Thus, the Bardo fits more appropriately within the view that such forms and ornament communicated particular cultural associations and assumptions about the "Orient" to Western audiences.

In terms of Diebitsch's two different exhibition submissions one might assume that the incorporation of an "arabesque" on Diebitsch's 1862 zinc vase is somehow an unremarkable, or perhaps an ironic, orientalist feature. Or, one could interpret it simply as historicism, like its Greek or Gothic Revival counterparts. Yet, there remains a certain complexity about the way in which Islamic forms were understood, studied, and ultimately utilized by practicing architects in the nineteenth century, certainly prior to the association of these forms with exoticism, entertainment, and empire. This inconsistency has been explored, for the most part, in the vast literature produced in the wake of Edward Said, which (among other things) notably linked Orientalism with the imperial-colonial project. In this context, however, the postcolonial critique leaves us with more questions than answers, especially after we leave the realm of the well-studied empires of France and Britain, which have garnered the most critical attention. In response to recent studies on the history of German Orientalism,[52] we can regard Diebitsch's work in a context that situates his work not within an Orientalist genre, but within the history of modern architecture. Diebitsch deserves to be understood as a modern architect and designer whose chosen "path" was, in the end, not the one "chosen" by later historians who shaped the narrative of the Modern movement. In this way he emerges as an active participant in a more complex and certainly less linear trajectory of modern architecture.

Conclusion

In this brief discussion of Diebitsch's career and examination of his entries for the 1862 and 1867 World's Fairs, my intention has been to advance the proposition that it was possible to produce a structure in the 1860s that can be considered simultaneously "modern" and "Islamic." I have refrained from labeling Diebitsch's work "Orientalist" for the particular reason that not only is his oeuvre not so easily classifiable, but also because such classifications do not do it justice and, in this case, are simply unhelpful. Diebitsch, like any other curious and ambitious architect, discovered an approach to architecture with which he identified and chose to pursue and adapt to the modern building industry. He was an independent practitioner who met, and was eventually hired by, a sovereign ruler of a country whose only colonial association at the time was the fact that it was still "technically" a province of the Ottoman Empire. In the end Diebitsch appears to be an architect who sought to have a successful practice designing in a style he believed was the answer to the most important question in nineteenth century architectural culture: *In what style should we build?*[53]

Diebitsch's use of "arabesques" and other "Islamic" motifs therefore should not be seen in a reductivist sense that interprets these motifs merely as orientalist whims that potentially reify an imperial worldview. Rather, they may be seen as evidence of a designer employing Islamic design motifs, in combination with new technologies, as part of a broader trend of a politically motivated modernization that had a particular meaning in both a German and international context. Diebitsch, as we have seen, was himself a product of the Berlin architectural tradition and a beneficiary of the government's industrial-political advocacy. Over time, he developed a distinctive, more abstract style, which challenged the entrenched Neoclassical and Gothic styles with a different approach that he believed was not only ultimately derived from the same sources as the Neoclassical and Gothic (the ancient Greeks), but that lent itself to modern materials and more efficient and advantageous means of production. Moreover, he recognized and developed the tendency he saw in Islamic architecture further toward abstraction by streamlining its details and mass-producing its elements. This significant work by Diebitsch—from his use of modern materials, methods of production, means of transportation, and challenge to Europe's entrenched styles—not only anticipated key features of the Modern movement in architecture, but his work ultimately complicates and challenges our understanding of what is considered "modern" design.

From his first international debut of the zinc vase in 1862, to his 1867 kiosk, which realizes architecturally the ideas nascent in the vase, Diebitsch's conflation of arabesque ornament with Western arts and industries represents his desire to modernize contemporary architecture. Thus, Diebitsch's original use of Islamic forms and ornament in the end represents a form of modernism that was subsequently conflated within a sea of more purely historicizing "Moorish" revivals. Despite the rift between

Khedive Isma'il and Diebitsch over the placement of the 1867 pavilion at the Exposition, the architect nevertheless returned to Cairo after the exhibition to dedicate himself for two more years to what would become some of his most important work and the fullest expression of his ideas, all owing to the patronage of the Khedive. However, it was the international forum created by the world's fairs that enabled his designs to appear in a context appropriate to the conception and reception of his oeuvre.

Notes

1 *Amtlicher Bericht über die Industrie- und Kunst- Ausstellung zu London im Jahre 1862, erstattet nach Beschluß der Kommissarien der Deutschen Zollvereins-Regierungen*, XVI. Heft. (Berlin: Verlag der Königlichen Geheimen Ober-Hofbuchdruckerei, 1864), 340.
2 The official entry is: "2033 Diebitsch, C. von, 4, Hafenplatz Berlin.—Zinc vases." *International Exhibition 1862, Official Catalogue of the Industrial Department* (London: Truscott Son & Simmons, 1862), 280. It is unlikely there were two vases as there is no evidence besides this single comment that there were any more than one vase. Also, the vase (like many of his smaller pieces) was probably cast in Berlin at the Ravené foundry. See Elke Pflugradt-Abdel Aziz, "Orientalism as an Economic Strategy: The Architect Carl von Diebitsch in Cairo," in Mercedes Volait, ed., *Le Caire-Alexandrie architectures européennes, 1850–1950* (Cairo: IFAO, 2001), 6.
3 *Amtlicher Bericht über die Industrie- und Kunst- Ausstellung zu London im Jahre 1862*, 340.
4 "Kunst-Chronik," *Die Dioskuren: Deutsche Kunstzeitung* 7, no. 17 (27 April 1862): 132.
5 For the full list of 1862 award winners see *International Exhibition, 1862. Medals and Honourable Mentions awarded by the International Juries with a list of Jurors, and the Report of the Council of Chairmen*, 2nd ed. (London: Geo. E. Eyre and Wm. Spottiswoode Printers, 1862), 340–341.
6 Arthur Channing-Downs, Jr., "Zinc for Paint and Architectural Use in the 19th Century," *Bulletin of the Association for Preservation Technology* 8, no. 4 (1976): 80–99. 82f.
7 According to an article in *The Builder*, an "M. Geiss of Berlin" was casting zinc "for architectural and decorative purposes not hitherto employed amongst us,—namely, *cast*. It appears that for seventeen years, zinc has thus been used in Berlin for architectural purposes, namely, for all exterior as well as interior ornamental parts of buildings, which, by casting, can be produced in the sharpest forms and are said to be at the same time capable of resisting all influence of weather." See "Cast Zinc in Decoration," *The Builder* (28 July 1849): 353.
8 Channing-Downs discusses the dramatic rise in use of zinc in Berlin in the 1830s and cites Karl Friedrich Schinkels description of its benefits. For Channing-Down's discussion of this, see Channing-Downs, Jr., "Zinc for Paint and Architectural Use in the 19th Century,"
9 "Kunst-Chronik," *Die Dioskuren: Deutsche Kunstzeitung* 7, no. 17 (27 April 1862): 132.
10 "the fine play of lines in Arabic ornament realized with a sure hand in an original way..." (*Amtlicher Bericht* cited in Fehle, s.108).
11 "Kunst-Chronik," *Die Dioskuren: Deutsche Kunstzeitung* 7, no. 17 (27 April 1862): 132.
12 Isabella Fehle. *Der Maurische Kiosk in Linderhof von Karl von Diebitsch: Ein Beispiel für die Orientmode im 19. Jahrhundert* (München: Kommissionsverlag UNI-Druck, 1987). And: Pflugradt-Abdel Aziz, "Orientalism as an Economic Strategy: The Architect Carl von Diebitsch in Cairo," 7.
13 For an analysis of the relationship between the Zollverein and Britain see John R Davis, *Britain and the German Zollverein, 1848–66* (New York: St. Martin's Press, 1997), 35ff.
14 *International Exhibition, 1862*, 340.
15 *Amlichter Bericht über die Industrie- und Kunst- Ausstellung zu London im Jahre 1862*, 340.
16 ["Dr. L. (zu Berlin)"] "Bauwissenschaftliche und Kunst-Nachrichten: Siebente Versammlung deutscher Architekten und Ingenieure zu Braunschweig vom 26. Bis 29. Mai 1852. Bericht von Herrn Dr. L. zu Berlin." *Zeitschrift für Bauwesen* 2, (1852): 334.
17 Ibid.

18 This idea that the Islamic style is appropriate for the modern age is conveyed in the records of Diebitsch's lectures described earlier and is also recounted and elaborated upon in: Pflugradt-Abdel Aziz, "Orientalism as an Economic Strategy: The Architect Carl von Diebitsch in Cairo," 3–22.

19 Matthew Digby Wyatt, "Orientalism in European Industry," *Macmillan's Magazine* 21, (1870): 552.

20 David Landes, *Bankers and Pashas: International Finance and Economic Imperialism in Egypt* (Cambridge, MA: Harvard University Press, 1979), 114f.

21 No one style dominated to the exclusion of others. Due to the diverse range of inhabitants and foreigners, including the viceroys themselves, Cairo buildings embodied many styles, especially prior to the 1860s. For a more thorough discussion of this see Mercedes Volait, *Architectes et architectures de l'Égypte moderne (1830–1950): genèse et essor d'une expertise locale* (Paris: Maisonneuve et Larose, 2005), 157–161.

22 Pflugradt-Abdel Aziz, "Orientalism as an Economic Strategy: The Architect Carl von Diebitsch in Cairo," 12–13.

23 Mohamed Scharabi, *Kairo: Stadt und Architektur im Zeitalter des europäischen Kolonialismus* (Tübingen: Ernst Wasmuth Verlag, 1989), 63.

24 Mohammad A. Chaichian, *Town and Country in the Middle East: Iran and Egypt in the Transition to Globalization, 1800–1970* (Lanham, MD: Lexington Books, 2009), 159–160.

25 The Khedive also commissioned the composer Giuseppe Verdi to write the opera *Aida*, which premiered in Cairo in 1871; cf. Mary Jane Phillips-Matz, *Verdi: A Biography* (London and New York: Oxford University Press, 1993), 570–573.

26 Carl Ohnesorge's journal and collection of letters was published later in his life by his children for his 70th birthday. He was Diebitsch's assistant while in Egypt and managed a number of his projects. See Carl Ohnesorge, *Orientalische Skizzen: Unsers Vaters Erinnerungen an sein Arbeiten un Wandern im Orient 1863–65 zu seinem siebzigsten Geburtstage dem 17. Juli 1908.* (Collected and presented by his children) (Magdeburg: A Wohlfeld, 1908), 5, 18.

27 Ohnesorge, *Orientalische Skizzen*, 5.

28 Among Diebitsch's clients were the famous Minister of Public works, and then Prime Minister of Egypt, Nubar Pasha (1825–99), the three-time Egyptian Prime Minister Sharif Pasha (1826–87) (the nephew of the illustrious general Suleyman Pasha al-Faransawi [1788–1860]), and his most famous client of all, the Khedive of Egypt, who, no doubt, gave Diebitsch the commission that best represents the apex of his work in Egypt—the Gezira Palace Pavilion.

29 Elke Pflugradt-Abdel Aziz, "Der Preußische Palast in Ägypten," in Wolfgang G. Schwanitz, ed., *125 Jahre Sueskanal: Lauchhammers Eisenguss am Nil* (Hildesheim and New York: G. Olms, 1998), 71.

30 Ohnesorge, *Orientalische Skizzen*, 5.

31 Pflugradt-Abdel Aziz, "Orientalism as an Economic Strategy," 10.

32 Ibid. Fahmy notes that it is highly unlikely that Sève achieved the rank of colonel in the French army before coming to Egypt but he does not prove it.

33 Ohnesorge, *Orientalische Skizzen*, 5.

34 Gabriel Guémard, "Le Tombeau et les 'armes parlantes' de Soliman Pacha," *Bulletin de l'Institut d'Egypte* 9, (1927): 72–73.

35 Ohnesorge, *Orientalische Skizzen*, 6ff.

36 Jürgen Osterhammel, *The Transformation of the World: A Global History of the Nineteenth Century* (Princeton, NJ: Princeton University Press, 2014), 710–711.

37 N.B.: Isma'il Pasha (1830–95, r. 1863–79), was initially the "Viceroy" of Egypt upon his accession to the throne. The title "Khedive" was bestowed upon him by the Ottoman Sultan through a *firman* (official decree or royal mandate) dated to 8 June 1867, which came just a week or so prior to the opening of the Egyptian Pavilion at the 1867 Expo in Paris and his trip there to participate in the celebration to open it. A detailed discussion of this can be found in: Georges Douin, *Histoire du règne du khédive Ismaïl* (Vol. 1, Les premières années du règne, 1863–1867) (Rome: Nell'Istituto poligrafico dello stato per la Reale società di geografia d'Egitto, 1933), 421f & 442 (for the *firman* itself).

38 Panayiotis Jerasimos Vatikiotis, *The History of Egypt,* 3rd ed. (Baltimore, MD: The Johns Hopkins University Press, 1985), 83.

39 Edmond About, *Le Fellah* (Paris: Hachette, 1883), 38. Cited in: Douin, *Histoire du règne du khédive Ismaïl,* 1.

40 Douin, *Histoire du règne du khédive Ismaïl,* 1–2.

41 Namely that of Egypt's representation at the 1889 Expo. See Zeynep Çelik, *Displaying the Orient: Architecture of Islam at Nineteenth-Century World's Fairs* (Berkeley, CA: University of California Press, 1992), 111ff.

42 It is also relevant to mention the fact that Austria was not represented at the Expo because they had just suffered a significant defeat at the hands of Prussia in the 1866 Austro-Prussian War.

43 This is not to be confused with the other *"Pavillon du Bosphore,"* which was set within the Ottoman exhibit at the same Expo.

44 Hippolyte Gautier, *Les Curiosités de L'Exposition universelle de 1867* (Paris: CH. Delagrave et Cie, 1867), 44. "On est tout étonné de trouver dans cet espace réservé aux expositions septentrionales, un élégant kiosque oriental aux murs revêtus d'éclatantes arabesques, surmontés d'un dôme en cuivre doré; c'est en effet un kiosque destiné aux rives du Bosphore, mais fabriqué à Berlin, et qui a passé par Paris avant de se rendre sur le sol où il doit définitivement s'élever."

45 Fehle, *Der Maurische Kiosk in Linderhof von Karl von Diebitsch,* 114–115.

46 Ibid., 155. Cornelia Köster and Wolfgang Schwanitz, "Kunsthistoriker, Architekten, Ingenieure und Manager: Ein Gespräch mit Elke Pflugradt-Abdel Aziz," in Wolfgang G. Schwanitz, ed., *125 Jahre Sueskanal: Lauchhammers Eisenguss am Nil* (Hildesheim and New York: G. Olms, 1998), 48.

47 Köster and Schwanitz, "Kunsthistoriker, Architekten, Ingenieure und Manager," 48. Also with regard to the architects of the Egyptian pavilions see Çelik, *Displaying the Orient,* 111. Çelik also associates Drévet with the name of another architect she identifies as 'E. Schmitz' but she only identifies him by the initial of the individual's first name and is only mentioned once.

48 Çelik, *Displaying the Orient,* 3.

49 This is directly evidenced by the work of both the supporters and detractors of Said such as Dipesh Chakrabarty (*Provincializing Europe,* 2000), Zachary Lockman (*Contending Visions of the Middle East,* 2009), Ibn Warraq (*Defending the West,* 2007) and Robert Irwin (*Dangerous Knowledge,* 2008) et al. However, a great deal of other work indirectly deals with many of these issues as well, for example Kwame Anthony Appiah and Mahmood Mamdani et al. to name just two.

50 *Revue générale de l'Architecture,* Vol. 26 (1868), 271f. Cited in: Stefan Koppelkamm, *Der imaginäre Orient: Exotische Bauten des achtzehnten und neunzehnten Jahrhunderts Europa* (Berlin: Ernst & Sohn Verlag, 1987), 143.

51 Çelik, *Displaying the Orient,* 123.

52 Namely: Suzanne Marchand, *German Orientalism in the Age of Empire* (Cambridge: Cambridge University Press, 2009), Nina Berman, *German Literature on the Middle East: Discourses and Practices, 1000–1989* (Ann Arbor, MI: University of Michigan Press, 2011), Ursula Wokoeck, *German Orientalism: The Study of the Middle East and Islam from 1800 to 1945* (London: Routledge, 2009), as well as Andrea Polaschegg, *Der andere Orientalismus: Regeln deutsch-morgenländischer Imagination im 19. Jahrhundert* (Berlin: W. de Gruyter, 2005), Sabine Mangold, *Eine "weltbürgerliche Wissenschaft" – Die deutsche Orientalistik im 19. Jahrhundert* (Stuttgart: Steiner, 2004), Todd Kontje, *German Orientalisms* (Ann Arbor, MI: University of Michigan Press, 2004), and Hodkinson and Morrison's edited volume of essays in *Encounters with Islam in German Literature* (Rochester, NY: Camden House, 2009).

53 See Heinrich Hübsch, *In welchem Style sollen wir bauen?* (Karlsruhe: C.F. Müller, 1828).

3 The Belgian reception of Italy at the 1885 Antwerp World Exhibition

Converging artistic, economic, and political strategies on display

Daniela N. Prina

Introduction

The multifaceted context of the relationships between Italy and Belgium, before and after the establishment of the Kingdom of Italy in 1861, has been the subject of renewed interest for historians who have investigated the broader cultural relationships between the countries as well as more narrow topics such as Belgian participation in Italian cultural events including the Venice Biennale.[1] The general picture offered by these studies is that Italy viewed Belgium as a sort of experimental "laboratory" from which to draw inspiration and to encourage or legitimize its own future, while Belgium considered Italy a developing country; a territory to conquer, whose own political, social, and economic identity was rarely acknowledged.[2] These views stem from a number of perceived differences between the two countries that developed during the nineteenth century. After achieving national independence in 1830, liberal Belgium rapidly adopted a stable institutional structure and reached a level of economic development that established the country among the more advanced European powers. In a pre-unified Italy struggling with industrial backwardness, the formation of a Belgian national state was therefore considered an aspirational model. Cultural exchanges—adopted on the Belgian side through the tradition of the *voyage en Italie*[3]—began to be forged around political, economic, and social issues. While earlier historiography has often studied these links in isolation, this essay will try to reunite the artistic, political, and social dimensions of these interactions through the examination of the debates that took place in Belgium and Italy during the second half of the nineteenth century concerning the relationship between art and industry, specialized training, and the quality of industrial products,[4] which were pressing issues across all of Europe at the time.[5] From an historical perspective, these debates are symptomatic of the positions that the governments of these young nation-states forged in relation to their artistic history and their cultural roots as they envisioned and launched processes of political and productive modernization.

Decorative arts, industry and the reforms in Italy and Belgium in the mirror of national exhibitions

In both Belgium and Italy, the reform and promotion of the decorative arts began in the first half of the century, prior to independence. In both, unification strengthened national awareness of the relationship between art and industry—an issue that remained central to the economies of each country. The problem of the union between art and industry, between quality and quantity, stood at the core of debates within the artistic, productive, and commercial sectors of each nation throughout the second half of the nineteenth century. A pivotal point of the discussion was the potential educational and operative utility of the arts, engendering attempts to establish new artistic organizations to educate consumers and promote standards of "good taste." This discourse is essential for the understanding of specific changes both in the economic growth of Belgium in the nineteenth century and in Italy's industrial development.

Conscious of its own slow industrial expansion, from the 1870s the Italian government carefully examined how foreign countries, and especially Belgium, were tackling the challenges posed by industrial growth.[6] Through industrial, political, and diplomatic relationships[7] as well as participation in international conferences, exhibitions, and world's fairs, Italy maintained close contact with several Belgian institutions, eager to confront economic realities and pursue reforms as effectively as possible.

After unification in 1861, Italy had to manage new and intense pressures stemming from the political and institutional changes that the young nation faced. Unification did not automatically bring economic integration, for Italy suffered serious financial problems as well as political, social, and economic fragmentation, a result of the diversity of the pre-unified independent states. However, in the fifteen years that follow the party known as "Historical Right" was able to build a centralized juridical and institutional framework and to solve the conflict between the State and the Church through the Law of Guarantees (1871).[8] During this time, the recently established nation-state, marked by a moderate liberal bourgeoisie of large landowners, industrialists, and military leaders in search of strong and rooted identities, tried to reorganize and merge into an integrated system the different local educational and productive sectors (such as ceramics, glassware, wood sculpture and wood carving, weaving, stone cutting, and many others) that had secured the wealth of the various independent Italian states.[9] Each Italian region had indeed developed one or more specific sectors of artistic production depending on the local availability of raw materials, establishing viable traditions based upon commissions from sovereigns and nobility, passed down from generation to generation.[10]

As part of this political project, Italian government emissaries were often sent to attend international meetings and events where the relationship between art and industry and educational reforms in schools of applied art were debated; others were tasked with fact-finding missions, aimed at studying and reporting on the progress made by foreign countries in the industrial and decorative arts, in order to examine which initiatives had been taken to improve both education and production.[11] While debates on art and industry occurred in several countries (including England and France), Belgium undoubtedly stood atop the list of countries of interest to the Italians, due to its renowned industrial capacities, artistic traditions, and the introduction of national educational reforms, initiated by an *ad hoc* commission known as the *Conseil de perfectionnement de l'enseignement des arts du dessin* (hereafter the *Conseil*), founded in 1859.[12] In 1880 Baldassarre Odescalchi (1844–1909),[13] a member of the Roman patrician society, connoisseur of architecture, active politician, and promoter of the capital's Museum of Industrial Art (founded in 1873),[14] was tasked by the Ministry of Agriculture, Industry, and Commerce (hereafter MAIC) to visit the South Kensington Museum in London.[15] During his mission, Odescalchi together with his secretary, Raffaele Erculei,[16] also visited France and Belgium. The visit to Belgium occurred during a special event, the 1880 Jubilee Exhibition held in Brussels, in a former military ground transformed for the occasion in a new green area named *Cinquantenaire* Park.[17] The exhibition building, a sprawling classically inspired winged structure culminating in the triumphal arch, flanked to each side by a colonnade, and surmounted at each end by a modern glass and iron vault, reflected the ambivalence of the Belgian state, divided between the support for modern industrial growth and the need for self-representation through didactic and easily recognizable architectural models (Figure 3.1). With this event, modelled on a world's fair and celebrating the 50th anniversary of Belgian independence, the government sought to strengthen a progressive vision of the youthful but already powerful nation, and to renew ties with its great artistic heritage. Available funds were mainly directed to a section of the exhibition known as the *fête industrielle*, which included two displays—one focused on contemporary industrial arts, the other on historical decorative arts—both accompanied by exhaustive catalogues.[18] Industrial and decorative arts were therefore presented at the Jubilee Exhibition as visual expressions of the past, present, and future of the nation, renewing and expanding upon a venerable Belgian artistic tradition. The connection between artistic research and its practical outcomes had already been promoted by the government in Belgium through the creation of institutions such as the Royal Commission for Monuments (1835) and the *classe des Beaux-arts* of the Belgian Royal Academy (1845), which encouraged studies of Belgium's national artistic heritage. The research published to illustrate the historical decorative arts presented at the Jubilee Exhibition stressed the relationship between tradition and modernity, emphasizing those elements that demonstrated the successful combination of art, industry, architecture, and national identity.

Figure 3.1 The exhibition building erected in the *Cinquantenaire* Park for the 1880 Jubilee Exhibition.

Source: Reproduced from Franz Herla, "*Exposition Nationale 1880. Album commemoratif*" (Bruxelles, Cie de Publicité et d'Emission, 1880) © Université de Liège.

A similar rhetoric animated the report published by Odescalchi and Erculei upon their return to Italy.[19] The report supported Italian educational, artistic, and industrial policies: in 1876 the government, led by the regionally fragmented liberal-conservative Historical Right, had been ousted and replaced by the Left Party, whose first measures were aimed at eradicating illiteracy.[20] The new government also tackled the development of artistic awareness among the working classes, starting with the renewal of the professional education system, and the reorganization of the Academies of Fine Arts—accused of an incapacity or unwillingness to relate their teaching to the cultural, social, political, and economic realities of post-unification Italy.[21] The Italian decorative arts displays at the 1878 world's fair in Paris had rested exclusively on a highly specialized production of luxury furniture and were criticized as lacking inventiveness and practicality.[22] This perceived failure at the Paris exposition was the pretext to look closely at institutional models in other countries that might be adapted and imported into Italy. Among the three countries studied by Odescalchi, the Belgian case was particularly interesting for Italy: although the two young nation-states were experiencing industrialization differently, they were both building a unified identity, expressed through high moral and artistic values. The national artistic tradition—the glorious fifteenth and sixteenth centuries in Belgium[23] and the Renaissance in Italy[24]—were seen as the foundation for a renewed artistic and industrial educational system created through laws and the establishment of new institutions. Moreover, the 1880 Jubilee Exhibition included plans for the eventual allocation of a portion of the buildings as permanent premises for a new museum of industrial arts (opened in 1889), which was intended to bridge the gap between the major and minor arts, applying the latter to industrial production.[25] Likewise, a Congress on education took place during the exhibition (Italy was represented by engineer and inspector for industry at the MAIC, Oreste Lattes), demonstrating the degree of attention that Belgium was paying to the impact that social and industrial policies were having on design reform efforts.[26]

These Belgian initiatives coincided with the Italian aim to recognize Rome,[27] the political capital, as the new national cultural center, following the example of other European nations.[28] Such efforts at centralization failed, however, to supersede Italy's vigorous and well-defined local cultural model. The plan to move all the political, economic, and cultural institutions to Rome with the creation of an "Italian Museum of Industrial Art" was hindered by local resistance and alternative initiatives. Indeed, in the same years, thanks to the protectionist action of the *Cairoli* and *Miceli* ministerial circulars (inviting local administrations to broaden the number of schools of applied arts by helping them with the government's subsidy of two-fifths of the annual expenses of the schools and leveling out the study programs), the educational institutions and local museums of applied arts of the major regional towns were strengthened, becoming active centers for the development of their own local productions.[29] As a result, between 1876 and 1882 five schools of applied arts were established in Venice, Naples, Milan, Palermo, and Florence. In Venice, Naples, and Milan, following the successful foreign example of the South Kensington Museum, these schools were connected with local industrial museums, thus responding to the government's desire to control and develop the industrial system through the education of artisans and workers.[30] This move allowed these schools to grow independently, and their productions were recognized at national exhibitions such as the *Esposizione Nazionale* in Milan in 1881,[31] and the *Esposizione Generale Italiana* in Turin in 1884,[32] dedicated to the progress of industry and to high production standards.

Cultural heritage and the development of industrial production

Organized in the picturesque park of the Valentino Castle, Turin's industry-centered exhibition was directed toward both education and entertainment. Among the variously designed pavilions and kiosks located in the gardens and illuminated by electric light, the sober *Padiglione del Risorgimento*—dedicated to the exposition of objects and documents illustrating Italy's unification process with long galleries displaying industrial productions and a theatre for cultural events and celebrations—represented focal points showcasing a modern Italy that could compete with more advanced European nations. The country's initiatives in many productive sectors were in turn noticed by Belgian visitors, who described the exhibition as "brilliant," adding that Italy "was a great country which had provided itself with a complete and perfected apparatus," and was "showing itself as an adversary to be considered in the near future."[33] The 1884 exhibition also demonstrated that one of the most important goals for late nineteenth-century Italy was to retrieve and reassert its artistic and cultural supremacy, a position Italy had not held among European neighbors since the seventeenth century.

Appreciated for its historical heritage, Italy wanted to build a strong national consciousness using art, in particular the scientific and philological study of the styles of the past, to shape a future national style in architecture and the decorative arts. At the Turin exhibition, particular attention was paid to the arts of the fifteenth century that were displayed in a special section called "Art History." Designed as the result of a thorough archaeological investigation by Alfredo d'Andrade,[34] a pioneer in the promotion and the teaching of industrial arts, this section, set up on the river banks in Valentino Park, offered an idealized reconstruction of a *Borgo Medievale* (Medieval Village) composed of faithful copies of existing buildings from the Piedmont and Valle d'Aosta regions (Figure 3.2). Thus, the area chosen to symbolize artistic unity was neither Florence nor Rome, renowned centers of Italian civilization in the Middle Ages and Renaissance, but the small towns on the slopes of the Alps. Conferring on these towns an artistic dignity equivalent to the larger cities implied a broader consideration of Italian artistic and cultural heritage. This exact scale-model reproduction functioned as a sort of three-dimensional *Dictionnaire Raisonné*, following Viollet-le-Duc's example,[35] aimed at revealing every aspect of medieval civilization through its architectural productions, decorative arts, and material culture. The village included a series of active workshops (pottery, blacksmithing, coppersmithing, weaving, carpentry, etc.) that enabled the visitor to observe the traditional phases of production of manufactured goods. The stylistic model of the fifteenth century extended, therefore, its formal coherence and unity of style to the products of art and industry, thus becoming a model for nineteenth-century architecture as well as for a modern vision of a qualitative industrial production.[36] The educational purpose of the *Borgo Medievale* was conclusively achieved when the village was donated by the Town Hall to the Civic Museums in Turin (founded with the intent to promote the history of art applied to industry, as well as to the fine arts) which subsequently transformed it into a separate branch of the Museum.

The valorization and protection of Italy's later medieval heritage was one of the consequences that the picturesque Artistic Section of the 1884 Exhibition had on the Italian cultural and political scene. With a Royal Decree of 29 November 1884, the first Royal Delegations for the conservation of monuments were formed under the auspices of the Ministry of Education,[37] and a permanent institution devoted to managing

Figure 3.2 Alfredo de Andrade, *Borgo Medievale* built for the 1884 Exhibition in Turin.
Source: © The Author.

and controlling the application of art to industry, the Central Commission for Artistic and Industrial Education (1884–1908; hereafter *Commissione*), was established with Royal Decree of 23 October 1884 under the guidance of the MAIC.[38] The *Commissione* was tasked with the improvement of industrial art museums, the organization of schools of applied art and, finally, the extension to primary schools of compulsory teaching of drawing. This move allowed Italy to be perfectly in step with parallel developments in the art-industry debate elsewhere in Europe.[39]

From Turin to Antwerp: The Italian participation to the first Belgian world's fair in 1885

Another consequence of Turin's exhibition was the acceptance of Belgium's invitation[40] to participate in its first world's fair, held in Antwerp in 1885.[41] After a series of disappointing performances at previous world's fairs, the favorable reception of the *Esposizione Generale Italiana* in Turin encouraged the Italian government to join the Belgian event, for it represented an ideal occasion to re-launch and re-brand Italian artistic and industrial production. Moreover, the government endeavored to coordinate its participation, in order to avoid the criticism the nation received at the Paris world's fair of 1878 and at the 1883 international exhibition in Amsterdam, where state sponsorship had not been fully granted.[42] Between 1878 and 1885 the Italian government had indeed considered it unnecessary to participate at international exhibitions, judging that a too assiduous attendance over a short period of time would have reduced these events' impact—for it was difficult to show progress and changes in industrial productions if such displays were organized too often. The government's abstention, however, could not prevent individual Italian manufacturers from participating in international exhibitions during this period. Therefore some government action was taken in order to assure the communication of essential information related to the exhibitions. In some instances the government underwrote the transportation or the installation costs of participation, though not as part of a national display strategy. This resulted in an often unbalanced presentation of Italian production, resulting in an unsuccessful critical response. Moreover, not being officially represented by their government, the Italian producers could not count on the presence of Italian jury members in the commissions responsible for awarding prizes.

With the rise of Italy's industrial production in 1885 the opportunity to participate in a new world's fair seemed particularly relevant to the government, and the great importance of Antwerp as a trading center offered the opportunity to broaden connections with other countries as well as to open new markets for national production.[43]

Like most European countries at the time, Belgium was affected by an economic recession, ultimately lasting from 1874 to 1895 following a period of strong economic growth in the third quarter of the century; however, Antwerp remained a leading trading center in Europe. Its importance had rapidly grown and was expected to increase due to the ambitious reorganization of the harbor, which straightened the quays for three and a half kilometers along the Scheldt, and included the construction of a new dock for inland navigation to the south of the city. The official opening of the new quays, completed in 1885, provided an unprecedented opportunity to organize a world's fair in Belgium's commercial and artistic capital. The exhibition was organized under the aegis of the State, but, in the spirit of Belgium's liberal political tradition, realized with private initiatives undertaken by three members of Antwerp's commercial bourgeoisie: civil servant Eduard Colinet, shipbroker and town councilor Frans Gittens, and architect Jan Laurent Hasse. By 1885, the exhibition site—thirty hectares freed up by the demolition of the old Citadel—was cleared out, and connected by tram lines to the harbor and the city center. Moreover, the year 1885 coincided with two important events: the recognition of King Leopold II's possession of the Congo Free State, publicized with a pavilion in the exhibition gardens,[44] and the triennial Fine Arts Salon in Antwerp, organized by the Société royale d'encouragement des Beaux-arts. The fair (opened from 2 May to 2 October 1885) thus promoted the new seaport, commerce and art, and identified Antwerp as a "Metropolis of Trade and Art,"[45] combining the modern and dynamic face of the city with its great historical and artistic past. This message was also conveyed by the architecture of the fair. Entrusted to Gédéon Bourdiau, who had also designed the glazed halls for Brussels' Jubilee exhibition in 1880, the monumental entrance gate, an ephemeral construction bearing symbols of maritime power, was covered with stucco and wooden plaques, whose iron skeleton as well as the structure of the Machinery Halls and Galleries alluded to local identity and the talents of Belgian industry[46] (Figure 3.3). The exhibition also included a garden designed by Louis Fuchs, complete with waterfall fountains and follies that hosted the tourist pavilions, displays organized by private companies, objects from the Congo Free State, and French and Portuguese colonial pavilions.

Situated inside the Work Hall, in the *Galerie Principale*, in a narrow but well-situated space of 3,778 square meters (an extra 411 square meters were acquired by the firm Ferro of Murano for the installation of a glassware workshop that resulted

Figure 3.3 Gédéon Bourdiau, Monumental entrance gate at Antwerp's Exhibition, 1885.
Source: Reproduced from René Corneli, Pierre Mussely, *Anvers et l'exposition universelle de 1885* (Anvers: Typographie Bellemans Frère), 111 © Artesis Plantijn Hogeschool. Koninklijke Academie voor Schone Kunsten Antwerpen, Antwerp.

in one of the most popular attractions of the fair), close to the Belgian, French, and Austrian sections, was the Italian section. Its entrance was lavishly decorated with Renaissance-inspired motifs by the renowned engineer Camillo Riccio, the designer of the 1884 exhibition buildings in Turin, whereas its internal display was refined and sober (Figure 3.4). The space allotted to the 665 Italian producers, mostly chosen from among those who garnered prizes at the Turin exhibition,[47] was restricted (each exhibitor was granted a site only 8 m wide by 4 m high) but skillfully arranged by Riccio. According to one writer, the Italian section was one of the rare ones in which "the fitting was done in a rational order, without any damage to the general view."[48] It was also designed to impress visitors, who immediately realized that they were "entering the country of classic arts."[49] As seen in Turin, the study and re-elaboration of historical models was used as an instrument to build a broad idea of modernity that could contribute to forging a national, cultural, and social identity, and give impetus to industrial production. The later Middle Ages and the Renaissance (until the 16th century, as revealed, for instance, in the furniture production presented by

Figure 3.4 Camillo Riccio, The Italian section at Antwerp's 1885 Exhibition.
Source: Reproduced from René Corneli, Pierre Mussely, *Anvers et l'exposition universelle de 1885* (Anvers: Typographie Bellemans Frère), 245 © Artesis Plantijn Hogeschool. Koninklijke Academie voor Schone Kunsten Antwerpen, Antwerp.

the manufacturer Zanetti), historical eras in which differences between major and minor arts were less pronounced than in later periods, became the privileged models for an extremely diversified industrial production of artistic quality that reflected beauty, harmony, and luxury. The products presented in Antwerp also revealed inspiration from older Italian past artistic traditions (Figure 3.5). Great attention had been given to the manufacture of silks, jewelry, ceramics, glassware, bronzes, and furniture. It was deemed a "triumph of art applied to industry."[50] Commentators remarked that Italy had improved not only its export of agricultural or mineral products (which until then had represented its chief strengths), but also its manufactured products.[51] Indeed, thanks to the strategic arrangement of the exhibition space, the variety of its products, the inclusion of a fully functioning glassware workshop, and the publicity it received, the Italian contribution to the Antwerp fair achieved remarkable success: its section was not only one of the most visited, but also one of the most rewarded.[52]

The prize system can be considered as an official recognition of the value of production; during Italy's past participation at world's fairs, the number of awards received had been disproportionately small given the number of participating exhibitors.[53] The juries of the various product sections were generally international groups of experts tasked with examining the products and guaranteeing a balanced and transparent evaluation of their technical and artistic quality. The comparison between

Figure 3.5 Jewellery by Giacinto Melillo exhibited in Antwerp, 1885.
Source: Reproduced from Théophile Fumière, *La section italienne à l'exposition universelle d'Anvers* (Brussels: Guyot, 1885) © Erfgoedbibliotheek Hendrik Conscience, Antwerp.

manufacturers of the same class of goods emphasized differences in industrial practices, thus becoming a useful instrument for producers in the re-examination of their approaches.[54]

At Antwerp's 1885 exhibition, organized into five main sections (teaching, industry, sailing/commerce, electricity, and agriculture) and nine merchandise classes, Italy had five jury presidents and a good number of vice-presidents in different groups, quotas which were in proportion to the size of the Italian contribution to the fair. The final report of the general commissioner of the Italian section, the Marquis Carlo Alberto Maffei di Boglio, who held the title of "extraordinary envoy and plenipotentiary minister" of the Italian King to the Belgian Government, stated that although it was impossible for the Italian delegation to have a member in every class of products to protect national interests (each country had the right to one jury member for every 40 exhibitors), Italy had received many awards even in the classes in which they had few or no representatives—for instance in the wine section, where the jury was mainly French.[55]

The exhibition resulted in a great number of gold medals and honorable mentions (570 of the 665 Italian exhibitors were awarded), many earned in classes in which Italy had not previously excelled (such as fabrics, passementeries, broidered laces). Reports and comments in Belgian official publications lavished praise for the outstanding performance of the Italian section; many commentators noted the progress made by Italian artistic industries and the advancements in educational practices.[56]

Théophile Fumière and the Italian decorative arts as a model for Belgian production

Indeed, while the general trend of the world's fairs had begun to shift from an initial emphasis upon decorative and industrial arts to technological innovation, industrial progress and the celebration of national power,[57] the exhibition in Antwerp continued to emphasize themes linked to educational policies and design reform. Although reforms in Belgium started early—the first commissions were created in 1852—they were initially hindered by the constraints of classicism, for the bodies of reforming institutions were mainly composed by academics (mostly painters) who did not want to modernize fine arts institutions with the creation of a specific training for artisans and workers, who, nonetheless, represented the majority of students at that time.[58] In Italy the artistic and industrial educational system was being reorganized in the 1870s and 1880s, and therefore Italy and Belgium shared a concern for these issues. Both countries aimed at healing the rift between fine arts and applied arts and improving their industrial production through closer collaborations between artists and manufacturers. Moreover, advancement in education was seen as necessary for artisans and workers to improve their social status; for government officials it represented an indispensable boost for national industrial progress, and a crucial element in maintaining the country's stable social relations (an issue deeply felt in Belgium after the political turmoil that had brought the Catholics to power following a long period dominated by the Liberals).[59] The Italian displays in Antwerp were therefore used as a benchmark by the Belgians, and even were the subject of a book, *La section italienne à l'Exposition d'Anvers*, by Théophile Fumière, a Belgian architect and author of numerous publications on the art-industry debate, exhibitions, and decorative arts education.[60] Fumière described the fair as the mirror of contemporary questions and social problems; by examining Italy's productions he intended to discern particular characteristics that were linked to current debates, both in Italy and Belgium. He associated the Italian revitalization of the arts with its recently accomplished unification, with the efficiency of a renewed educational system, and with the support of its industrial museums. He praised the "liberal government, friend of the instruction"[61] for its active role in the creation of institutions dedicated to artistic and industrial education, anchored in the vigorous Italian regional model, which allowed for the development of a variety of styles and craft practices. Fumière considered Italy's choice to concentrate mainly on its arts industries as perfectly in step with modern times, with industrial production being the only fertile ground for original design creations in contemporary Europe. His analysis reveals a crucial moment in which Italian productive districts, following local craft and artistic traditions, were beginning to undertake the necessary technical and economic modifications towards modernization, in the framework of a political project aimed at monitoring their productive and formal processes. Although Fumière analyzed all the decorative productions of the Italian section (furniture, glasswork, bronzes, goldsmith, etc.), he broadly, and not without reason, focused in particular on two of them: education and ceramic production. At that time, Fumière was also a designer at the Boch Frères ceramic manufacture in Tournai: he had therefore a specific interest in the subject, and we can suppose that the Italian display provided him with new ideas and inspiration to apply to the Belgian firm.[62]

Promoted by the MAIC, the collective exhibition of the industrial schools was met with a positive response from the public as well as by the international juries, who praised the rapid developments in these institutions and awarded the MAIC two honorary diplomas and sixteen gold medals to the schools (including the arts and crafts schools of Vicenza, Foggia, Biella, Bologna, and Naples, and the schools of industrial art of Florence, Venice, Turin, and Rome) as well as eight silver medals. Following foreign examples, the schools of industrial art were each linked to a museum, in order to strengthen the economic and cultural project of the government. That project was going to be further broadened thanks to the newly formed *Commissione*, represented in Antwerp by one of its members, architect Camillo Boito,[63] one of the most relentless advocates for the reorganization of the educational system.

In addition, samples of earthenware were sent to Antwerp by several schools,[64] thus endorsing ceramic production as one of the most suitable media for technical and formal experimentation (as well as one of the most economical, hence the great interest for a country like Italy, in its early phase of industrialization). At the exhibition, Italian ceramics proved to be one of the most successful areas of the entire section. Fumière reported on the technical modifications of ceramic products as well as on their originality and the creative process, stimulated by the presence in Italy of a network of industrial museums (which, at that moment, was lacking in Belgium). His in-depth study of and praise for the modernization of the Italian ceramic industry identified two main productive tendencies: the renewed use of ancient models by the manufacturers Pio Fabri (Figure 3.6), Castellani, Molaroni, Cantagalli (Figure 3.7), and Antonibon, and the elaboration of new decorations and forms, in the works of Schioppa & Cacciapuoti, Sansebastiano & Moreno, Fratelli Cacciapuoti, and Richard (Figure 3.8). The resurgence of Italian artistic pre-eminence in this sector

Figure 3.6 Ceramic by Pio Fabri exhibited in Antwerp, 1885.
Source: Reproduced from Théophile Fumière, *La section italienne à l'exposition universelle d'Anvers* (Brussels: Guyot, 1885) © Erfgoedbibliotheek Hendrik Conscience, Antwerp.

Figure 3.7 Ceramic by Cantagalli exhibited in Antwerp, 1885.

Source: Reproduced from Théophile Fumière, *La section italienne à l'exposition universelle d'Anvers* (Brussels: Guyot, 1885) © Erfgoedbibliotheek Hendrik Conscience, Antwerp.

Figure 3.8 Ceramic by Richard exhibited in Antwerp, 1885.

Source: Reproduced from Théophile Fumière, *La section italienne à l'exposition universelle d'Anvers* (Brussels: Guyot, 1885) © Erfgoedbibliotheek Hendrik Conscience, Antwerp.

was the result of a long process that had developed following unification.[65] In-depth research had contributed to the technological growth of the sector and the strategic importance of government initiatives and support was also demonstrated by a subsidy granted to the government official and scholar Giuseppe Corona, who published an artistic and technological study of Italian ceramics and their contemporary development. His work, entitled *L'Italia Ceramica* (1880), was also presented—and awarded—at Antwerp's exhibition.[66]

The Italian system, based on the development of education, artistic, historical, and technological studies, appeared as an effective model that confirmed current tendencies in Belgium, where a new educational organization in academies and design schools was being tested (a new study plan compiled by the *Conseil* was finalized in the years 1874–1876 and the creation of a Higher Institute in Antwerp under state control was expected by 1886)[67] and where plans to create a national museum and new institutions devoted to the training of artisans and workers were in development.[68] The Italian example also found corroboration in Belgium's own vital and well-defined local cultural model, for it attempted to monitor industrial growth through a more centralized project which simultaneously maintained a regional dimension, expressed by the diversity of artistic schools and practices disseminated throughout the national territory. Moreover, Italy's capacity to reinvent itself by investing in art manufactures was highly praised by Belgian observers, along with its recent prosperity in a period marked by economic crisis, thanks to the consolidation of its traditional industrial apparatus—extile, food industry, and luxury goods—as well as through the government's investments in the construction of a stronger railway system which favored the rise of heavy industry.[69]

Not surprisingly, Italy was also present in the fine arts section at the Antwerp exhibition. Located in a separate pavilion, Italian art did not stimulate the same interest provoked by its decorative arts exhibition. Indeed, contemporary Italian art could not match the heights achieved by its great Renaissance and Baroque masters. Fumière dedicated an entire section of his book to the subject, insisting more on Italy's great past artistic tradition than on the noteworthy but not exceptional current school. His opinion reflected current debates in artistic and academic circles: even the customary *voyage en Italie* had lost much of its appeal in the eyes of young Belgian artists and architects who looked for new trends in other European capitals, especially Paris.[70] The focus on the artistic importance of Italian productions had demonstrably shifted from the fine to the decorative and industrial arts, thanks to the exhibition as well as to the extensive publicity through newspapers and other print media, where Italy was represented as a virtuous example of hard work resulting in an artistic and industrial renascence.

Towards the new century

After the exhibition, both Italy and Belgium began a series of initiatives aimed at the development of decorative arts training. In Belgium, two major projects—Brussels' Decorative Arts School (1886)[71] and the Royal Museums of Decorative and Industrial Arts (1889)—were promoted by liberal burgomaster and reformer Charles Buls.[72] However, contrary to Buls' expectations, the two institutions were not created interdependently within the framework of a coherent plan conceived to enhance decorative and industrial arts. Resulting from the merging of already existing artistic and plaster casts collections, the Royal Museums of Decorative and Industrial Arts struggled from the very beginning to achieve their goals. Belgium's artistic policies, aimed at renewing its institutions in order to give the image of a young and dynamic country, were also nurtured by more conservative impulses, still present in artistic and governmental bodies. In the establishment of the new museum, the latter prevailed over the educational ambitions promoted by reformers and manufacturers. Therefore, collecting treasures of decorative art, supporting mass education, and preserving the national heritage of the nation, eventually took precedence over creating a specialized institution for the exclusive training of craftsmen and artisans.[73]

As a result, Brussels' Decorative Arts School began to form its own collection under the direction of architect Jean Baes, who at that time was searching for more progressive technical and aesthetic approaches, in harmony with alternative educational ideals to counter obsolete academic training.[74] Similar innovative practices—for instance those based on the geometrical analysis and abstract representations of nature, in particular of the vegetal world—had also been developed by other Belgian institutions such as the design schools founded in the 1860s in Brussels, and by the Catholic neo-gothic Saint Lucas Schools, which contributed to the advent of Art Nouveau[75] (Figure 3.9).

Figure 3.9 "Feuilles de chène," a model used at the Saint Lucas Schools in Ghent, 1882.

Source: Reproduced from Joseph Marès, *Modèles gradués pour servir d'exercices préparatoires à l'étude du dessin à main levée, Série D 2ème cahier,* (Bruges: Desclée, De Brouwer & Cie, s.d.), pl.16 © Erfgoedbibliotheek Hendrik Conscience, Antwerp.

Belgium's progress in design was thus the result of the interaction of a dual political strategy—partly centralized, partly local—that ensured, even if belatedly, the completion of an educational organization of the kind already achieved in Italy, where the dialectic between a national and regional approach had built the backbone of the decorative arts system. Its achievement led to the recognition and development of a common set of educational principles and to the creation of a new artistic expression.

In the meantime, Italy continued to look at Belgium as an example of effective political and social organization, and as a model for liberal initiatives promoting the economic benefits of the industrial arts. Belgian initiatives were observed and studied through the prism offered by its exhibitions—for instance the *Grand concours de l'Industrie*, an international event held in Brussels in 1888 on the site of the Jubilee exhibition,[76] and, also, through investigations conducted by the newly formed *Commissione* and its members, which continued to look at foreign applied arts and architecture, in order to corroborate or improve the strategies undertaken by both public institutions as well as private enterprise. During the 1890s, investigations into Belgian artistic and industrial education were published in the magazine founded by the *Commissione*, titled *Arte Italiana Decorativa e Industriale* (hereafter *AIDI*).[77] Examples of the Italian interest in Belgian production include the religious artifacts of the Saint Lucas Schools—seen by Boito in their entire splendor in 1885 in Antwerp and considered to be a qualitative and successful economic enterprise to be imitated in Italy. Indeed, a series of exhibitions of sacred art objects was held in 1896 (Orvieto) and 1898 (Turin), and the creation of a company for the production of artistic artifacts for religious use was proposed by long-time member of the *Commissione* Odescalchi to cardinal Mariano Rampolla del Tindaro, secretary of State under Pope Leo XIII. Odescalchi's proposal aimed to involve the Roman Museum of Industrial Art—still functioning under precarious conditions—as the headquarters and operative section of the company. This proposal was reassessed by Boito in the preface of the Orvieto exhibition catalogue, but was never realized.[78]

At the end of the century, new artistic directions roused the artistic and industrial world, and the *Commissione*'s members turned to new issues: Camillo Boito, for instance, reported on foreign publications on floral art and initiated a series of articles on Art Nouveau.[79] Essays focusing on 'the new industrial art' were penned for *AIDI* by Alfredo Melani,[80] professor at Milan's school of applied arts. This development occurred through the accomplishment of the most urgent tasks of the *Commissione*, which had identified a broad selection of Italian models and had adopted a set of rational principles aimed at the development of "good taste." The *Commissione* thereby produced an almost spontaneous engagement in new artistic and original ideas, thus paving the way for the advent of a "new style" in the industrial arts and architecture, oriented, like in the best Belgian examples, toward the synthesis of past styles and nature. The International Decorative Arts exhibition in Turin in 1902 (Figure 3.10) embraced a cultural and economic change begun earlier and offered another occasion for Italy and Belgium to meet on an international stage, to rekindle a shared dialogue based upon the renewal of artistic and architectural language, and to reassess, in a changed context and under the auspices of the new century, the future role and value of industrial and decorative arts.[81]

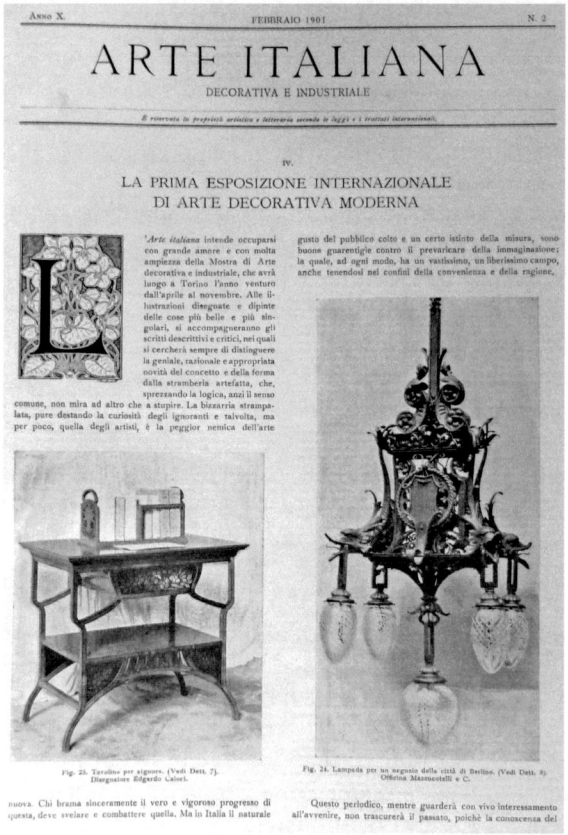

Figure 3.10 A page from the article "La prima esposizione internazionale di arte decorativa moderna."

Source: Reproduced from *Arte Italiana Decorativa e Industriale*, n. 11 (Milan: Hoepli, February 1901), 1 © Biblioteca di "Storia ed analisi dell'architettura e degli insediamenti," Politecnico di Torino, Turin.

Notes

The research for this essay was supported by a grant from the Fondation Nationale Princesse Marie-José.

1 See Michel Dumoulin, "Hommes et cultures dans les relations Italo-Belges, 1861–1915," *Bulletin de l'Institut Historique Belge de Rome*, t. LII, (1982): 271–565; Michel Dumoulin and Herman Van der Wee, *Hommes, Cultures et Capitaux dans les relations italo–belges aux XIXe et XXe siècles* (Rome: IHBR, 1993); Sabina Gola, *Un démi–siècle de relations culturelles entre l'Italie et la Belgique (1830–1880)* (Rome: IHBR, 1999); Andrea Ciampani, Pierre Tilly, and Vincent Viaene, "Italia e Belgio nell'Ottocento europeo. Nuovi percorsi di ricerca, Atti del convegno internazionale di Roma," *Rassegna storica del Risorgimento* III, suppl. (2002): 1–184, and Martina Carraro, "I Belgi e la Biennale: premesse e protagonisti del primo padiglione nazionale ai giardini (1895–1914)," (PhD diss., Università Ca' Foscari Venezia, 2010).
2 Dumoulin, "Hommes et cultures," 272–273.
3 See the essays in Ciampani, Tilly, and Viaene, "Italia e Belgio," Michel Dumoulin, "Découvertes de l'Italie par les belges aux 19e et 20e siècles ou la psycho-géographie

d'un malentendu," in *Hommes, Cultures et Capitaux*, ed. Dumoulin Michel, Van der Wee Herman (Rome: IHBR, 1993), 115–136, and Christine Dupont, *Modèles italiens et traditions nationales. Les artistes belges en Italie (1830–1914)* (Rome: IHBR, 2006).

4 For the debate in Italy and Belgium see Annalisa B. Pesando, *Opera vigorosa per il gusto artistico nelle nostre industrie. La Commissione centrale per l'insegnamento artistico industriale e il "sistema delle arti" (1884–1908)* (Milan: Franco Angeli, 2009); Annalisa B. Pesando and Daniela N. Prina, "To Educate Taste with the Hand and the Mind. Design Reform in Post-unification Italy," *Journal of Design History* 25, no. 1 (2012): 32–54; Claire Leblanc, ed., *Art et industrie: les arts décoratifs en Belgique au XIXe siècle* (Brussels: MRAH, 2004); Claire Leblanc, ed., *Art nouveau & Design. les arts décoratifs de 1830 à l'expo 58* (Brussel: Racine, 2005), Daniela N. Prina, "L'unité des Arts avant l'Art Nouveau. La réforme de l'enseignement artistique et industriel en Belgique pendant la deuxième moitié du XIXe siècle" (PhD diss., Politecnico di Torino & K.U. Leuven, 2009).

5 For the European debate see the bibliography in Prina, *L'unité des Arts avant l'Art Nouveau*, 319–362.

6 See for instance the report on the state of industrial and vocational schools in Belgium in: "Notizie e documenti sulle scuole commerciali e popolari in Italia e all'estero," *Annali dell'Industria e del Commercio* 10 (1879): (151–277).

7 For a recent survey on the diplomatic relationships between the two countries, see Ugo Colombo Sacco di Albiano, *Un omaggio a oltre 150 anni di amicizia Italo-Belga attraverso luoghi e protagonisti della diplomazia* (Rome: Servizi Tipografici Colombo, 2014).

8 See Carlo Ghisalberti, *Storia costituzionale d'Italia 1848/1948* (Bari: Laterza, 2000), 87–154.

9 Sandra Pinto, "La promozione delle arti negli Stati italiani dall'età delle riforme all'Unità," in *Storia dell'Arte Italiana*, vol. 6 ed. F. Zeri (Torino: Einaudi, 1982), 791–1079. In 1861 the Ministry of Public Education (hereafter MPE) supervised all the literary and artistic institutions, including Fine Arts Academies, while the Ministry of Agriculture, Industry and Commerce (hereafter MAIC) supervised the technical and professional schools (Royal Decree of 28 November 1861, n. 347).

10 Titti Carta, "Artigianato," in *Enciclopedia Italiana – VI, Appendice* (Milan: Treccani, 2000), 703; App. III, i, 139.

11 See for instance the report commissioned by the MPE to Naple's University Professor Alberto Errera, *Studi sull'istruzione primaria, industriale professionale e commerciale nel Belgio* (Rome: Tipografia Eredi Botta, 1880).

12 For the institution's history and aims: Daniela N. Prina, "Design in Belgium before Art Nouveau: Art, Industry, and the Reform of Artistic Education in the Second Half of the Nineteenth Century," *Journal of Design History* 23, no. 4 (2010): 329–350.

13 Baldassarre Odescalchi was a long-time member of the *Commissione*, and an expert on foreign applied art museums: Baldassarre Odescalchi, *I musei d'arte e d'industria in Italia: considerazioni e proposte* (Rome: tip. Eredi Botta, 1880). For a biographical note see "Baldassarre III Odescalchi," in Sofia Crifò, *Raffaello Ojetti architetto nei primi cinquant'anni di Roma capitale* (Florence: Polistampa, 2004), 72–75.

14 The problematic vicissitudes of the Roman Museum of Industrial Art are illustrated in: *Del M.A.I.: storia del Museo artistico industriale di Roma* (Rome: ICCD, 2005) and *Del M.A.I.: storia del Museo artistico industriale di Roma: Collezioni d' arte antica. Inventari 1876, 1884, 1956* (Rome: ICCD, 2011).

15 Rome, Archivio di Stato, Fondo Odescalchi, 11 c F 1 (44), *Letter from the Minister of Agriculture, Industry and Commerce, section 2a N° 14953, to Baldassarre Odescalchi, 25 august 1880*.

16 Raffaele Erculei was first secretary (1874–1885) and later director (1885–1898) of Rome's Museo Artistico Industriale. For a biographic note: Maurizio Donati, "Profili di Raffaele Erculei, segretario (1874–1885) e primo direttore del M.A.I. (1885–1898), e di Alberto Gerardi, ultimo direttore (1932–1959)," in *Del M.A.I.*, 93–106.

17 See *Exposition nationale de 1880. Catalogue officiel* (Brussels: Mertens, 1880). For a comparative analysis of the event, see Jean–François Constant, "Entre mémoire et avenir. La nation aux expositions nationales de Bruxelles et de Montréal (1880–1884)," in *Vivre en ville. Bruxelles et Montréal (XIX–XX siècles)*, ed. Serge Jaumain and Paul–André Linteau (Brussels, 2006), 351–371.

18 Théophile Fumière, *Les arts décoratifs à l'exposition du Cinquantenaire belge* (Brussels: Guyot, 1880); *Exposition Nationale, IV section, Industries d'Art en Belgique antérieures au XIXe siècle. Catalogue Officiel* (Brussels: Vanderauwera, 1880).

19 Baldassarre Odescalchi and Raffaele Erculei, *Il movimento artistico industriale in Inghilterra, nella Francia e nel Belgio e istituzioni intese a promuoverlo* (Rome: tip. Eredi Botta, 1880).

20 The government of the Left covered the period between 1876 and 1896. See Giampiero Carocci, *Agostino Depretis e la politica interna italiana dal 1876 al 1887* (Turin: Einaudi, 1956); Mario Bendiscioli, "La sinistra storica e la scuola," *Studium* 4 (1977): 447–466.

21 For the debates on Italian academies see the bibliography in Pesando, *Opera vigorosa*, 313–341.

22 See *Exposition universelle. Paris. 1878. Rapports du jury international. Groupe III. - Classes 17 et 18. Rapport sur les meubles à bon marché et les meubles de luxe, ouvrages du tapissier et du décorateur* (Paris: Imprimerie nationale, 1880) 22, and Anna Pellegrino, *La città più artigiana d'Italia. Firenze 1861–1929* (Milan: Franco Angeli, 2012), 123.

23 For the uses of neo-Renaissance in Belgium see Alfred Willis, *Flemish Renaissance Revival in Belgian Architecture (1830–1930)*, (PhD diss., Columbia University, 1984); Piet Lombaerde, "A la recherche d'une identité historique: l'architecture néo-Renaissance à Anvers au 19ème siècle," in *Le XIXe siècle et l'architecture de la Renaissance*, ed. Frédérique Lemerle, Yves Pauwels, and Alice Thomine-Berrada, (Paris: Picard, 2010), 165–178; Benoît Mihail, "Un mouvement culturel libéral à Bruxelles dans le dernier quart du XIXe siècle, la 'néo-Renaissance flamande'," *Revue belge de philologie et d'histoire* 76, no. 4 (1998): 978–1020.

24 See in particular: Rossana Pavoni, ed., *Reviving the Renaissance. The Use and Abuse of the Past in Nineteenth-Century Italian Art and Decoration* (Cambridge: University Press, 1997).

25 Daniela N. Prina, "Belgian Decorative Arts in the Later Nineteenth Century. Needs for a National Museum and Debates Surrounding Didactic Collections in Brussels," *Journal of the History of Collections* 24, no. 2 (2012): 257–274. Brussels had also an Industrial Museum established in 1826. See Musée Royal de l'Industrie. *Catalogue des collections de gravures, de dessins, de modèles, etc., formées dans le but de propager les applications des arts aux diverses industries* (Brussels: Dehou, 1876).

26 *Congrès international de l'enseignement 1880* (Brussels: Hayez, 1880).

27 Italy's capitals were: Turin (1861–1865); Florence (1865–1871) and Rome from 1871.

28 "Il Museo Italiano d'arte industriale," *Annali dell'Industria e del Commercio* 2 (1879).

29 See *Annali dell'Industria e del Commercio*, 6,10,13 (1880).

30 For an overview on Industrial Museums in Italy, see Fiorella Bulegato, *I musei d'impresa. Dalle arti industriali al design* (Rome: Carocci, 2008) and Monica Amari, *I musei delle aziende: la cultura della tecnica tra arte e storia* (Milan: Franco Angeli, 2011).

31 On Milan's exhibition see Ilaria M.P. Barzaghi, *Milano 1881: tanto lusso e tanta folla: rappresentazione della modernità e modernizzazione popolare* (Cinisello Balsamo: Silvana Editoriale, 2009).

32 On Turin's Esposizione Generale Italiana see Umberto Levra and Rosanna Roccia, *Le esposizioni torinesi. 1805–1911. Specchio del progresso e macchina del consenso* (Turin: Archivio storico del Comune di Torino, 2003).

33 *L'exposition Nationale à Turin, Extrait du Bulletin de l'association des ingénieurs sortis de l'Ecole spéciale de Gand*, Archivio Centrale dello Stato (hereafter ACS), Fondo MAIC, 60A (temporary).

34 See Maria Grazia Cerri, Daniela Biancolini Fea, and Laura Pittarello, ed., *Alfredo d'Andrade. Tutela e restauro* (Florence: Vallecchi, 1981); Annalisa B. Pesando, "Un inedito d'Andrade: innovatore nell'insegnamento delle arti decorative," *Bollettino S.P.A.B.A.*, LII (2004): 265–286.

35 Eugène Emmanuel Viollet-le-Duc, *Dictionnaire raisonné de l'architecture française du XIe au XVIe siècle* (Paris: B. Bance - A. Morel, 1854–1868).

36 See Elena Dellapiana, "Il mito del medioevo," in *Storia dell'architettura italiana. L'Ottocento*, ed. Amerigo Restucci (Milan: Electa, 2005), 411.

37 See Mario Bencivenni, Riccardo Dalla Negra, and Paola Grifoni, *Monumenti e istituzioni. Parte I. La nascita del servizio di tutela dei monumenti in Italia 1860–1890* (Florence: Alinea, 1987); Id., *Monumenti e istituzioni – Parte II. Il decollo e la riforma del servizio di tutela dei monumenti in Italia 1880–1915* (Florence: Alinea, 1992).

38 On the work of the *Commissione*, see Pesando, *Opera vigorosa*; Pesando, Prina, *To educate*.
39 In 1881 both France and England had set up a board of inquiry mandated to analyze the state of technical and artistic instruction. See Report of the Royal Commission on technical instruction, First report, France (London: Eyre and Spottiswoode, 1882), Second report, Vol. I, Foreign Countries and United Kingdom, Minutes of Evidence; Vol. II, The Continent and America, (London: Eyre and Spottiswoode, 1884), and Antonin Proust, *Commission d'enquête sur la situation des ouvriers et des industries d'art* (Paris: Quantin, 1884).
40 "Esposizione Internazionale di Anversa. Circolare diretta alle Camere di commercio del Regno," *Bollettino di notizie commerciali* s. II 1, no. 6 (1884): 172–173.
41 On Antwerp's exhibition see René Corneli and Pierre Mussely, *Anvers et l'exposition universelle de 1885* (Antwerp: Bellemans, 1886); Mandy Knockaert and Jan-Lodewijk Grootaers, ed., *De panoramische droom: Antwerpen en de wereldtentoonstellingen 1885, 1894, 1930* (Antwerp: Antwerpen 93, 1993).
42 "Esposizione Internazionale di Anversa," *Bollettino di notizie commerciali* s.II 1, no. 9 (1884): 283–285.
43 See the report *Il commercio Italiano e la piazza di Anversa* by Professor Edouard Heinzmann–Savino in ACS, Fondo MAIC, 60A (temporary).
44 King Leopold II allowed the objects he was collecting to create a Congolese museum (later opened in 1898 in Tervueren) to be displayed at the exhibition. The fair therefore constituted also an instrument of propaganda for the colonial and economic politics of the King.
45 Mandy Nauwelaerts, "The dream of the Metropolis. Antwerp and the World Exhibitions," in Knockaert, Grootaers, eds., *De panoramische droom* (Antwerp: Antwerpen 93, 1993), 66.
46 Corneli and Mussely, *Anvers et l'exposition universelle*, 137.
47 See the letter dated 2 September 1884 and signed by Tommaso Villa, chief of the executive committee of Turin's 1884 exhibition in ACS, Fondo MAIC, 62 (temporary).
48 Corneli and Mussely, *Anvers et l'exposition universelle*, 236.
49 Ibid.
50 Ibid., 235.
51 Ibid., 236.
52 "Relazione del Regio Commissario (marchese Maffei di Boglio) sui premi ottenuti dagli espositori italiani," *Bollettino di notizie commerciali* s.II 11, no. 35 (1885): 648–664. A minute of this report with supplementary information is in ACS, Fondo MAIC, 74 (temporary).
53 Anna Pellegrino, "L'Italia alle esposizioni universali del XIX secolo: identità nazionale e strategie comunicative," *Diacronie* 2, no. 18 (2014): 12.
54 For instance in 1862, during London's world fair, the technical director of the Doccia manufacture, Paolo Lorenzini, wrote a personal letter to the marquis Ginori, owner of the factory, to inform him that it was imperative to renew their production system and the design of their porcelains. The Doccia manufacture was indeed still successful but needed to be updated in order to be in step with the advancements of the other producers. See Archive of the Museo delle Porcellane di Doccia, Document n. 2525, in Sandra Buti, *La Manifattura Ginori. Trasformazioni produttive e condizione operaia (1860–1915)* (Firenze: Olschki, 1990), 24.
55 Same as in note 52.
56 Positive reviews for decorative arts are expressed in Anatole Bamps, "L'exposition universelle d'Anvers," *Revue Internationale* no. 8 (1885): 223–224; precision instruments were appreciated in Victor Van Tricht, *L'exposition universelle d'Anvers, revue scientifique* (Brussels: Vromant, 1885) 82–88; schools were praised in: *Rapports presentés à l'administration communale de Bruxelles* (Brussels: Baertsoen, 1885) 117–139.
57 Pellegrino, *L'Italia alle esposizioni*, 9.
58 See Prina, *L'unité des arts*.
59 Emiel Lamberts, Jacques Lory, *1884: un tournant politique en Belgique* (Brussels: FUSL, 1984). Belgian Socialist Party was born in 1885.
60 Théophile Fumière, *La section italienne à l'exposition universelle d'Anvers* (Brussels: Guyot, 1885); Id., *Les arts décoratifs à l'exposition du Cinquantenaire belge* (Brussels: Guyot, 1880); Id. *Des expositions et de l'enseignement des arts décoratifs. Leur développement en France et leur avenir en Belgique* (Brussels: Guyot, 1882); Id. *L'exposition d'Amsterdam et la Belgique aux Pays-Bas* (Brussels: Guyot, 1883).

61 Fumière, *La section italienne*, 20.
62 Unfortunately the archives of the manufacture were partially destroyed in 1985 and we do not have much information on Fumière's role at Boch Frères. The author wants to thank Mme Vanessa Bebronne, curator of Keramis in La Louvière, for providing detailed information on the Boch Frères manufacture.
63 On Boito see Guido Zucconi, *L'invenzione del passato. Camillo Boito e l'architettura neomedievale* (Venice: Marsilio, 1997); Elena Dellapiana, "Camillo Boito (1836–1814)," in *Storia dell'Architettura Italiana*, ed. Amerigo Restucci (Milan: Electa, 2005) 590–605.
64 The list of products sent by the schools is in ACS, MAIC, 62 (temporary).
65 See Elena Dellapiana, *Il design della ceramica in Italia, 1850–2000* (Milan: Electa, 2010).
66 Giuseppe Corona, *L'Italia Ceramica* (Rome: Tip. Eredi Botta, 1880). A subvention to Corona was also granted for Antwerp's exhibition: Letter of Minister of Agriculture, Industry and Commerce Bernardino Grimaldi to Maffei, 7 september 1885, in ACS, MAIC 66 (temporary).
67 Daniela Prina, "From Centralization to Local Policies: Design Reform Dynamics in Belgium and the Creation of Antwerp's Higher Institute (1830–1914)," in *Tradition, Transition, Trajectories: Major or Minor Influences?* ed. Helena Barbosa and Anna Calvera (Aveiro: Universidad de Aveiro, 2014) 557–562.
68 Prina, *L'unité des arts*.
69 Valerio Castronovo, "La storia economica," in *Storia d'Italia vol. 4 Dall'unità a oggi* (Torino: Einaudi, 1975), 5–129, and Luciano Segreto, "Storia d'Italia e storia dell'industria," in *Storia d'Italia. Annali. L'industria*, ed. Franco Amatori, Duccio Bigazzi, Renato Giannetti e Luciano Segreto, (Torino: Einaudi, 1999), 7–21.
70 Christine Dupont, "Les artistes belges en Italie" in Ciampani, Tilly, Viaene, "Italia e Belgio," 125–138.
71 *École des arts décoratifs. Eléments de l'enseignement. Organisation* (Bruxelles: Mertens, 1879). Jean Paul-Midant, "La création de l'École des arts décoratifs et l'enseignement de l'architecture à l'Académie des Beaux-arts," in *Académie de Bruxelles. Deux siècles d'architecture* Id., (Brussels: AAM, 1989) 309–323.
72 On Buls, see Marcel Smets, *Charles Buls et Les principes de l'art urbain* (Liège: Mardaga, 1995).
73 See Prina, "Belgian decorative arts," 266–269.
74 Jean Baes, "Conseil de perfectionnement de l'enseignement des arts du dessin. Réponse aux questions posées par le ministre de l'Agriculture, de l'Industrie et des Travaux publics," *L'Emulation* (1893), n. 5, col. 65–7; n. 6, col. 81–5; n. 7, col. 97–103; n. 8, col. 113–20.
75 See Prina, "Design in Belgium"; on the St Lucas Schools, see Jan De Maeyer, *De Sint-Lucasscholen en de neogotiek 1862–1914* (Leuven: KADOC, 1988); Wilfried Wouters, *Van tekenklas tot kunstacademie. De Sint- Lucasscholen in België 1866–1966* (Kortrijk-Heule: UGA, 2011).
76 The Italian section was successful but the profits were lowest when compared to 1885. See the minute of the report of 28 September 1888 written by Antonio Monzilli to Vittorio Ellena, undersecretary at the MAIC, ACR, Fondo MAIC, 88 (temporary).
77 Raffaele Erculei, "L'insegnamento artistico–industriale in Europa. Belgio" *AIDI* 5 (1894): 42–43.
78 Letter by Odescalchi to the Cardinal Rampolla, 10 December 1895, Archivio di Stato, Fondo Odescalchi, 11c F4 (14); Raffaele Erculei, "L'esposizione di arte sacra in Orvieto" *AIDI* 2 (1897): 20; Camillo Boito, "L'industria delle suppellettili sacre," in *Oreficerie, stoffe, bronzi, intagli all'esposizione di arte sacra in Orvieto*, ed. Raffaele Erculei (Milan: Hoepli, 1898), VI–IX.
79 See for instance: Camillo Boito, "Cenni di recenti pubblicazioni sul nuovo stile e sullo stile della natura," *AIDI* (1901): 83–84; 96–98; and Id. "Il professore M. Meurer e i suoi studi sulle piante e sull'ornamento," and M. Meurer, "Lo studio ornamentale della pianta," *AIDI* (1901): 53–57.
80 Alfredo Melani, "L'Arte industriale nuova. Un'occhiata all'estero," *AIDI* (1901): 51–52; 65–67.
81 Rossana Bossaglia, *Torino 1902: le arti decorative internazionali del nuovo secolo* (Milan: Fabbri, 1994).

4   A Danish spectacle

Balancing national interests at the 1888 Nordic Exhibition of Industry, Agriculture, and Art in Copenhagen

Jørn Guldberg

To the committee it is evident that the efforts made by practically all civilized nations to lend to the products of craft a more full stamp of beauty and taste, necessarily have to be taken up here, because such efforts bear with them the promise of a higher stage of culture and a happier life for the individual, and in addition, they hold out prospects of more profitable sale of products, and finally because they, in particular, are directed towards supporting the smaller enterprises in order to lend to them the power of resistance to the emerging large-scale manufacturing industry. What is crucial is to lift up the level of the whole people's understanding of beautiful forms and decoration and to make possible a livelier collaboration between the arts and handicraft.

> The exhibition committee in an address to the Lord Mayor
> and City Council of Copenhagen, November 11, 1886

The Nordic Exhibition of Industry, Agriculture and Art—The Great Nordic Exhibition—of 1888, was arranged by the Copenhagen Industrial Association as one of three projects, all intended to accelerate the process of industrialization in Denmark, and more specifically, to stimulate the growth of domestic applied art, or "kunstindustri" ["art industry"] as it is called in the Scandinavian languages. The first project was achieved when the Association launched the journal *Tidsskrift for Kunstindustri* [Journal of Applied Art], in 1885. The second was realized with the 1888 Nordic Exhibition, and the third reached fruition with the founding in 1890, and subsequent public opening of The Danish Museum of Decorative Arts in 1895.

The 1888 exhibition served as the official celebration of three anniversaries: the centenary of the abolition of serfdom among the peasantry in 1788; the fiftieth anniversary of the Industrial Association; and the twenty-fifth anniversary of the reigning king, Christian IX.

The exhibition

The Great Nordic Exhibition was open for four and a half months during the summer of 1888. It attracted close to one and a half million visitors at a time when Copenhagen's population was about 320,000. A little more than 6,000 exhibitors (companies, workshops, and individual manufacturers, and artists) presented their products. The vast majority of exhibitors were Scandinavian; Denmark alone was responsible for about two-thirds of the exhibitors, and Danish manufacturers and artisans dominated the prestige section, the display of applied art in the main exhibition hall.[1]

The exhibition quickly outgrew initial expectations. According to the original plan, the exhibition was to be organized in six parts, with fifteen main sections divided into forty subsections, in accordance with a simple system of classification. However, the amount and character of items submitted to the various special committees made it necessary to revise the plan. For instance, an independent subsection for machinery, engines, and heavy devices was created by separating large-sized items from the initially rather broadly defined main section for industry, and a great number of female exhibitors were moved from the main section for domestic industry, which was housed in a side building, to a separate subsection for textiles as part of the main section for applied art in the main building.[2]

The Great Nordic Exhibition was situated in what would become the future city center of Copenhagen. The exhibition grounds occupied approximately 40 acres, with an additional 16 acres added from the Tivoli amusement park, whose facilities formed an integral part of the exhibition complex. A souvenir lithograph of the exhibition grounds (Figure 4.1) shows, to the lower left, the main building, with its dome, transepts and oblong, central hall with protruding bays to either side. To the right is the building complex that housed the previous Nordic Exhibition of 1872, later adopted as the home of the Industrial Association. The view is framed by a band with interlaced decoration based upon old Nordic ornamentation, popular at that time, with portraits of the king (top center), the president (top left) and the three vice-presidents for, respectively, industry, agriculture, and art.

The 1888 exhibition's main building stood where the present town hall was built, beginning in 1892. Both the 1888 main building and the new town hall were designed

Figure 4.1 Souvenir lithograph of the exhibition grounds.
Source: Photo courtesy The Royal Library, Copenhagen.

by the architect Martin Nyrop, who was responsible for all except one of the "official" buildings of the Nordic exhibition: the pavilions housing applied art, fine art, machines and tools, gardening, fishery, agriculture, a working dairy, and others.[3] In addition, a number of smaller pavilions and spectacular monuments were erected by private companies, such as the Carlsberg and Tuborg breweries. The latter had a giant beer bottle built with an elevator inside that transported visitors to an observation deck at the top, offering them a panoramic view of the city.

In a general review of the section of applied art, the secretary general of the exhibition committee, Camillus Nyrop, a historian of industry, technical schools and a number of individual trades, stated boldly that the Nordic Exhibition attained the status of "world exhibition in its own right."[4] This is surely an overstatement, for several European countries did not participate and some of them were not invited at all. Nonetheless, a handful of non-Nordic countries joined Denmark, Norway, and Sweden in the main building, which housed the sections displaying applied art of the exhibition. Exhibitors from Russia (a total of 225), Italy (63), Germany (58), France (42), England (9), and the city of Budapest (1) contributed to the Copenhagen exposition, which also saw a private entrant: in the same year he began to edit and publish the journal, *Le Japon Artistique*, Sigfried Bing was allotted floor space and some vitrines in a prime location near the café in the main building. As a result, Bing was able to exhibit both a giant Buddha statue in bronze and a selection of small Japanese items and prints from "La Porte Chinoise," his shop in Paris. Bing's display was considered by many reviewers a pleasant surprise and the true sensation of the exhibition. During the exhibition, he presented examples from his collection of Japanese decorative arts to the committee planning the future museum of applied art.

At the time of the opening on May 17, only the sections of France and Russia were in place, and Russia continued to add recently produced objects throughout the exhibition until its closure on October 2, which is reflected in the relatively high number of exhibitors from that country. A considerable part of the Russian contribution, however, consisted of objects from the tsar's private collection and from court-owned factories and workshops. Finland, at that time a grand duchy under the Russian tsar, was given the opportunity to be included in an independent, although small, subsection of the Russian presentation.[5] The Nordic Exhibition was not complete until July 7, when the German section was installed, and that display also included many items from royal porcelain and metal manufactories and from the Prussian court's own collection of decorative objects in china, gold, and silver. In general, many of the items in the non-Nordic sections were historical objects, resembling a museum display more than a showcase of contemporary production. Some of the exhibits in the Russian, French, Italian, and German sections could even be characterized as antiquarian; this partially applies to the Norwegian section as well.[6]

Both in his general review and in the official exhibition report he authored, Nyrop complained that the foreign committees had not chosen more recently manufactured objects, especially within the field of applied art, to showcase their nations' current states in relation to taste, creativity, and productive competence. In general, Nyrop expressed disappointment that he was unable to identify a consistent national style in any of the participating nations' sections in the main exhibition building. What he encountered was a confusing mixture of geographical (national, regional, local) and historical styles, crisscrossing individual countries' displays. All the figurative porcelain on display was related more or less explicitly to the Meissen Rococo style or to antique vase shapes and/or surface decoration; wood carvings were inspired by Italian Renaissance or Baroque techniques and patterns; and the idioms of historical wood carving were transferred to cast iron. It should be emphasized that neither Nyrop nor other Scandinavian or European commentators on the Nordic Exhibition adopted an anti-historicist stand; the need to develop a contemporary style, an expression of a "new age" was hardly introduced. Rather, the subject for discussion in Denmark during the 1880s was to identify a historical source for the renewal or reinvention of a genuine national and specifically *Nordic* style.

Political, organizational, and ideological contexts

As mentioned earlier, the Great Nordic Exhibition functioned as a triple celebration, with competing agendas and differing political and ideological implications; the three anniversaries were represented in the exhibition's three main concerns: agriculture, art and industry, and the monarchy. However, what united the members of the organizing committee was a nationalistic ideology nurtured by Denmark's traumatic defeat by the Prussian-Austrian army in 1864, a defeat that resulted in the loss of one third of Danish territory. Demonstrating a well-organized agricultural economy, a prosperous and artistically oriented industry, and celebrating the king as a symbol of national unity, the exhibition helped to construct a new political self-image as a "Small State," compelled to find its own way in the world, defeated and deserted as it was by its neighbors to the south, east, and west. In a cantata written to the 1872 Nordic Exhibition, the poet H.P. Holst introduced a saying that neatly summarized this new nationalistic élan: "What is outwardly lost must be inwardly regained." The saying is undoubtedly the most cited sentence in popular, as well as scholarly, accounts of Danish national policy since the 1860s as it turned to domestic resources for its organizational, educational, and inventive structure and capacities. Thus, the dominant nationalistic ideology during the period of planning, accomplishment, and evaluation of the Nordic Exhibition of 1888 was a strange mixture of a paralytic Lilliputian mentality, on the one hand, and self-confidence and self-containment, on the other.

In the press, the term "jubilee festival" was coined to characterize the exhibition. The agricultural part of the exhibition was not simply a tribute to rural life in the countryside. The centenary of the abolition of serfdom was framed by the recent creation of a specific kind of agrarian capitalism, known as the cooperative movement, for which Denmark is recognized. This process transformed the agents of agrarian economy and culture from peasants into farmers, resulting in a cooperative business organization operating at a highly efficient technological and industrialized level.

The Copenhagen Industrial Association was created as a semi-political body in 1838 by leading manufacturers, businessmen, scientists, academics, political leaders, and a handful of master artisans. Its first president was the geologist and professor at the University of Copenhagen, Johan G. Forchhammer. Among the founding members was its lifelong supporter, the first Danish "designer," German-born architect and professor at the Royal Academy of Arts, Gustav F. Hetsch. In addition to supplying leading manufacturers in Copenhagen with drawings of decorative objects and furniture, Hetsch published collections of designs for craftsmen as well as several essays promoting the applied arts, on commercial as well as cultural grounds. A classicist by conviction, Hetsch strongly believed in both improving export rates and in raising public taste by means of disseminating "universal" classicist doctrines of beauty among leading manufacturers and the workforce.

The main political aims of the Association were: international commerce, the introduction of a secondary school system offering professional training of the workforce, and abolition of the country's medieval guild system. The Association was lobbying in pursuit of these aims when a Parliamentary Act on Freedom of Trade was prepared and passed in 1857, anticipating full adoption by 1862, and it succeeded in the foundation of its own technical school called Teknisk Institut (The Technical Institute) as early as 1843. After a number of revisions, this school was transformed into Teknisk Selskabs Skole (The Technical Society's School) in 1876. The school's director was appointed as inspector general of all technical schools in Denmark. By 1890 there were about 100 schools spread all over the country.[7]

In its official invitation, published in May 1838 with its appeal to take part in the founding of a new association, the group of initiators outlined a number of urgent activities upon which the success of the future Industrial Association would depend. Among the activities were: "Public display and diffusion of excellent domestic products; disposal of the objects displayed by lottery, and the publishing of news from abroad as well as drawings of foreign artefacts that might be useful to us..."[8] This last mentioned point was meant literally: useful objects from other countries should be copied for the benefit of Danish consumers. However, in relation to the role of "design," that is, the growth of applied art, the purpose was to pursue two interrelated aims: on the one hand, to further industrial development by removing the traditional trade system with all its hampering regulations, restrictions, and monopolies, and, on the other, to set a standard for well-designed industrial artifacts by means of general, public education and supplementary, professional training of young craftsmen. At that time, in Denmark as elsewhere, secondary "artistic" or "theoretical" education outside the workshop meant instruction in drawing techniques. The Technical Institute founded in 1843 offered classes in drawing held on weekday evenings and on Sundays. After the amalgamation of the Technical Institute into the Technical Society's School, a special educational program in applied art was introduced, in 1882.[9]

The Industrial Association was primarily concerned with industrial policy; however, its constituency included both businessmen and intellectuals with strong nationalistic aspirations. Both groups of members stressed the essentially philanthropic and patriotic character of the Association's endeavors. And, by the mid-1880s, the Association, due to its newly (1883) formed partnership with the otherwise competing Handicraft Association (a society of master artisans founded in 1840), had a considerable influence in shaping public and professional discourse on industrial policy, particularly in regard to manufactured goods during the period of transition from a traditional to a modern society that the Nordic Exhibition of 1888 epitomized.

The Great Nordic Exhibition was first and foremost a celebration of an event, the end of serfdom, occupying a paradigmatic status in the national Danish narrative. By marking its own anniversary, the Industrial Association was also able to act as a semi-official body representing the whole system of manufacture in the country. Finally, the festive celebration of King Christian IX as a symbol of national unity and identity changed—at least to a certain extent—the general attitude toward an otherwise unpopular king. German rather than Danish by birth, in 1863 he was selected among several candidates only after a highly politicized process involving several courts throughout Europe.[10]

The end of serfdom and establishment of free trade were part of a liberal, yet hardly democratic, policy. In fact, the words "democracy" or "democratic" appear nowhere in the many official guides, announcements, and reports accompanying the exhibition as well as more critical reviews and comments in the press, not even in the social democratic newspaper. The reason for this omission is quite simple. During the 1880s and until 1894 Denmark was governed by a small group of the country's richest landed proprietors who ruled the country by means of a series of so-called "provisional laws," the implementation of which ignored the majority of the members of the lower chamber of the parliament, the *Folketinget*, founded in 1849. The head of government was the wealthy landowner J.B.S. Estrup. This undemocratic ruling was sanctioned by the king, who took advantage of his formal, royal privilege of appointing members of the cabinet. In 1885 a number of civil, democratic rights (e.g. freedom of speech and assembly) were restricted, and the right-wing government established its own paramilitary force and intelligence agency to confront any rebellion throughout the country.

The rallying point of the ruling faction was the so-called "defense question," endorsing the fortification of Copenhagen to be financed via the budget. The significance of this conservative defense policy was visible at the exhibition, where special sections were set up for presentations of army and navy weaponry and technical equipment, and military officers were appointed as members of several of the special juries of the exhibition's different sections.

While the king served as the protector of the Nordic exhibition, the richest landowner in Denmark, Count C.E. Krag-Juel-Vind-Frijs, became president of the exhibition committee. Another count, and one of Estrup's peers, C.F. Danneskjold-Samsø, represented the agricultural section as a vice-president, while Phillip Schou, director of The Royal Porcelain Factory, member of the Royal Academy, and artistic director of Bing & Grøndahl ceramics, served as vice-presidents of industry and applied art. A sign of how the organizers handled the delicate political situation is that Estrup, the head of state, was given absolutely no role in connection with the national exhibition.

The buildings: national romanticism in wood

The organizers of the Nordic Exhibition had two principal ambitions. They wanted the official buildings to signify a national or pan-Scandinavian style of building, and to demonstrate the cultural and economic benefits of quality-based small-scale industrial production as an alternative to mass production. The latter was seen as unsuitable for relatively small economies such as Denmark and other Scandinavian countries.

Accordingly, prestige was linked to the principal buildings: the main building, and the buildings housing the sections of fine arts and fishery, all of which included references to Nordic styles of building. As temporary buildings, they were all in wood.[11] The only building not designed by the chief architect, Martin Nyrop, was the hall of fishery, designed by his contemporary and colleague, Erik Schiødte. Schiødte's building adopted the dragon head motif, known from the stems and bows of Viking ships, in the roof's ridge. All of the exhibition buildings included the reworking of details referencing stave churches and other regional wooden architecture, and old Nordic ornamentation and traditional decorative patterns, such as intertwining snakes, filigree motifs, ribbon like ornaments, and color schemes surviving in the domestic handicrafts.

Figure 4.2 Entrance of the main building.
Source: Photo by Juncker Jensen. The archive of DTU Library. The Technical University of Denmark.

In the main building, Nyrop transferred a number of elements from medieval churches and interiors into decorative details (Figure 4.2). The arch crowning the entrance was inspired by a golden altar from Jutland, and the entrance to the Hall of Fine Arts was flanked by carved wood panels replicating stone carvings found in another church in the same area. Nyrop also used a color scheme derived from what were, at that time, believed to be the traditional colors used on house facades and in the domestic crafts and industries. The presence of ancient motifs in the wooden buildings and a number of other architectural details triggered debates during the 1890s about a revival of Bronze-, Iron-, and Viking-age ornamentation as a possible means of reshaping or constructing a regional Nordic style. In this respect, the architecture accomplished its mission.

Around 1890 most Danish critics advocated a "new" style based upon the idiom of North German or Dutch Renaissance precedents known as the "King Christian IV Style," characteristic of a number of public buildings and castles erected during the reign of that king (1588–1648), in Copenhagen and Northern Zealand. One of the strongest arguments, however, pointed in another direction. It came from the architect Alfred Raavad, who would later join Daniel Burnham's staff in planning the Chicago World's Columbian Exhibition of 1893. In the Industrial Association's journal Raavad published an essay entitled "A National Style," in the year of the Nordic Exhibition. He argued for a contemporary, national style to be built upon the old Nordic tradition of ornamental decoration as it had survived in the villagers' style and feeling for form. His recommendation was not to imitate the old patterns and schemes, but, rather, to cultivate their formal principles, thus making it possible for architects and decorative and applied artists to have at their disposal stable, generalized forms that subsequently might be adapted to various specific contemporary constructive or decorative functions.[12] In fact, Raavad presented in this essay one of the very first formulations of the principle of the "type," predating by 20 years the German Werkbund's program of *"Typisierung,"* expounded by Herman Muthesius, which was to be elaborated further, in theory as well as in practice, by leading protagonists of the Modern movement in the 1920s (e.g. Le Corbusier, Walter Gropius, Marcel Breuer, and, among Scandinavians, Poul Henningsen [PH] and Gregor Paulsson). Raavad foregrounded contemporary Norwegian woodcarving and filigree work in metals and glass as positive examples, as well as the Swedish architect Frederik Wilhelm Scholander, whom he noted for his combination of classical ordering on building façades and decorative ornaments in the old Nordic style to frame doors and windows.[13]

The exhibits and their presentation

Inside these buildings, exhibition designers employed markedly different strategies in displaying various categories of objects, from fine arts to foodstuffs. This was both to appeal to visitors and a means of expressing the character of the item type.

Of the applied arts in the exhibition's main building, contemporary critics and commentators agreed that the exhibition of ceramics was the most interesting and also the most representative (and favorable) of Denmark's contributions. Therefore, the display of ceramics is a useful example of the strategies used in the design and arrangement of the exhibition.

There is much evidence that many manufacturers saw the Nordic exhibition as an occasion for promoting specific artistic and commercial ventures, particularly in the medium of ceramics, especially porcelain—a field of special interest to larger, export-minded manufacturers. The Royal Porcelain Factory, established 1775 (today's Royal Copenhagen), occupied the place of honor in the central hall. Its podium was situated at the very front of the hall, just below the stairways leading from the open area beneath the dome toward the central hall (Figure 4.3).

Figure 4.3 The central hall in the main building with the section of applied art, seen from the entrance.
Source: Photo by Juncker Jensen. The archive of DTU Library. The Technical University of Denmark.

The Royal Porcelain Factory showed a broad selection of its products, from dinner sets without decoration or with simple, stenciled patterns to highly wrought decorative ware including vases with Japanese-inspired decorations.

Philip Skou, director (from 1872) of the Alumnia faience works that took over the Royal Porcelain Factory from the state in 1882, became the managing director of both companies. He was also president of the Industrial Association as well as president of the Joint Committee of the Industrial and Handicraft Associations at the turn of the century (see above). One of three vice-presidents of the committee of the Great Nordic Exhibition, Skou was *de facto* in charge of that organizing body. An innovative industrialist and trained as an engineer, Skou directly engaged in product development for his firm, and his professional training contributed to his firm's experiments with ceramic production technologies such as firing, mixtures of clay substances, and glazing.

In 1884 Skou went a step further and added artistic matters to his responsibilities. It is likely that he initiated a design policy that much later became a part of Scandinavian design culture, that is, the engagement of an artist as a member of the company's management. Within the later historiography of Scandinavian design, one of the standard, discursive elements has been the Scandinavian tradition of engaging in-house designers, not as consultants, but as regular members of the managing staff.[14] That same year, Skou engaged the architect Arnold Krog, first, to revive the firm's blue, fluted china that had been in a state of decline for decades, and, second, to develop new decorative patterns for vases, bowls, and jars. In collaboration with Skou, Krog reinvented the special technique of under glazing by which the decorations were painted on the items before the final firing. This procedure not only protected the painted decoration from direct wear and tear, but also gave the colors more depth and brilliance. Travelling across Europe prior to his engagement with the Royal Porcelain Factory, Krog had, like many others, been introduced to the Japonisme attracting the artistic community, and in his work for the factory he combined the Japanese idiom with Danish landscape motifs along with Danish flora and fauna. This new, naturalistic approach to decoration was a sensation with many observers. A year after his engagement, Krog was appointed "artistic manager" as an acknowledgement of his skills, and from 1890 onwards his title was "artistic director".

The outcome of Krog's efforts as displayed in 1888 was celebrated by Danish and foreign critics, and, at the Paris Exposition Universelle the following year, the Royal Porcelain Factory received several prizes for its products. A notable export success followed, and by 1900 the factory had its own outlets in New York, London, and Paris and later in other American and European cities.

The Royal Porcelain Factory's policy relative to the approaching Nordic Exhibition may serve as an example of how both large and smaller companies approached their respective displays. The ceramic industries were in the forefront of engaging artistic assistance, either by appointing painters, sculptors, or architects as managing members of the design office or by temporarily hiring experts as consultants. The second "big business" in Denmark's production of porcelain and faience, Bing & Grøndahl, also wanted to prominently promote and display its wares, partly in order to compete with the Royal Porcelain Factory. The company engaged the painter Pietro Krohn as master decorator and later art director. Krohn, who was appointed the first director of the Danish Museum of Decorative Arts in 1893 two years prior to its opening, designed the extravagant "Heron Set" dinner and coffee service, so called for its dominant motif, for the 1888 exhibition. One reviewer ironically stated that the "Heron Set" constituted a dissertation on the bird's behavior, including its standing postures and wing movements, suggesting that Krohn's interpretation had led to aesthetic exaggerations.[15]

The same reviewer, art historian and critic F. J. Meier, was, on the other hand, mostly enthusiastic in his comments on the display of another Copenhagen based ceramic works, Peter Ipsen's Widow's Terracotta Factory. The Ipsen factory produced imitations of classic and traditional ceramic works. Its repertoire included copies of forms representing the entire ceramic heritage. In his review, Meier stated:

> Here we find a rich selection of the firm's works, from small, inexpensive and plain merchandise to huge and costly decorated vases and reproductions of Thorvaldsen's statues, close to the originals in size. It is probably the richest *collection of vases* to attract the eyes of the spectator. The vases are to be seen in all sizes, from quite small to very tall, from simple to richly decorated, in all styles: archaic Greek, pure Greek, Italian Renaissance, Rococo, Lewis XVI's style and a few decorated in Persian, Egyptian and Japanese style. In particular, attention must be drawn to the outstanding series of reproductions of pieces in the Munich collection of vases.[16]

Noteworthy in the above quote is the absence of any critical distance from the practice of copying foreign and old styles. On the contrary, Meier seems impressed by the Ipsen factory's ability to provide a wide range of ceramic objects in an eclectic array of period styles. As Meier mentions, in addition to vases and vessels Ipsen produced miniatures of celebrated sculptures and reliefs; for example, scaled-down replicas of neo-classicist sculptures by Bertel Thorvaldsen (Figure 4.4).

Figure 4.4 The display of a selection of ceramic works from P. Ipsen's Widow's Terracotta Factory, Copenhagen.

Source: Photo by Juncker Jensen. The archive of DTU Library. The Technical University of Denmark.

The factory engaged the architect Vilhelm Koch to design a series of three vases featuring the king's 25 years of regency, the centenary of peasant freedom, and Fredensborg Castle, the king's residency in northern Zealand. Each vase was about three feet tall; the first mentioned was executed in the Rococo style, the second in the style of Louis XVI, while the third adopted the Italian Renaissance style.

Again, this simple-minded approach to the issue of style imitation in combination with an absence of artistic inventiveness is characteristic of a great number of exhibitors at the Nordic Exhibition. This was still an era of revivals, which design historians have disparagingly labeled *historicism*; the idea of new, contemporary forms reflecting the industrial age was hardly acknowledged. Few consumers expected such new forms at the time, nor could critics blame producers for manufacturing goods that exerted popular commercial appeal.

There were, however, alternative approaches to both design and display at the Nordic Exhibition. As a reaction to the planned Nordic exhibition, an independent group of artists (architects, painters, and graphic artists) formed a working community with a master potter who had his workshop on the outskirts of Copenhagen. This group called itself "The Association of Decorators" and has been seen in the literature as a Danish parallel to the Arts and Crafts movement. There is no evidence that the group had any knowledge of the British movement (the Arts and Crafts Exhibition Society was established only in 1888), nor of the first manifestations of Art Nouveau on the continent. Nevertheless, many of the group's ideas were in line with the Arts and Crafts movement, including the revival of craft and the use of local materials; traditional techniques, forms and decorative patterns; and a predilection for medieval forms and presumed production methods. Like Nyrop in his design of the main building, the artists sought to reinvent and conventionalize a "national" color scheme for glazing. The favored colors were dark blue and earthy, brownish tones. Another typical feature of the group's ceramic objects is their preference for rather simple or rough shapes and outlines almost to the point of appearing rather primitive and amateurish. In some cases modernized Viking-age motifs were used as decoration, and in other cases decorations had a narrative touch, referring to mythological themes, Nordic folk song motifs, or fairy-tale characters. A few of the ceramic objects derived their forms from animals; rather than relying upon "heroic" or heraldic animals such as lions or eagles, however, they were drawn from local flora and fauna. In other words, the Association of Decorators aimed at construing a vernacular tradition in response to "industrial" ceramics that in their eyes were based on foreign influences, exemplified, for instance, by the blue fluted china that the Royal Factory had introduced around 1790 or the Ipsen Factory's style pastiches (Figure 4.5a). In addition to their own ceramic works, The Association of Decorators presented a few furniture pieces designed in the 1850s by the romantic painter Lorenz Frölich for his own home, as well as a selection of embroideries with botanical motifs created by popular textile artist Ida Hansen.

However, the display of the Association of Decorators was met by criticism in newspaper reviews. The objects on display were for the most part, so the critics claimed, unique, one-of-a-kind pieces.

Accordingly, they were art works, rather than prototypes adaptable for production in series of manufactured goods (Figure 4.5b). On the other hand, leading art critic and historian Karl Madsen resolutely argued in favor of the Association's innovative approach using the ceramic medium for works that were to be considered art rather than utensils. A Danish tradition of studio ceramics was born.[17]

Figure 4.5a Vilhelm Koch, The King's Vase. P. Ipsen's Widow's Terracotta Factory.
Source: Drawing by C. Mortensen.

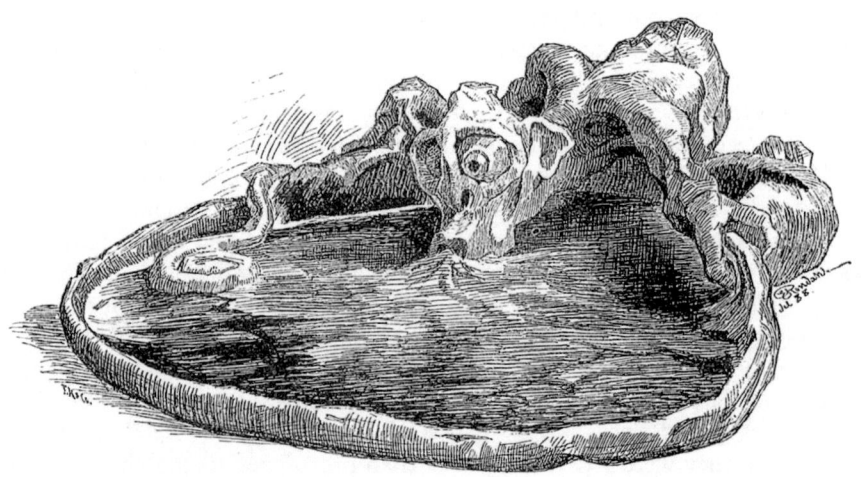

Figure 4.5b Suzette Skovgaard, (Association of Decorators), The troll who sucks out the lake.
Source: Drawing by E. Romdahl. Both drawings are from *Tidsskrift for Kunstindustri* [The Journal of Applied Art], 1888).

As revealed by photos of the Royal Factory display, each company was allotted one or, in rare cases, a couple of free-standing units, the actual space depending on the type of objects (Figure 4.6).

It was a principle that prevailed throughout the central hall of the main building that the stands should be approachable and visible from all four sides rather than laterally against the walls. One result of this approach was that visitors experienced the central hall as both crowded and labyrinthine.

The arrangement of stands was a challenge for exhibition designers. As with the displays of ceramics, the basic principle of presentation for most manufacturers was to display as much as possible of their product lines. Very often, this meant that the limited space only allowed one example of each product to be included in order to leave space for more displays of unique, presentation pieces. Most common for relatively small objects was the exhibition designer's demonstration of the broad range of wares offered by the manufacturer. Its impression was mediated by the compact arrangement of items of different sizes, shapes and colors, providing an overall impression of abundance and variety. In other words, the focus was the sheer number and variety of the merchandise, rather than demonstrating the benefits of mass production of identical or serially manufactured products.

But with other types of products the exhibition organizers utilized a very different approach to display. The exhibition of packaged goods including fruit juice, beer, alcohol, preserved vegetables, refined salt and sugar along with non-edible products such as candles, envelopes, and razor blades, demanded a more sophisticated strategy. These types of products invited designers to work out solutions that stressed an unlimited supply of identical items. This resulted in displays that downplayed the actual physical or functional specifics of individual articles, but at the same time visually communicated the product's ubiquity through packaging and repetition. In many cases the solution was architectural.

Figure 4.6 The display of ceramics from The Royal Porcelain Factory.
Source: Photo by Juncker Jensen. The archive of DTU Library. The Technical University of Denmark.

Designers used models and schemes taken from exotic, often Middle Eastern and Asian, building structures. More than one reviewer remarked on the spectacle of pagodas, mosques, temples, castle gates, porticoes, and arcades constructed of arrangements of bottles and cans (see Figure I.2). Such patterned display was also used to underline the multiplicity of, for instance, a single part of an agricultural implement or a carpenter's tool.

What characterizes this exhibition strategy is the de-emphasizing of the singular item or specimen. Instead, the fanciful aesthetic of the stand design itself becomes the attraction, compensating for the triviality of packaged salt, a lead pipe, or a wooden ladle. A kind of attraction, even power, accrues to ordinary products of everyday use, the result of the imaginatively designed identical units that function as bricks in the construction of a dream-like architecture. The stand of the Copenhagen company Ludvigsen & Hermann, for example, featured a decorative arrangement of toilet traps and other sanitary equipment arranged as radiating archivolts of a monumental entranceway framing an industrial porcelain bathtub and urinal (Figure 4.7). Unlike the displays of ceramics, glass, and other materials, which demonstrated *diversification* as part of the manufacturer's capacity, the issue here is *diversion* by means of an aesthetic that invests the banal with architectural grandeur.

The various strategies and techniques of presentation in the prestige sections of applied art and light industry emphasized the "natural" bond among the physical, functional, and marketable properties of the objects or substances on display.

Figure 4.7 The Copenhagen company Ludvigsen & Hermann's exhibition of water traps and sanitary objects.
Source: Photo by Juncker Jensen. The archive of DTU Library. The Technical University of Denmark.

Objects such as tableware and furniture, products that appeal to consumers' personal identity or social aspirations, were shown as discrete, individual items. The product lines exhibited by, for instance, the Royal Porcelain Factory, included both the prestige pieces of the blue fluted service and expensive wares, such as bowls and vases, along with cheaper plain and undecorated pieces; the latter being targeted toward the lower middle classes. As noted by reviewers, it seems that the factory's and other manufacturers' displays were designed to attract attention from different groups of customers, hence the emphasis on diversity and the customer's ability to choose among alternatives representing a broad range of historical styles and differences based upon cost.[18]

By contrast, all the humble, yet necessary, commodities, remedies, and ingredients of everyday life and of housekeeping, including both durable goods such as boots and brushes and replaceable items such as food or razor blades, did not give consumers much room to express their tastes and define themselves in their choices. Although these items are products of industry in the same way as, say, tableware is, the form of such things was dictated almost exclusively by practical, functional concerns, or by tradition. Only very few objects in this category would, at that time, be expected to manifest any aesthetic concern on the part of the manufacturer. And since chromolithographic poster or print advertisement had not yet reached Denmark in 1888, the solution most exhibitors preferred in order to transcend the constraints of presentation was to endow the stand itself with a striking presence that the items or substances did not possess themselves.

Thus, by presenting these unimpressive things as parts of an aestheticized structure, the enchantment of the arrangement refers to both the possibility of producing, by industrial means, multiple identical items, and to the magical power of the individual entity.

The second part of the industrial products display was installed in the hall of machinery. As a category, "machines" comprised heavy, stationary engines designed for the factory floor, as well as mobile, mechanical tools and implements typical of the different trades and small-scale industries. In addition, machine-made products were, in some cases, on display together with the parts of the machinery. This was the case, for example, with canvas aprons shown next to a powered sewing machine or stacks of rope grouped with a rope maker's wheel. In such cases, however rare they were, a metonymic relation of product to the mechanical means of its production conveyed a simple narrative. In most cases, however, the machinery itself "told a story" about magnitude, weight, power, and, speed; for instance, the number of operations performed per minute. Small signboards, with such minimalistic, decontextualized information were added to the exhibit. In a number of cases, machines and engines were "working"; or more precisely, they were simply running (Figure 4.8).[19]

Thus visitors had no possibility of directly experiencing a causal relationship between the machines and what they produced. The effect placed technology in the realm of mystery and enchantment. All the components of factory-based industrial production—dirt, heat, smell, physical exertion, and, at least to a certain extent, noise—were absent in this *ersatz* display of industrial scenery. It is not surprising that it was the socialist press and a few left-liberal minded Copenhagen newspapers that raised critical comments on this issue. For instance, in the coverage of this part of the exhibition the social democratic newspaper, expressed its reservation thus: "...the exhibition cannot be seen as [a] shiny, newly brushed hat, whereby society is featured

Figure 4.8 The hall of machines with a row of steam engines from competing firms.
Source: Photo by Juncker Jensen. The archive of DTU Library. The Technical University of Denmark.

as a blameless gentleman, saluting all the other nations. As finery jollying up the head of the existing state of things it is humbug."[20] This statement was not only about the exhibition as such. It was, at the same time, addressing the liberal and conservative press and its celebration of industrial leaders and enterprising merchants as acting on behalf of public society.

Conclusion

In comparison to other, contemporary world's fairs, the Great Nordic Exhibition only had a few examples of human showcases.[21] Denmark's status as a minor colonial power was manifested only by the presence of one pavilion, for the West-Indies (as the Virgin Islands were called until the three islands were sold to the United States in 1917), situated close to the building complex of the agricultural sections. During the exhibition, this pavilion, known as the West-Indian House, was inhabited by a St. Croix family who opened their temporary residence to the public and demonstrated the handling of sugar canes. Canes were brought to the family's hut, where a simplified process of crushing, pressing, boiling down, and dehydration took place. Catalogues and reports make no mention of objects or products from another colony, Greenland, and the names of exhibitors do not in all cases reveal what they produced or imported.[22]

The 1888 Nordic Exhibition did include a number of working shops and mini factories (e.g. a goldsmith's shop and a chocolate factory). However, the real attraction, not only of the agricultural section, but of the exhibition in general, was the working dairy (Figure 4.9).

Figure 4.9 The working dairy with the newest equipment.
Source: Photo by Juncker Jensen. The archive of DTU Library. The Technical University of Denmark.

Here, the city dwellers had the opportunity to observe from second-story galleries surrounding the dairy floor how milk was processed, the raw milk being delivered from a cow barn next to the dairy and in the vicinity of the capital. More importantly, however, it served as an example of a modern, mechanized, and efficient production facility rather than juxtaposing machines with finished products or the decontextualized movement of machines *sui generis*.

The working dairy, as a snapshot of the potentials of the "new" agrarian economy, along with Arnold Krog's ceramics for the Royal Porcelain Factory, can be seen as anticipations of future developments. Both the agrarian sector and the branch of ceramic industry represented by the Royal Factory were, as indicated above, dependent upon industrial means of production, and both targeted export markets with their products: bacon, butter, and ceramics.[23]

The popularity of the dairy and of the agricultural section in general should be seen in relation to the revolution taking place within Danish agriculture in the 1880s. In the 1870s massive imports of American grain caused a dramatic drop in prices in the European market. Danish farmers, beginning with owners of larger and middle-sized farms, shifted from grain to animal production, introducing a novel cooperative system, first initiated in the dairy industry in 1882 (700 were in operation by 1890), and followed by bacon factories opened in 1887 and then others in the beef processing industry as well as in the production and processing of feed. In this respect the 1888 Nordic Exhibition indeed marked an anniversary, celebrating the 1788 liberation of the Danish peasant who, 100 years later, was an independent farmer who now associated with his equals by establishing a modern, collective, and democratic agrarian

economy with its own financial system (the Cooperative Banks), a culture of secondary education (The Folk High Schools, beginning in 1844), and systems of local
cooperative retailing.[24] This culture with its member-owner structure was, for about
a century, characteristic of Danish society. Asked about what, in one word, the identity of Denmark was, cultural historian Palle Lauring once answered: "Denmark is
a Co-op."[25]

Both the dairy and bacon businesses (and the other cooperative initiatives) increased
the demand for specialized technology and in this way the agricultural sector in effect
stimulated industrial development and in many cases functioned as a prime mover
of industrialization due to the growing demand for efficient farming implements and
machinery for processing food stuffs. Historians have debated the question of when
Denmark became an industrial rather than an agricultural nation. Some industrial
historians argue that the decisive step was taken only after World War II, when the
contribution of industry to the gross national product exceeded that from the agricultural sector, while another group of historians claims that the 1888 Nordic Exhibition
eventually symbolized the very beginning of the era of transition that also saw the so-
called "change of political system" in 1901 by which real political democracy was introduced. The change of system was caused by an ever stronger pressure from a group
of liberal-minded farmers and independent urban tradesmen and workshop owners
claiming the constitutional right to execute political power according to the group's
majority share of seats in the legislative chamber of the parliament, the *Folketing*.
More generally, the grassroots character of the many cooperative initiatives and a
plethora of popular self-governing organizations across the country as well as the
many professional societies and associations established during the last quarter of the
ninetieth century created channels for articulating demands for more expert services,
supplies of equipment, implements, specialized tools etc. to facilitate manufacturing
processes and cater to leisure activities. This development is linked to both a demand
for influencing the political course of the country and a demand for an increase of
productivity and administrative efficiency; demands that eventually could only be met
by developing an industrial economy.

In terms of party politics, this large group or movement of liberals was by 1870
organized into The United Left (left meaning liberal in this context), and the group
gained support from urban intellectuals. The growing strength of the workers
movement is reflected, first, in the founding of a Social-Democratic party in 1871,
and, secondly, in the national Workers Union entering a general agreement in 1899
with the Danish Employers Organization leaving once and for all settlements and
regulations concerning wages, work conditions, the shop steward system, holidays,
etc. to direct collective bargaining between the two central organizations and their respective sub-sections. Internationally, this is often referred to as the "Danish Model"
(sometimes called "The Scandinavian Model") of the labor market.[26]

The 1880s and 1890s have traditionally been difficult to interpret historically because of the complexity of political, economic, and cultural tensions and controver-
sies that characterize these decades. In many respects, the Great Nordic Exhibition
represents an attempt to balance competing interests. The head of the right wing
government, J.B.S. Estrup, a landed proprietor, played absolutely no role at all in the
organization of the exhibition, even if the exhibition's president, Count C. E. Krag-
Juel-Vind-Frijs, was of his peers. Estrup's name appears nowhere in the official
literature, nor was he mentioned in the numerous newspaper reports or magazine

features. Neither did he play a role in the exhibition's festive ceremonies. This indicates the problematic nature of the political situation the organizers faced.

The class Estrup represented was not particularly enthusiastic about the democratic changes taking place in the agrarian economy. The cooperative movement with its "self-help ideology" represented a challenge to the older, paternal, and landed economy, and, in turn, to a political culture that was based upon a socially divided parliamentary system. In fact, the political groupings that made up the cooperative movement's constituency had won the majority of seats in the "lower" chamber of the parliament, the *Folketinget*, back in 1872. But the "upper" chamber, the *Landstinget*, simply refused to adopt the *Folketinget's* decisions and legislative initiatives. In return, the *Folketinget* did not vote for the prime minister's proposed Budget Acts. As noted earlier, the effect of this was that Estrup ruled by provisional laws and essentially suspended all democratic political processes.

The Nordic Exhibition of 1888 reveals the contours of a new alliance between medium-light industry and cooperative farmers, who increasingly would demand more sophisticated equipment and machinery for cooperative farming. The economic developments of the last quarter of the nineteenth century included several subgroups; among these a more radical and social-liberalist group of intellectuals with a strong hold in the urban middle class and, to a certain extent, also in the class of small farmers. This group's anti-militarist, pro-European, anti-nationalist, and outspoken democratic aspirations had support from the majority of the cultural elite, and it was a representative of this elite, Georg Brandes, a controversial literary historian and cultural critic, who coined the phrase that still serves as a description of the cultural dynamic of the period 1870–1900 in Denmark: "the breaking through of the modern." In its way, the displays of the 1888 Nordic Exhibition reveal the social, political, and economic complexities of this era, while also contributing to shaping an emerging and modern Danish identity.

Notes

1 About 5,600 exhibitors were from Scandinavian countries, around 90% of the total. Just over 4,100 exhibitors were from Denmark. Of around 2,500 Scandinavian exhibitors displaying applied arts—furniture, ceramics, and wares in gold, silver, glass, etc.—about 1,800 were Danish.

2 Camillus Nyrop, *Den nordiske Industri-, Landbrugs- og Kunstudstilling: Officiel beretning* (Copenhagen: The Exhibition Committee, 1890), 119–125. [The Nordic Exhibition of Industry, Agriculture, and Art. Official Report]. This report is the most important source of information about all kinds of statistics relating to all aspects of the exhibitions. Recently, a review of the huge 1888 exhibition archive in the National Archive has been published, see Nicolai Falberg Jensen, "Den nordiske Industri-, Landbrugs- og kunstudstilling i København 1888—en introduction til arkivet og dets anvendelsesmuligheder," *Erhvervshistorisk Årbog* 1 (2015), [The Nordic Exhibition of Industry, Agriculture, and Art in Copenhagen 1888—an introduction to the archive and its use]. Jensen's indirect account of data concerning the exhibition differs in some details from Nyrop's original report, in particular in relation to the ordering and hierarchy of parts, sections and subsections, but Jensen seems to be right in his well-documented and accurate account. In short, the 1888 exhibition included main sections for applied art from non-Nordic counties, Nordic applied art (the largest section), a section for home industry, public health, and educational matters. The main sections 6 to 12 included the various subsections for agriculture (including dairying and a dairy show, fishery, forestry and gardening. Then there were two sections for machinery (one with agricultural tools, implements, and farming machines) and one for means of transport. Finally there were three sections for the fine arts (painting and graphic art, sculpture, and architectural design).

3 The "official" buildings covered a total of c. 58,000 square meters; with the main building alone covering c. 15,000 square meters.

4 Camillus Nyrop, "Kunstindustrien på den nordiske udstilling" ["Applied art at the Nordic exhibition"], *Tidsskift for Kunstindustri* [Journal of Applied Art] 4, (1888), 264. Camillus Nyrop was the architect Martin Nyrop's cousin.

5 On Finnish nationalism in this period, see also Chapter 7 in this volume, Bart Pushaw, "Finland at the 1900 Exposition Universelle."

6 Randi Gaustad, "'Dette eiendommelige norske', Norge på den Nordiske industri-, landbrugs- og kunstudstillingen i København 1888" [The uniquely Norwegian, Norway at the Nordic Exhibition], in *Den stora nordiska utställningen i Köbenhamn 1888*, ed. Severi Parko (Helsinki: Nordic Forum of Decorative Arts, 1989), 47–51.

7 The establishment of these schools parallels developments in other European countries at the time. See also Chapter 3 in this volume, Daniela Prina, "The Belgian reception of Italy at the 1885 Antwerp World Exhibition," p. 53.

8 The National Archive, Industriforeningen i Koebenhavn. Generalforsamlingen.

9 This program, along with The Women's School of Drawing and Applied Art, established in 1875, subsequently formed the roots of what is today's School of Design, which has been a section of the Royal Academy of Arts in Copenhagen since 2011. Accounts of developments within the technical education in Denmark are to be fund in, among others; Camillus Nyrop, *Bidrag til Dansk Håndværker-Undervsinings Historie* [Contributions to the history of craftsmen's education in Denmark] (Copenhagen: The Technical Society, 1893), esp. pp. 189–236; Verner Rasmussen, "De tekniske skolers historie" ["The history of technical schools"], *Uddannelseshistorie* [History of Education] 3 (1969), 7–41; Ida Juul, "De danske håndværker-uddannelser i krydsfeltet mellem det europæiske og det nationale" ["The system of of Danish craftsmen's education in the cross field between the European and the national"], *Uddannelseshistorie* 47 (2013), 60–80.

10 With the death of the childless king Frederik VII in 1863 the succession was broken. The only heir to the throne from the Oldenburg dynasty that had reigned since 1448 was a distant female relative, and since the duchy of Holstein was under German jurisdiction in such matters, a female heir to the throne could not be accepted. The Russian tsar, Nicolai I, was a descendant of the Oldenburg dynasty and therefore in a position to claim rights to the throne or, at least, the right to play a decisive role in pointing out a suitable member of another dynasty to be the future head of the Danish monarchy. Prince Christian of Glücksburg was accepted by the tsar, and diplomatic activities paved the way for the so-called London Treaty of 1852 (formally ending the "First Prussian War" (1848–51)). The treaty was co-signed by England, Russia, France, Prussia and Austria, and it installed the Glücksburg dynasty as successor to the Danish throne. The Glücksburg name is referring to a castle situated in the duchy of Schlewig, south of what since 1864 and the defeat in the "Second Prussian War" has been the border between Denmark and Germany. Christian IX was nicknamed the "father in law" of Europe since his daughters and sons were installed in several European courts; the son, Wilhelm was crowned as king George I of Greece, while his grandson Carl was elected king of Norway under the name of Haakon VII. His daughter Alexandre married the later king Edward VII of England, and another daughter, Dagmar, married the later Russian tsar, Alexander III. His son followed him as Frederik VIII of Denmark. Other daughters married princes and dukes in Germany and England.

11 Whereas wood was associated with traditional architecture in Denmark, in other Nordic countries stone took on similar nationalistic overtones. See Chapter 7 in this volume, Pushaw, "Finland at the 1900 Exposition Universelle," p. 133.

12 Alfred Raavad, "En national stil," *Tidsskrift for Kunstindustri* 4, (1888), 25–33.

13 The inclusion of Norwegian wood carving and filigree work in the Nordic Exhibition 1888 is discussed by Gaustad *op.cit.*, and Scholander's ideas of architecture and the Nordic heritage is covered by Swedish art and design historian, Bo Grandjean in his *Frederik Wilhelm Scholander och drömmen om Renässancen* (Stockholm: Nordiska Museet, 1979).

14 This rhetoric crystallized in the 1950s. See, for instance, the introductory text by Gotthard Johansson (at that time, president of the Swedish Society of Arts and Crafts and Industrial Design), in the catalogue, *Design in Scandinavia*. A travelling exhibition of the same name that toured to 24 places in the USA and Canada, 1954–57. Among numerous publications repeating the standard comment, see, for instance, Jeremy Aynsley, *Nationalism and Internationalism, Design in the 20th Century* (London: Victoria and Albert Museum, 1994), especially the chapter titled "The case of Scandinavia," 42–45.

15 F. J. Meier, "Noget om dansk Keramik på Udstillingen" ["On Danish ceramics at the exhibition"], *Tidsskrift for Kunstindustri* 4, (1888), 80.

16 Ibid., p. 85.

17 See, Karl Madsen, "Dekorationsforeningen," *Tidskift for Kunstindustri* 4, (1888), 145–154; Mirjam Gelfer-Jørgensen, *Influences from Japan in Danish Art and Design 1870–2010* (Copenhagen: Arkitektens Forlag, 2013), p. 132ff.; Lars Dybdahl, *Dansk keramik 1850–1997* [Danish ceramics], (exhibion catalogue) (Lyngby: Sophienholm, 1997), esp. 17f.

18 See, for example, Meier, *op.cit*, 78, 83. As noted above, stylistic novelty effectively had no commercial appeal; historical revival styles were the most saleable.

19 The machines produced nothing; there was no input, nor an output, and the source of steam power that made the machines run was placed outside the halls in a separate building from where the steam was transported in subterranean tubes to the running machines and engines.

20 *Social-Demokraten*, May 6, 1888, p. 2.

21 Cf. Paul Greenhalgh, *Ephemeral Vistas. The Expositions Universelles, Great Exhibitions and World's Fairs, 1851–1939* (Manchester: Manchester University Press, 1988), especially, Chapter 4, "Human Showcases," pp. 82–111.

22 At that time, no Danish manufacturer would declare if a commodity, for instance, bottled thrane oil, came from Greenland, the Faroe Islands or elsewhere.

23 Today the vast majority of former independent dairies have been shut down or, in larger units, subsumed under the Danish-Swedish concern Arla Foods, while most of the bacon factories are now parts of the giant Danish Crown corporation. Both are still cooperative organizations, but assume a de facto role of monopolies. And the Royal Porcelain Factory is now a part of Royal Copenhagen, a company structure that includes a number of former independent firms (the Royal Porcelain Factory, Bing & Grøndahl, Georg Jensen's and A. Michelsen's precious metal smithies, Holmegaard's glass works). Since 2013, Royal Copenhagen has been owned by the Finnish corporation Fiskars and has become a part of a family of firms that include Iittala, Rörstrand, Raadvad and Gerber.

24 By 1896 these local organizations were united into the Danish Cooperative Wholesale Society, and an umbrella organization, The Danish Cooperative Societies, was set up in 1899.

25 Lauring's statement is widely cited, and it was used as an intro to another popular historian's (Paul Hammerich) television series, "Old Denmark," which was broadcast by the Danish Broadcasting Company, in the early 1970s. The television series was later transformed into a publication in six volumes, but nowhere is the original source mentioned. An indication of the multiplicity of old and new references (in Danish) to Lauring's statement is easy to find on the internet: "Danmark er en brugsforening."

26 Among the events marking the change of system, mention should be made of the abolition of the system of privileges of the feudal nobility by 1919, that is to say, conversion of entailed estates into fee simple, thus removing legal obstacles to subdividing large tracts of privately held land. This was followed by the re-distribution of the land of some estates to smallholders, which contributed considerably to extend and intensify cooperative structures and culture. The smallholders were supported politically by both the Social Democrats and by a faction of the left liberals within the United Left, who by 1905 had founded a party called the Radical Left (meaning non-socialist social liberals) that was supported by both the urban and landed intelligentsia, such as newspaper editors and journalists, university professors, school masters, and representatives of the Folk High School movement.

5 A neoclassical translation

The Hôôden at the 1893 World's Columbian Exposition

Hannah L. Sigur

The Hôôden, Japan's pavilion at the 1893 World Columbian Exposition, survives mainly as drab photos that barely hint at its impact on contemporary visitors.[1] Though dwarfed by the White City, the 4,050 square foot structure was an exposition highlight. *Inland Architect and Building News* vividly described its unpainted exterior of "warm and delicate cream color;" ambulatories and verandas, exterior walls and roof trusses in the distinctive textures of coarse-grained cypress, fine fir, and knotty cedar; and projecting roof with an "upper surface the dark natural color of copper hand-made tiles." Embraced by a garden on the north point of a nearly empty Wooded Island in a large lagoon, accessible only by boat or by bridge, its isolation and exotically elegant rusticity made an arresting parity with the neoclassical grandeur of the nearby Court of Honor around the rectangular Grand Basin. It was stunningly lovely (Figures 5.1 and 5.2).

Three exquisite interiors climaxed in its central "Jodan-no-ma," where movable walls (*fusuma*) painted with pines and phoenixes (*hôô*) "emblematic of power and glory,"[2] shimmered beneath two hundred and seventy panels with the hôô motif on gold ground coffered by gold lacquer and gilded metal.[3] American periodicals

Copyrighted by C. D. ARNOLD.   FRONT VIEW OF THE HŌ-Ō-DEN.

Figure 5.1 Kuru Masamichi, The Hôôden, World's Columbian International Exposition, 1893.

Source: Reproduced from *The Hôôden (Phoenix Hall), An Illustrated Description of the Buildings Erected by the Japanese Government at The World's Columbian Exposition Jackson Park, Chicago* (Tokyo: Ogawa, 1893), n.p. Photograph: C.D. Arnold.

Figure 5.2 View of the north end of the Wooded Island, showing the Hôôden, the Women's Building, the Illinois Building, the Government Building, and the circular wing of the Fisheries Building. Starks W. Lewis, Amateur, Brooklyn.
Source: NY Brooklyn Museum Archives, Goodyear Archival Collection (S03_06_01_image 2257).

describe brilliant blues, yellows, blacks and reds "...so distinctly associated in our minds with Japan."[4] Opulent, glinting in the light, this interior intended to overawe. "It is probable," declared a reviewer, "that even Japanese visitors may be astonished to see such a revival of ancient decoration"[5] (Figures 5.3 and 5.4).

To stand as a nation of note through architecture had been an ongoing motivation over the course of Japan's presence in the exposition movement. The Japanese had witnessed the 1862 London exhibition and actively participated in the 1867 Paris Exposition Universelle. The latter had introduced national pavilions to showcase the received wisdom that the products of each country vying in the great palaces owed their distinctive qualities and relative competitiveness to a national character distillable to a viewable quintessence. From this point, the movement had matured from grand yet straightforward competitions of technology and manufacturing into grandiloquently visualized ideological statements necessitating a structural presence as a centerpiece of identity. At that event, an Edo businessman, the shogunal Tokugawa clan, and their bitter rivals the Shimazu of Satsuma all featured buildings. The outsized success of the "Satsuma prince's hunting pavilion" may have provided an object lesson for the Shimazu, who were prominent in the 1867 coup that ended the shogunal system and united the nation politically. The new Meiji regime enthusiastically embraced the expositions as a matter of policy. Its debut at the next major fair, the 1873 Vienna Weltausstellung, featured a garden complex with scale-model shrine, temple, and market structures. It followed at the 1876 Philadelphia Centennial with a teahouse/bazaar and residence that stood apart on the grounds. But neither made the dignified statement of a true national pavilion.

The 1893 World's Columbian Exposition, to date the grandest and most comprehensively ideological in architectural vision, offered the perfect opportunity to advance Japan's cultural, economic, and political distinction through the placement and form of a true national pavilion. In its success, the Hôôden was an apotheosis not

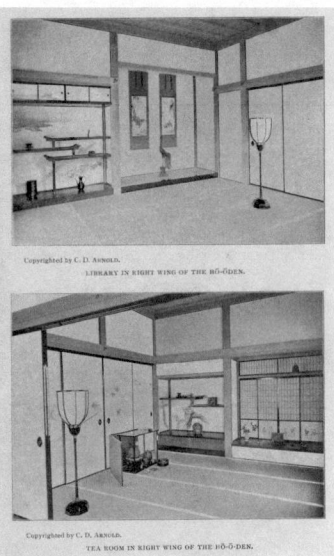

Figure 5.3 Interior of the Left Wing ("Fujiwara Epoch"); Library in Right Wing of the *Hôôden* (top); Tea Room in the Right Wing of the *Hôôden* ("Ashikaga Epoch").

Source: Reproduced from *The Hôôden (Phoenix Hall), An Illustrated Description of the Buildings Erected by the Japanese Government at The World's Columbian Exposition Jackson Park, Chicago* (Tokyo: Ogawa, 1893), n.p., Photographs: C.D. Arnold.

Figure 5.4 "Central Hall": "Jodan-no-ma" (bottom); Kon-no-ma (top left), Sho-Sai (top right).

Source: Reproduced from *The Hôôden (Phoenix Hall), An Illustrated Description of the Buildings Erected by the Japanese Government at The World's Columbian Exposition Jackson Park, Chicago* (Tokyo: Ogawa, 1893), n.p., Photographs: C.D. Arnold.

only in expression of Japanese identity, but also, in the masterful negotiations that brought it to life, of resolute "underdog" nations of its day. The case can be made that symbolically it pointedly reinterpreted the Euro-centered biases of exposition dogma. One stemmed from the implicit link between a country's relative standing and authority over its image. Japanese dissatisfaction had peaked with the French, who imposed fanciful "Japanese" display facades and ornamental gateways designed by local architects inside the Main Buildings at the 1867 and 1889 Expositions Universelles. The latter fair also featured a "Japanese House" in the Histoire d'Habitation Humaine, a pseudo-anthropological assemblage of "representative" housing of various cultures. It offended because of its location among "underdeveloped countries," rather than in "proximity to the countries of the West," and because the displays conveyed the impression that Japan was "a recently opened, naïve country." That some fairgoers even likened Japan to a part of China was deemed "beyond deplorable." For Japan to "present to the world the reality of the everlasting great empire of the East"[6] henceforth required absolute control over every detail.

Second was the truism of Western nations' innate political, technological, cultural, and economic superiority predetermined by progress from Hellenic origins to Roman civilization to the Italian Renaissance, culminating in the current Industrial Age. This saw expression architecturally in Beaux Arts neoclassicism.[7] Beginning with the Weltausstellung, visual reference to the Greco-Roman orders and an organizational syntax of symmetry and geometry characterized many expositions, particularly in the New World. In Japan, where the coup had ended not merely a regime, but even the philosophical and intellectual basis on which it had rested, progressive intellectuals paradoxically embraced this biased ideology in the drive to modernize, appropriating its visual analogue of neoclassicism for civic structures.[8] This chapter will show that in the context of the expositions, the Hôôden expressed neoclassicism "in Japanese," proof that Japanese civilization, like a rediscovered lost tribe, shared the West's idealized heritage and thus enjoyed a natural and inevitable rise as a world economic and political power.

Finally, from the 1855 Exposition Universelle onward, the exposition movement highlighted the eighteenth-century theories of Batteaux, d'Alembert, and Winckelmann of "fine art" as a moral and aesthetic apogee.[9] "Civilized" nations reified and adulated their identities with both ancient treasures and new works. Dovetailed with the accepted wisdom of Greco-Roman origins, these usually appeared in prominently located Galleries of Fine Arts, secular temples of commandingly conservative neoclassical design even when the overall architectural scheme took other directions.[10] This imbued not merely dignity, but implied that as humanity's purest expression, a nation's fine art was the ultimate symbol of its superior Hellenic origins. Finding the notion vague and inconsistent yet noting its signature status, Meiji government and intellectual circles—fiercely competitive and determined to be on total parity with Western nations—immediately embraced it.[11] As Japanese language lacked even an equivalent word, in time for the Weltausstellung debut they had linguists contrive a cognate: *bijutsu*, "fine art"—and its extension, *bijutsukan*, "fine art museum."[12] And importantly, they adapted its associated dogmas, according fine art a focal point in the new ideology of national unity. Two decades later, the same officials planning the Hôôden concurrently were directing their deepest attention to plans on the drawing boards for permanent museums in the ancient capitals of Nara and Kyoto. Pointedly designed by Japanese architects, these inserted into Japan's cultural heartland neoclassical structures to display heritage artifacts re-identified as national fine art.[13]

But if the presence of fine art in a nation denoted preeminence, it also highlighted inconsistences at the heart of the ideology of globalism and a resulting network of shoals to navigate at the fairs. One was its hyper-idealization as a vague unfungible essence revealed by style and subject, untainted by any practical function, which raised a society "above the level savagery."[14] Second were stipulated "pure" categories: painting, sculpture, architecture, and related media. Third was inconsistency: Selected heritage works graced art galleries solely as distillations of a nation's quintessence. New fine art competed for medals, cash prizes, and buyers—a conflation with commodity. In fine art Japan had always fared poorly; works by living artists in particular deemed either inferior to Western verisimilitude or stylistically inauthentic. Meanwhile Japanese textiles, ceramics, and fine crafts in the Manufactures and Liberal Arts palaces always took top prizes, yet classification as utilitarian—lower in the production hierarchy— rendered even the best unsuitable as suitable icons of high civilization.[15] Japan's exposition commissions lobbied to eliminate this illogical barrier over the course of several fairs. They attained partial headway in 1893 when judges at last conceded that the categories were "largely artificial."[16] Tour-de-force ceramics, bronzes, embroideries, and enamels finally competed at the top level re-ranked as fine art. Yet even as Japan's mass-market products led a journalist to enthuse, "that European and American manufacturers will have to reckon that country in [the] future as a serious competitor,"[17] the revisions failed to displace the status value of conventional works in the Art Gallery, where a new "objective" rating system effectively eliminated those by living Japanese. A lengthy *New York Times* article on Japan's entries blared, "High standards of merit established by the [Exposition] Judges [were] so rigorously enforced as to cause discontent—not a single picture in oils [was] accepted and few watercolors or drawings."[18] Nothing had really changed.

The Imperial Commissioners charged with Japan's national pavilion and its fine art at the exposition, Kuki Ryûichi (1852–1931), Okakura Kakuzô (Tenshin) (1862–1913), and Yamataka Nobutsura (1842–1907) were cosmopolitan, ardent nationalists who straddled political, economic, intellectual, and art worlds in a manner inconceivable today. Kuki and Yamataka had served in prior fairs and in 1878 were founding members of the Dragon Pond Society (Ryûchikai), established to encourage and commoditize modern, yet identifiably Japanese, fine art. Their activities facilitated politicized theories Okakura was then refining as Founding Director of the Tokyo Academy of Art (Tôkyô Bijutsu Gakkô)—whose faculty and top students executed the paintings inside the Hôôden.[19] All participated at the highest levels in an ongoing government program to interweave Japan's exceptional material heritage into the new imperial ideology. Their embrace of the ideological aspect of fine art, clear from the national museums that they then were planning, suggests that one intent for the pavilion in Chicago was to recast Japan's fine art along Japanese cultural contexts in a way its uninformed viewers would comprehend as "authentic" and therefore valid. To gain acceptance as a "civilization" on its own terms would resolve the discrepancy between Japan's lower hierarchical status and obvious international rise, with its corollary *entitlement* to economic, social, and political primacy.

The Commissioners initially had envisioned three separate accurate-to-era buildings representing successive apogees in Japanese civilization. Customary of the didactic tableaux that since 1867 had been part-and-parcel of major fairs, they were to be furnished with ornaments, objects, and an occasional role-player or mannequin. Yet these were not to show "usual" life as typical of such displays, but that of Japan's glorious past: a colorful *shindenzukuri* hall of the twelfth-century imperial court; a *shoin* study and teahouse in the restrained taste of the fifteenth-century Ashikaga warlords—a cultural apogee and also correlating with Columbus's 1492 discovery of America; and an opulent palace shoin of the Tokugawa shoguns' eighteenth-century zenith and the setting for the grand paintings. It was chosen partly for the paradoxical reason that that regime had in 1854 accepted the first trade treaty with the United States.[20] Warmly regarded by Americans, in Japan that highly disadvantageous pact had been a major factor in the coup and remained a very sharp thorn in the new government's side.[21] Indeed, in Japan such structures stood as symbols of a decrepit, deservedly defunct past.[22]

As ultimately realized in Chicago these three architectural forms provided the interior references. But the surmounting of their overall stylistic and material incoherence, and even an inadvertent undertone of political inferiority, owed to architect Kuru Masamichi (1855–1914). Both training and temperament led him to an inductively creative leap: recognizing that historical reference, location at the fair, and above all exterior appearance could be made to accord with neoclassical theories of rational order and balance. The result was a dignified, entirely distinct, symbolically powerful whole in complete unity with the exposition ideology. In arguably equalizing Japanese visual identity in relation to acknowledged powers, it provided an elevating context for the interior *fusuma* wall paintings and heritage furnishings.

Kuru began work in April 1892, initially dismissed by the Pavilion Committee as "a contract architectural work supervisor."[23] He had never designed or supervised traditional wood construction.[24] As a pupil at Tokyo Imperial University's elite architecture program he had lived and studied in a neoclassical building in conditions of total indoctrination to the Beaux Art theories necessitated by the government's civic modernization program.[25] A central axiom held that a "civilization" had "architecture." This meant masonry, ideally stone—interpreted as a material whose intractability compelled the "honesty" and rationalism first expressed in the Classical Orders.[26] It dismissed wood structures as "carpentry." Yet expositions demanded "authenticity" of national pavilions, which in Japan's case meant wooden structures. At the International Congress on Architects concurrent with the Exposition, a colleague presented an address by their former professor Josiah Conder that invoked equivalency with classicism to contend that Japan's building heritage was nonetheless architecture. Conder averred that the archipelago's seismic and climactic intractability likewise inspired "rational" and "honest" progression, and therefore innately,

> The wooden style of Japan is based upon rules of proportion almost as studied and refined as those of the Greeks, and it displays constructive and applied decoration of a high order; so that on the score of artistic qualities...[it] can hardly be denied the title of Architecture...every constructive and decorative detail which it possesses is the natural and logical outcome of the most effective use of timber... none of these structural or ornamental qualities would possess rationality or stability if executed in masonry.

The English Conder possessed eminence; trained at University of London and in the atelier of William Burges, a Soane Medal winner, he had come to Japan in 1877 as a professor of architecture at Tokyo Imperial University and recently retired. His praise and this equivalency reflected favorably upon the Hôôden, and by extension on the objects within—newly designated as "fine art," they now resided in a building deemed "architecture"; the "mother of all arts."[27]

Technically Kuru was superfluous to the job at hand because a master carpenter oversaw construction for Japan's official complexes at every fair. Along with the buildings themselves, traditional garb and methods entertained and assured foreigners of "authenticity" to highly positive effect (Figure 5.5). In 1893 the honor fell to a lead artisan for the new Imperial Palace, Oda Senkichi of Japan Civil Engineering Co.,[28] who in a familiar pattern directed a crew to construct and disassemble the pavilion in Japan, and then reassemble it in Chicago. Yet Kuru maintained a conspicuous, continuous presence, arriving before Oda and remaining through construction. At the dedication he stood at the podium, delivered a brief address in Japanese, and placed the building key in the Imperial Commissioner's hands.[29] Once home, he spoke at the Architectural Institute, and was honored at professional events.

His inclusion surely owed in part to the value of his credentials to the pavilion's claim as "architecture," critical to its own status, that of its contents and thus of the nation. But he also brought an unusual asset. His graduation thesis had addressed a conundrum of his training—how to reconcile native buildings against Beaux Arts evolutionary theory, which began with Egypt, through Greece, Rome, the Italian Renaissance, and culminated in Western Europe.[30] Kuru's "History and Theory of Japanese Architecture," the first to nativize its concepts, began in prehistory and referenced the classic literature on which the government based its new Shinto Emperor cult, such as the eleventh-century *Ôkagami (The Great Mirror)* and the eighth-century *Kojiki (Record of Ancient Matters)*.[31] His binding of Japanese architectural history to nationalist ideology within a Western progressive and rationalist academic frame brought an intellectual following for a number of years. It is thought that Okakura chose Kuru based upon his lecture at the Tokyo Academy of Art on "The Art of Architectural Decoration," that is, native motifs.[32]

Figure 5.5 Japan's dedication.
Source: Reproduced from *The Dream City, A Portfolio of Views of the World's Columbian Exposition* (St. Louis: N.D. Thompson Publishing Co., 1893), n.p. https://upload.wikimedia.org/wikipedia/commons/1/1d/Hooden.jpg.

For Chicago, unbowed and undeterred by the class-based aristocratic arrogance of Kuki, Okakura, and Yamataka, Kuru dismissed their idea that "Japan" should appear as a complex of three traditionally accurate buildings. Bringing honed aesthetic education to bear, he observed that a Tokugawa grand shoin, Fujiwara shinden-zukuri and Ashikaga villa shoin and tea room shared little structural or historical relation; simply grouped they did not meet the expressed aim of "architectural magnificence."[33] "It's just as if it were sashimi garnished with candy," he railed, "it proceeds by any means possible as long as it can be tacked on, and is truly uninteresting..."[34] His criticism persuaded Kuki to propose three period interiors inside one structure of singular importance, modeled on the opulent Rokuonji (1397), better known as Kinkakuji, or "Golden Pavilion." With shinden on each of its first two floors and a "Chinese" *kara-yô* on the top, it offered a perfect historical precedent of combined architectural styles. Western fascination with the exotic regality of its gilded exterior only added appeal.[35]

But fair officials rejected Kinkakuji as competitively tall with the White City. This no doubt suited Kuru, who thought it a poor choice because flat Jackson Park did not match the historic structure's hilly setting. But the Wooded Island location inspired him to imagine another possibility. Secured over the protests of other nations resentful of the obvious distinction bestowed, and of Frederick Law Olmsted who feared for his landscape design, this untrammeled setting had met a critical stated goal—the "expression of architectural beauty"[36] by Japanese standards. Efforts to clinch it were well underway by December of 1891, when Daniel Burnham wrote the Imperial Consul General requesting that he come as soon as possible to Chicago to discuss the matter, followed by a rather defensive letter to Olmsted protesting innocence on the suggestion but concurring with the Japanese on the compatibility.[37] But Kuru's Beaux Arts training would have predisposed him to consider Japan's pavilion in the context of the entire fair, including the island's large size and the selected site's central placement between the neoclassical Court of Honor and Art Gallery on the South Pond, circumstances of visual and symbolic parity. Kuru also recognized that the bankside situation approximated that of an exquisite, culturally defining vestige of Japan past: the graceful Byôdôin, poised on an artificial peninsula to cast its reflection on a pond (Figure 5.6). Built in 1053 by the Fujiwara clan in its glory days as a villa, then repurposed as a Buddhist temple, it is a highlight of Uji, a locale redolent of ancient imperial Yamato. To the Japanese its shape recalled a great phoenix alighting, hence the sobriquet Hôôdô, "Phoenix Temple." But Kuru almost certainly recognized that its columned, symmetrical wings flanking a central structure coincided with classical form—in short, the Byôdôin matched Western criteria of high aesthetic status and beauty on Japanese terms. Over resistance from Okakura and Yamataka, he convinced Kuki of its suitability as a model.

Aligning with Beaux Art theories of architectural perfection in which visual harmony supplants authenticity, he adapted the Hôôdô's alate form, reconceiving the Japanese pavilion as three distinct buildings linked by detached columned ambulatories. He then cloaked and manipulated their disparities: Ignoring chronology, he placed at center the Tokugawa shoin, grander than the Ashikaga villa and Fujiwara shinden. He added Fujiwara-style railings to the Tokugawa and Ashikaga buildings and changed Ashikaga shutters (*mairado*) to Fujiwara lattice (*shitomi*). He roofed all with copper plating authentic to none. The suffix *-do* a religious designation, he emended as *-den*, hence Hôôden, translated as "Phoenix Hall." Formal elegance dwarfing modest scale,

Figure 5.6 Byōdō-in Phoenix Hall, Uji, Kyoto prefecture, Japan, 1059.
Source: Photograph: Oilstreet2, converted to gray scale from color original. https://commons.wikimedia.
org/wiki/File:Byodoin_Phoenix_Hall_Uji_10-1R.jpg.

the symbolic bird of newly risen Japan now also arose as a weapon of identity aesthetics to equal the White City's monumental neoclassicism. One unanswerable question is whether Kuru had such parallels in mind with respect to Kuki and Okakura's goals for the furnishings and fusuma wall paintings—both now "fine art"—the purpose for which it was perfectly suited.

Laconic vagueness belied and surely obscured to readers the significance of Okakura's exposition booklet essay to the elaborate brocade of such concerns. Written in English, he describes Japan's art as beginning in A.D. 629, when Chinese influences expelled "that purity and simplicity which were distinctively Japanese" and introduced Buddhism, with its "traces of Indian, Greek and other western schools."[38] That "art" began only with rejection of the indigenous in favor of Western-tinged alien forms was considered not merely plausible, but even a necessary dogma; similar allusions pepper period commentary by influential native and foreign thinkers. The tableaux, which featured objects from imperial collections, supplied a key element.

As head of the Education Ministry in the 1870s Kuki had enjoined Okakura and many others in a massive, ongoing project to catalogue Japan's material heritage. The initiative entwined with exposition goals as a basis for displays and manufacturing models.[39] But it shortly took a starring role in the drive to unify, consecrate, and empower Japan's historically regionalized population by turning the emperor, relegated by long tradition to impotence, into the supreme living deity of State Shinto. A "politico-faith," State Shinto held that Japan's identity resided in these rulers as successive links in an unbroken generational chain into the mythological past to Amaterasu omikami—Sun Goddess and daughter of the celestial birthers of the islands. Emperor and Japan were one, immortal and godly. The possessions of dismantled fiefs and religious communities, denatured of ancient divisive associations, transformed into shared *heritage*: *Japanese* objects of the *imperial* state and its beneficence, and aestheticized as *fine art*.[40] A procession of objects stretching to the early Yamato kings now assumed roles in Meiji's lineage, a material chronicle to unify the population and meet foreign notions of nationhood. The government formalized this tectonic shift of identity and function by repurposing an old feudal term: *kokuhô* "national treasure."[41] An 1872 inventory of the eighth-century treasure house of Emperor Shômu revealed items from the Mediterranean among his intimate possessions, linking one of the last politically potent monarchs to the ancient West.[42] Kokuhô provided impetus for the Beaux Arts imperial museums planned for the historic capitals. In the distant Neoclassical White City, the Hôôden and its contents stood as their contextual twin.

The mural-like fusuma of the Hôôden's interiors climaxed in the dramatic Tokugawa Room (Figure 5.4). Tokyo Academy of Art Chief Professor Hashimoto Gahô, whose *Landscape (Sansui),* like other Japanese entries, seems to have won no prize in the Fine Art competition, led with a painting of phoenixes (Figure 5.7). It featured the Academy's official style, bearing the evocative name *nihonga* "Japanese Painting." A contemporary concoction of heritage aesthetic lineages, nihonga was deliberately designed to address a problem of both symbolic and functional identity, highlighted by the expositions and pithily expressed by the influential Ernest Fenollosa:

> It is well understood…that Japan's future position in the world's art cannot be established by throwing away her special gifts of pure and delicate design, in the quixotic desire to compete with France and America in the field of realistic oil painting. Neither can she fall back listlessly upon the fame of her past achievements.[43]

But in essence conceived as a riposte to Western art ideals, nihonga entwined incompatible notions and unrealistic goals. The Academy faculty of old-school luminaries set an aesthetic standard, yet were directed to not mold an olio of young talent to one, atelier style according to tradition, but to the Western model of individual masters. The stylistic blend was to materialize and celebrate the newly unified, contemporary

Copyrighted by C. D. ARNOLD.

HŌ-Ō (PHŒNIXES) AT PLAY, BY HASHIMOTO-GAHŌ.

Figure 5.7 Hashimoto Gahô, *Hô-ô-(Phoenixes) at Play.*
Source: Reproduced from *The Hôôden (Phoenix Hall), An Illustrated Description of the Buildings Erected by the Japanese Government at The World's Columbian Exposition Jackson Park, Chicago* (Tokyo: Ogawa, 1893), n.p., Photographs: C.D. Arnold.

Japan, instruct Westerners and Japanese alike to appreciate Japanese culture as unique, yet also be a globally relevant commodity. The Hôôden, venue for nihonga's debut had to embody the vital political actuality and commercial intentions of "modern Japan": an ancient imperial civilization whose distinctive products for the international market were the world's best.

A brainchild of dual parentage, nihonga reflected the convictions and advocacies of Okakura and Fenollosa, his mentor, who together had earlier conceived the curriculum of the Tokyo Academy of Art.[44] Though by no means unique in inserting Western aesthetics into Japan's intellectual matrix, this self-promoting American enjoyed for a time connections that made him among the most influential of foreign invitees, allowing him to give voice to resurgent conservatism. Ironically, government progressives had hired him to educate on Emersonian, Darwinian, Hegelian, and Spencerian philosophies of political economy, from 1878 at Imperial University and from 1886 at the Ministry of Education under Kuki.[45] While reformist, these deeply didactic modes of thought emphasizing order, hierarchy, discipline, tradition, and propriety found a fertile home in Japan's Neo-Confucian soil, in which the emperor cult was being nurtured as the magnetic nucleus of national identity. In fact deeply illiberal, Fenollosa espoused Ruskinian repudiation of the Industrial Revolution, romanticizing a rusticated past, and convinced that art reflected a society's current values and paved the path of its future. He averred that Japan's "purity" survived only in the rapidly vanishing remnants of pre-Meiji life. But ultimately unable to accept ebbing importance before his own theories of Japanese cultural supremacy, or his dependence on the bilingual and more broadly educated Okakura, Fenollosa returned in discontent to the United States in 1890.[46] In 1893 he served only nominally as advisor in selecting paintings for the Art Gallery competition. Okakura, whose career was in bloom as Tokyo Art Academy's Founding Director and as a Commissioner for the fair, made these ideas the basis for theories of Japan at the pinnacle of an Asian hierarchical cultural order.[47] In this activist role, he persuaded colleagues that revision of fine art classifications be the first priority,[48] and saw Japan's national pavilion as the perfect opportunity to promote nihonga as combined commodity, political statement, and declaration of identity.

The self-aware cultural nationalism of the Hôôden's nihonga paintings were Okakura's avowal of victory in the sparring of pro- and anti-Western factions in Japan's art world that had played out in the demise of the Technological Art School (Kôbu Bijutsu Gakkô). A vanguard of modernization founded in 1876 with Italian faculty, its impetus had been to teach the objective pictorial methods essential to modern technologies; hence its incorporation, along with the architecture program, into Tokyo Engineering College (Kôgaku-ryô), later Tokyo Imperial University. But its graduates joined others trained abroad, and aligned with intellectuals in a growing community espousing the new media and objective realism for the creation of "fine art." Dubbed yôga, "Western Painting," it became progressives' calling card. It deeply unsettled those who saw its supplanting of traditional methods and ideals of a subjective "higher" reality as undermining of Japanese society's fundamental philosophical groundings. Then at his height of influence, Fenollosa endorsed this view of yôga as a harbinger of unfettered contamination, unleashing a full-blown attack in print, from bully pulpits, and as co-organizer with Kuki and Yamataka of the traditionalist Kangakai (Painting Appreciation Society) and Ryûchikai.[49] Engulfed by a wave of hostility, the Technological Art School closed in 1883. In its wake Tokyo Academy of

Art opened in 1889, its nihonga curriculum a manifesto of repudiation.[50] It has been noted that the simultaneous debut next door of the Tokyo School of Music (Tôkyô Ongaku Gakkô) made a statement of national identity and cultural arrival.[51] But this was probably unconscious, as a pair they showed the irresolution surrounding what should constitute a modern Japan: The Academy of Music was an entirely Western program housed in a building the junior architect for which, coincidentally, had been Kuru Masamichi.

Paradoxically, at the heart of the inferences materialized by the Hôôden lay a conviction of Greek DNA in Japan's cultural genome. Showing "Hellenism" as germinal to classical Yamato culture "explained" the modern nation's rise as inevitable according to the West's own ideological rationale and a means to resolve the conceptual problem of integrating Western empirical thought, critical to modernization, with the new government mythology of Imperial godhead. "Proof" of these broadly incompatible notions, and their union, centered on the revered temple complex of Horyûji, established by the fabled founder of Buddhism in Japan, Crown Prince and Regent Shôtoku Taishi (574–622).[52] An 1888 publication had linked this monastic complex, not incidentally the world's oldest extant wooden structures, to ancient Assyria, an "Asian" civilization predating that of Europe.[53] But coincident to the 1893 fair a rising star of Tokyo Imperial University's architecture community, Itô Chûta, initiated an illustrious career with *Hôryûji kenchikuron (A Theory of the Architecture of Hôryûji)*. He interpreted the entasis of certain structural columns as vestigially Greek, and thus Japanese civilization as a rediscovered branch of the West's accepted defining lineage, separate but equal.[54]

In the same spirit, Horyûji's most famous carved image, the gilded wood Guze Kannon, had excited Fenollosa to the highest praise he could imagine: "...It was the aesthetic wonder of this work that attracted us most," he wrote, "From the front the figure is not quite so noble, but seen in profile, it seemed to rise to the height of archaic Greek art"[55] (Figure 5.8). This inspired his tortuous theory of "Japanese Greco-Buddhism," which saw Horyûji's era as the first in which two waves of supreme Mediterranean elements arrived via China, the second being the fifteenth-century feudal Ashikaga epoch. "Indeed, this Japanese naturalization of far-away Greek types so parallels the mediaeval unconsciousness of the classic tradition that remotely conditioned its work," he opined, "as to justify us in adopting for this style, if not for the Greco-Buddhist Art as a whole, 'the Buddhist Gothic.'"[56] Then, applying the Spencerian social theories of natural selection to which he ascribed, he concluded that as only Japan had avoided invasion, Japanese painting retained aesthetic vigor while that of China had withered, leaving Japan Asia's beacon of Greek ideals. He regarded as its ideal the works of the Kano school, the dominant taste of samurai aristocrats since Ashikaga times.[57]

In short, this made Hashimoto Gahô, among the last of the Kano, a natural choice to lead Tokyo Academy of Art's faculty, and nihonga a "dialect" in the visual language of modern Hellenes. Partnered with Kuru's neoclassically inspired configuration for the Hôôden in Chicago, it neatly fit into the exposition movement's ideological cosmos that entwined architecture and fine art as the identity aesthetics of political stature and entitlement. As a menu of what Tokyo Academy of Art could offer the well-appointed Western home, the interior paintings merely laid claim to the same muddied distinction between art and commodity allowed by contemporary art in the Exposition's Art Gallery. Okakura's booklet lists each painting and its artist. Yet not

Figure 5.8 Guze Kannon (7th Century).
Source: https://commons.wikimedia.org/wiki/File:GUZE_Kannon_Horyuji.JPG Tokyo Bijutsu Gakko [Public domain], via Wikimedia Commons.

a single contract of sale seems to have resulted. Moreover the message of which they comprised such an important part was futile by being completely indiscernible to its intended viewers, who came largely unquestioning of the exposition movement's dogmatic inconsistencies, in thrall to the sumptuous alien aesthetics, and conditioned to trust the veracity of whatever greeted them.

Indeed, primed by the reaction to the Satsuma pavilion in 1867, Japanese officials from the Meiji regime's debut in 1873 onward had seized upon a Western obsession with "authentic" as a tool. Long isolation had made Japan backward, but also guaranteed a reliably uninformed viewership abroad. In Chicago as elsewhere, gullible fair officials yielded to demands for total control in the interests of "authenticity," while all involved toyed with fact, confident in the knowledge that details of material, execution, garb, and exotic taste made "truth" out of omission and legerdemain: Kuru placed aesthetic unity over chronology and structural accuracy. Okakura's instructional essay spanning 629–1550 cited as apogees the Tempei, Fujiwara, and Ashikaga eras.[58] It did not coordinate with and even elided the historicity of the tableaux that visitors saw: the Tempei, dominated by a noble family whose exercise of marriage politics controlled a succession of emperors; the Ashikaga, and then the Tokugawa of 1600–1868, the two most transformational of military clans whose governments pointedly overshadowed the royal court—but according to Fenollosa's and Okakura's Ruskin-tinged Romantic theorizing showed "Greek" stylistic lineage as Japan's "Gothic" age. The nihonga paintings featured thematic and stylistic pastiche. Had critics and ordinary visitors alike recognized themselves as dupes, all the hyperbolic indignation typical of the time would have been brought to bear. But no evidence suggests that anyone had any cognizance of the complicated message much less its sleight of hand.

In fact, one can say that the paintings for all their glamour were completely overshadowed by the Hôôden complex itself, and to the extent that they succeeded at all owed to their role as elements of an exotic whole. Given their ingrained social superiority and aesthetic conservatism, it is unlikely that Kuki, Okakura and Yamataka understood what Kuru had accomplished in service to their nationalist ardor. Yet as their expectations for the paintings flew over heads in Chicago, Kuru's vision proved compelling as tool and metaphor of the state. To succeed the Hôôden did not need recognition as a Beaux Arts expression. It needed only that fairgoers subliminally respond to its symmetrical, ordered dignity and grace in order to suspend the biases embedded in the architecture of exposition complexes—a dogma of supremacy the Commissioners themselves ironically embraced. The complex's interlocking of aesthetics with ideology changed the standard for Japanese identity at the fairs. Exterior elegance henceforth superseded interior displays in importance. It also set a precedent: every subsequent major exposition saw an architect with top credentials over a traditional master carpenter in the design of Japan's national pavilions and compounds.

The Hôôden's success came at a cusp in histories of Japan, the expositions, and those behind its creation. In 1893 the government reopened the reviled unequal treaties, attaining in 1894 a genuinely reciprocal international relationship.[59] Its victory that year in the Sino-Japanese War inaugurated an imperialist profile that would intensify in coming decades.[60] Kuru designed Japan's pavilion for the 1904 St. Louis exposition but then vanished into the obscurity of a workaday career as engineer at the Ministry of Education. Kuki, Okakura, Yamataka, and Fenollosa shortly went their separate ways, three in deep acrimony.[61] A gift to Chicago, the Hôôden remained in Jackson Park, an apogee never repeated even as Japan continued its rise, and as its national pavilions continued to be grand. Its mysterious destruction by fire in 1943 may have been a measure of the success with which it wove a story of history and culture as convincing as it was loose with fact. For so many, the Hôôden *was* "Japan."

Notes

1 Two *ranma* (transoms) have been restored and are on view at the Art Institute of Chicago. The *fusuma* (wall paintings) of the Tokugawa Room have recently been rediscovered.
2 "The Interior Treatment of the Hooden at the Fair," *The American Architect and Building News* 41, no. 923 (September 2, 1893), 143.
3 P. B. White, "Japanese Architecture in Chicago, Part II," *The Inland Architect and News Record* 20, no. 6 (January 1893), 61.
4 "The Interior Treatment of the Hooden at the Fair."
5 White, "Japanese Architecture in Chicago, Part II," 60.
6 "Opinions on Participation in Foreign Exhibitions, Report on the Pris (sic), France Exposition Universelle" and "The Report," Article 4 Section 20, The Hôôden" in Masahiro Mishima, "1893 nen Shikago bankokuhaku ni okeru Hôôden no kensetsu ikisatsu nitsuite" ("On the Circumstances of the Construction of the Hôôden in the 1893 Chicago World's Fair"), *Nihon Kenchiku Gakkai ronbun hôkokushû (Transactions of the Architectural Institute of Japan)*, no. 429 (November 1991), 151–163, 155.
7 See, for example Henry Van Brunt, "Architecture at the World's Columbian Exposition" *Century Illustrated Magazine*, August 1892 and "The Personal Equation in Renaissance Architecture" reprinted in William A. Coles, *Architecture and Society: Selected Essays of Henry Van Brunt* (Cambridge: Harvard University Press, 1969), 150–157.

8 On Japan's adaption of neoclassicism see William H. Coaldrake, *Architecture and Authority in Japan* (London/New York: Routledge, 1996), Chapter 9.

9 Best illustrated by the 1867 Exposition Universelle, where planners placed it at the very center of the vast oval Main Building as the apex of an ascending hierarchy of production.

10 Apart from 1867, the art gallery was always neoclassical regardless of the general architectural theme. Vienna, all major American fairs after 1893, and those in Australia were entirely Beaux Arts campuses.

11 For example, Nishi Amane's "Bimyôgaku Setsu," "The Theory of Aesthetics," 1877. See Michael Marra, "Introduction," in *Modern Japanese Aesthetics, a Reader* (Honolulu: University of Hawai'i Press, 1999), 2.

12 See Alice Y. Tseng, *The Imperial Museums of Meiji Japan: Architecture and the Art of the Nation* (Seattle/London: University of Washington Press, 2008), 37, 173–176 and Dôshin Satô, *Modern Japanese Art and the Meiji State: The Politics of Beauty,* trans. Hiroshi Nara (Los Angeles: Getty Publications, 2011), 67–72.

13 See Tseng, Chapters 4 and 5 on each of these two museums.

14 Walter Smith, *The Masterpieces of the Centennial International Exhibition, Vol. I, Industrial Art* (Philadelphia: Gebbie & Barrie, 1875), 3.

15 For an overview of planning and exhibits overall, see Ellen P. Conant, "Japan 'Abroad' at the Chicago Exposition, 1893," in *Challenging Past and Present: The Metamorphosis of Nineteenth-Century Japanese Art,* ed. Ellen P. Conant (Honolulu: University of Hawai'i Press, 2006), 257–263.

16 "Japanese Art at the World's Fair," *The Art Amateur: A Monthly Journal Devoted to Art in the Household,* August, 1893: 29, 3. Also, see the example of Suzuki Chokichi, "The Juni-no-taka Bronzes," *The New York Times,* December 28, 1893, 8. The quest ultimately bore fruit in 1904 in St. Louis, with abolition of all art categories.

17 "Interior Decoration at the World's Fair," *The Art Amateur; A Monthly Journal Devoted to Art in the Household* 30 (March, 1894): 4.

18 "Japanese Art at Chicago" *The New York Times,* April 23, 1893, 7.

19 Other equally urbane officials negotiated concessions these men sought as key to the Hôôden's message.

20 See Masahiro Mishima, "Hôôden no keitai to sono seiritsu yôin ni tsuite" (The Factors Surrounding the Form of the Hôôden") in *Nihon Kenchiku Gakkai ronbun hôkokushû (Transactions of the Architectural Institute of Japan)* 434, (April 1992), 107–116, IV–3.

21 Failure to renegotiate brought down the Foreign Minister in 1887. Negotiations, reopened in 1893, succeeded first with the British in 1894. See W.G. Beasley, *Japanese Imperialism 1894–1945* (New York: Oxford University Press, 1987), 33.

22 Coaldrake, *Architecture and Authority in Japan,* 209 and Watanabe Toshio, "Vernacular Expression or Western Style? Josiah Conder and the Beginning of Modern Architectural Design in Japan," in *Art and the National Dream,* ed. Nicola Gordon Bowe (Dublin: Irish Academic Press, 1993), 43–52, 48.

23 Mishima, "1893 nen Shikago bankokuhaku ni okeru Hôôden no kensetsu ikisatsu nitsuite," 1991, 159.

24 He assisted the senior architect for the wood-constructed 1890 Sôgakudô, Tôkyô School of Music, which was Western but for its roof tiles. Hiroshi Watanabe, *The Architecture of Tôkyô, An Architectural History in 571 Individual Presentations* (Stuttgart/London: Axel Menges, 2001), 64–65.

25 Don Choi, "Shaping the Architect at the Imperial College of Engineering," Paper presented at the Society of Architectural Historians 65th Annual Conference, Detroit, MI, April 18–22, 2012.

26 On the symbolic role of stone see Coaldrake, Chapter 9, 212–222.

27 Josiah Conder, "The Condition of Architecture in Japan," reprinted in *The Japan Weekly Mail,* September 30, 1893, 392–394. Architecture as a fine art or simply as engineering was debated everywhere École des Beaux Arts theories were taught. In Japan, by the first decades of the twentieth century an "aggressively engineering-oriented group of architects exerted tremendous influence." See Jonathan M. Reynolds, *Maekawa Kunio and the Emergence of Japanese Modernist Architecture* (Berkeley: University of California Press, 2001), 19–20.

28 Mishima, "1893 nen Shikago bankokuhaku ni okeru Hôôden no kensetsu ikisatsu nitsuite," 1991, 160.
29 "In Japan's Temple." *Chicago Daily Tribune*, April 1, 1893.
30 See Don Choi, "The End of the World as They Knew It: Architectural History and Modern Japan," in *Seeking the City: Visionaries on the Margins, Proceedings of the 96th ACSA Annual Meeting* (Washington: ACSA Press, 2008), 738.
31 Mishima, "1893 nen Shikago bankokuhaku ni okeru Hôôden no kensetsu ikisatsu nitsuite," 1991, 160.
32 Tseng, *The Imperial Museums of Meiji Japan*, 126.
33 Mishima, "1893 nen Shikago bankokuhaku ni okeru Hôôden no kensetsu ikisatsu nitsuite," 1991, 155.
34 Kuru Masamichi, "On the Hôôden Exhibit for the American Exposition," *Journal of the Engineering Society*, October, Meiji 26 [1893] in Mishima, "1893 nen Shikago bankokuhaku ni okeru Hôôden no kensetsu ikisatsu nitsuite," 1991, 156.
35 Kinkakuji inspired a tea trader's association building at the 1904 exposition, and the government's reception pavilion at the 1915 fair.
36 Mishima, "1893 nen Shikago bankokuhaku ni okeru Hôôden no kensetsu ikisatsu nitsuite," 1991, 155.
37 Letter to Takahira dated December 29, 1891; Letter to Olmsted dated February 5, 1892.
38 Kakudzo Okakura, *The Hôôden (Phoenix Hall), An Illustrated Description of the Buildings Erected by the Japanese Government at The World's Columbian Exposition Jackson Park, Chicago* (Tokyo: Ogawa, 1893), 13.
39 Sato, 162, and Christine M.E. Guth, "Kokuhô: From Dynastic to Artistic Treasure," *Cahiers d'Extrême-Asie* 9, (1996), 313–322, 315.
40 See Julie Christ Oakes, "Contestation and the Japanese National Treasure System," (PhD diss., University of Chicago, 2009), 70, for a wonderful analysis of this.
41 On the term, formally adopted in 1897, see Guth (1996), 313–322.
42 See Seiroku Noma, *The Arts of Japan: Ancient and Medieval* (New York: Kodansha, 1978), 81 and Penelope Mason, *History of Japanese Art,* second edition revised by Donald Dinwiddie (Upper Saddle River, NJ: Pearson Prentice Hall, 2005), 58.
43 Ernest Francisco Fenollosa, "Contemporary Japanese Art: With Examples from the Chicago Exhibit," *Century Illustrated Magazine* 46, no. 4 (August 1893), 577.
44 See Chapters 1 and 4 in Victoria Weston, *Japanese Painting and National Identity: Okakura Tenshin and His Circle* (Ann Arbor: Center for Japanese Studies, University of Michigan, 2004).
45 Fenollosa was among the earliest and most influential sources of Hegel. See John Clark, "Okakura Tenshin and Aesthetic Nationalism," in *Since Meiji, Perspectives on the Japanese Visual Arts 1868–2000*, ed. Thomas Rimer (Honolulu: University of Hawai'i Press, 2012), 212–256, 230 and Kaneko Toshiya, "Cultural Light, Political Shadow: Okakura Tenshin (1862–1913) and the Japanese Crisis of National Identity, 1880–1941," (PhD diss., University of Pennsylvania, 2002), 51–70.
46 Kôjin Karatani attributes their estrangement to Okakura's politicization of what Fenollosa cherished as pure aesthetic superiority. "Bijutsukan to Shite no Nihon," "Japan as Art Museum: Okakura Tenshin and Fenollosa," *Hihyô Kûkan (Critical Space)* 2, no. 1 (1994), in Michael F. Marra, ed., *A History of Modern Japanese Aesthetics* (Honolulu: University of Hawai'i Press, 2001), 43–52, 46.
47 Toshiya, "Cultural Light, Political Shadow," 70.
48 Mishima, "1893 nen Shikago bankokuhaku ni okeru Hôôden no kensetsu ikisatsu nitsuite," 1991, 141–163.
49 Formed in 1881 and 1879 and included members of the Kano school, who had been displaced by initial rejection of past models but once again rose to prominence. The Kangakai focused on ancient works, the Ryûchikai on art commerce. Michael F. Marra, *Modern Japanese Aesthetics, a Reader* (Honolulu: University of Hawai'i Press, 1999), 71, text, and footnote 20, and Toshiya, "Cultural Light, Political Shadow," 67.
50 See Noriko Murai, "Okakura's Way of Tea: Representing Chanoyu in Early Twentieth-Century America," *Review of Japanese Culture and Society* 14 (2002), 60–77, 69.

51 Toshiya, "Cultural Light, Political Shadow," 108, Tseng, *The Imperial Museums of Meiji Japan*, 84, Fred G. Notehelfer, "On Idealism and Realism in the Thought of Okakura Tenshin," *Journal of Japanese Studies* 16, no. 2 (Summer, 1990): 326, and Kôjin Karatani, "Japan as Art Museum: Okakura Tenshin and Fenollosa," in *A History of Modern Japanese Aesthetics,* edited by Michael F. Marra (Honolulu: University of Hawai'i Press, 2001), 44.

52 On Greco-Buddhism and its ideological significance, see Stanley K. Abe, "Inside the Wonder House: Buddhist Art and the West," in *Curators of the Buddha*, ed. Donald S. Lopez, Jr., (Chicago: University of Chicago Press, 1995), 63–106.

53 See Stefan Tanaka, *Japan's Orient, Rendering Pasts into History* (Berkeley: University of California Press, 1993), 53.

54 Choi, "The End of the World as They Knew It: Architectural History and Modern Japan," 739, 740.

55 Ernest F. Fenollosa, *Epochs of Chinese and Japanese Art*, vol. I (New York: Frederick A. Stokes Co., 1912), 67.

56 Fenollosa, "Contemporary Japanese Art," 147–148. As proof, he noted that Gothic Revivalist architect Ralph Adams Cram (1863–1942) had "independently" come to the same conclusion.

57 The above, with a few expansions, abridges analysis by Toshiya, "Cultural Light, Political Shadow," 63–65. Although Fenollosa's books were written years later, the perspectives were formed in Japan.

58 Okakura, *The Hôôden (Phoenix Hall)*, 3. What he referred to as Tempei is now transliterated as Tempyô. The modern dates for all these periods are slightly different.

59 Beasley, *Japanese Imperialism 1894–1945*, 33.

60 Okakura's ideas matured into the *Awakening of the East*; *The Ideals of the East With Special Reference to the Arts of Japan*, and *The Awakening of Japan*, nationalist statements in English written 1901–1906. The *Book of Tea*, treating the aesthetic surrounding the beverage as an allegory of Japanese culture, he dedicated to close friend, artist John La Farge. Posthumously reinterpreted in Japanese by ultranationalists, his works resulted in his virtual canonization in the lead-up to World War II. Toshiya, "Cultural Light, Political Shadow," 298.

61 Fenollosa and Okakura were fully estranged. Okakura dismissed by Kuki due to erratic behavior and an affair with Kuki's ex-wife. Disgraced, Okakura took a position at the Museum of Fine Arts Boston formerly held by Fenollosa, who had lost it due to an affair of his own but spent his last years in Japan, writing his books.

6 Paris, 1900

The *Musée Centennal du Mobilier et de la Décoration* and the formulation of a nineteenth-century national design identity

Anca I. Lasc

A poster advertising the opening in 1899 of *Paris en 1400* with its *Cour des Miracles* (Court of Miracles) at 100, Avenue de Suffren survives in the collections of the Musée Carnavalet (Figure 6.1). Designed by Eugène Colibert, a student of Eugène Emmanuel Viollet-le-Duc and famed creator of the *Grande-Bastille* reenactment at the 1889 Paris world's fair, *Paris en 1400* was among the spectacular attractions prepared for the 1900 Exposition Universelle. It included a large-scale reconstruction of a fifteenth-century Paris neighborhood complete with streets, buildings, and fortifications, and was meant to give an accurate impression of life under Louis XI (reigned 1461–1483). To the restaurants, shops, taverns, cabarets, dairies, concert halls, and even whole streets from the period, Colibert added a restoration of the Court of Miracles as it had been vividly described by Victor Hugo in his popular 1831 novel *Notre-Dame de Paris*. Colibert hired a team of jugglers, charlatans, minstrels, illuminators, engravers, jewelers, knights and potters to perform on site, and appointed dedicated actors to stage the removal of Esméralda and the fall of Quasimodo from the towers of a mock Notre-Dame.[1] For the price of one franc per person,[2] visitors "would have lived for a few hours in the history of five centuries earlier, whose evocation would [thus] have been complete."[3]

Paris en 1400 with its *Cour des Miracles* was not the only themed immersive environment at the 1900 Exposition Universelle. Reenactments like this have been seen by historians as manifestations of "a desire to escape the present by studying the past (and especially the French past)" in nineteenth-century culture.[4] While several immersive displays were on view at the 1900 fair, not all historicizing displays were purely nostalgic or escapist. A key example of this is the display of French period rooms known as the *Musée Centennal du Mobilier et de la Décoration* (Centennial Museum of Furniture and Decoration). This chapter argues that the *Musée Centennal* leveraged history to promote contemporary decorative arts. The organizers and leading manufacturers involved with the *Musée Centennal* attempted to showcase the "originality" of nineteenth-century French home goods in a period characterized by blatant historicism by "rebranding" traditional period styles as a sequence of theatrical and immersive displays that adopted a format similar to that of other themed environments at the exhibition.[5] In the case of the *Musée Centennal*, historical reconstructions could be used as confident assertions of aesthetic continuity to bolster and inform modernity. The *Musée Centennal* displays were thus comparable with immersive attractions dedicated to mass entertainment. However, unlike the *Cour des Miracles*, they revived the recent rather than the distant past. In so doing, they

Figure 6.1 Exposition de 1900: Paris en 1400, la Cour des Miracles.
Source: © Musée Carnavalet/Roger-Viollet/The Image Works.

asserted a new confidence in the country's contribution to the world of decorative arts, design, and manufacturing both before and in the aftermath of its humiliating defeat in the Franco-Prussian War and amid increased competition from abroad. They also situated the work of progressive designers such as Émile Gallé and Louis Majorelle as the culmination of a historicist aesthetic, which was more palatable to a public long-accustomed to Classical, Medieval, Renaissance, Louis XIV, Louis XV, and Rococo revivals.[6] The *Musée Centennal* thus associated the decorative arts and their constructed interior settings with a form of immersive entertainment in Empire, Restoration, Louis-Philippe, and Napoléon III interiors that offered an impression of aesthetic continuity under the guise of nationalism.

The chapter begins with an overview of the *Musée Centennal du Mobilier et de la Décoration* as it was presented to visitors at the 1900 Exposition Universelle. It continues with a brief account of the popular archeological and ethnographic displays at the fair, situating these in a longer tradition of themed environments characteristic of nineteenth-century exhibitions. The chapter follows with a description of some of the most prized interior decorating ensembles created for the 1900 fair by a variety of public and private institutions eager to demonstrate their achievements in the realm of decorative arts and design. Finally, it returns to the *Musée Centennal* to examine its pivotal role in bringing together the similarly disparate French contributions to turn-of-the-century decorative arts and design in the context of spectacle and entertainment.

The *Musée Centennal du Mobilier et de la Décoration*

As the 1900 Paris Exposition Universelle marked the end of one era and the beginning of the next, the French faced the challenge of demonstrating their country's century-long accomplishments within the realms of art and manufactures along with its present-day achievements. Fair officials demanded that each French product section add a retrospective or "centennial museum" to its contemporary displays.[7] Among those included were the Centennial Museum of Section 12 (Photography), the Retrospective Museum of Section 14 (Geographic and Cosmographic Maps and Appliances), the Retrospective Museum of Section 15 (Coins and Medals), as well as the Retrospective Museum of Section 67 (Stained Glass Windows).[8] Sections 66 (Fixed Decoration), 69 (Furniture), 70 (Carpets), 71 (Mobile Decoration), and 97 (Bronzes), joined forces to create a single Centennial Museum of Furniture and Decoration (*Musée Centennal du Mobilier et de la Décoration*) under the leadership of François Carnot, president of the Union Centrale des Arts Décoratifs and head of the installation committee.[9] Situated in the middle of the French sector on the Esplanade des Invalides, which connected the newly built Pont Alexandre III with the famed Hôtel des Invalides (Figure 6.2), the Centennial Museum was not only centrally located but also "one of the main elements of success of the French exhibition" at the fair.[10] Lavishly

Figure 6.2 "L'Esplanade – Vue Prise de l'Hôtel des Invalides."
Source: Ludovic Bachet, ed., *Le Panorama: Exposition Universelle 1900* (Paris: Librairie d'art, 1900).

ESPLANADE DES INVALIDES — LES PALAIS DE LA DÉCORATION ET DU MOBILIER

Figure 6.3 "Esplanade des Invalides – Les Palais de la Décoration et du Mobilier, MM. Larche et
        Nachon, architectes; M. Esquié, architecte."
Source: Ludovic Bachet, ed., *Le Panorama: Exposition Universelle 1900* (Paris: Librairie d'art, 1900).

decorated buildings dedicated to all aspects of interior design lined both sides of the
Esplanade, with French decorative art displays on the western side and foreign ones
on the eastern side.[11] Visitors could find the *Musée Centennal* on the western side of
the Esplanade, on the ground floor of architect Pierre Esquié's Palace of Decoration,
Furniture, and Miscellaneous Industries and by the rue de Constantine (Figure 6.3),
tucked between the French national furniture displays to the north and bronze displays to the south.[12]

The Museum covered a surface 80 m long (north/south axis) and 20 m wide (west/
east axis) and was divided into two sections brought together under the unifying
design of architect Jacques Hermant: a "museum" section and a "period room"
section, the latter of which included a series of themed rooms arranged according to
recent political eras. The construction featured a white-colonnaded, green-draped
façade interspersed by gold and green valances deemed by organizers to be in the
Empire Style[13] (Figure 6.4). Its two parts took advantage of the natural light available in the building. The museum section per se stretched from north to south along
the central passageway and included a chronological display of objects arranged by
period. The periods here ranged from the reign of Louis XVI, "the veritable departure point of a movement whose end point is the new art [*l'art nouveau*]," to the
present, with the final display dedicated to works by Jules Chéret, Émile Gallé, and
Louis Majorelle that served as "a link to contemporary exhibitions."[14] Instead of
using current revival terminology—Neoclassical, Egyptian Revival, Gothic Revival,
Renaissance Revival—the organizers substituted a fresh set of labels, suggesting the

existence of a *Style Directoire* ("an imitation of the antique and especially of objects recovered from the excavations at Herculaneum and Pompeii"[15]), *Style Empire* (a mix of *Style Directoire* and *Style Louis XVI* with Egyptian and Oriental additions[16]), *Style Restauration* (with Romantic affiliations[17]), *Style Louis-Philippe* (two styles in one: the Empire Style and an adaptation of the Gothic style to contemporary tastes that would bring about the collapse of the Classical style[18]), and even *Style Second Empire* ("a mix of all the others"[19]).[20] While the Centennial Museum sought to present the coherence of national taste from the era of Louis XVI until the present, climaxing with what organizers would call *Le Style Troisième République*, difficulties arose in making such a claim. In the case of the *Style Empire*, for example, organizers struggled to differentiate Italian from French contributions, determined as they were to present contemporary achievements as the culmination of a century-long national progression towards a new interior decorating style. The last section [*travée*] illustrated the history of decorative arts in France during the Third Republic and up until 1889, where fashion returned to the "rare pieces from the past" appreciated under Louis-Philippe's reign. Once items of Gothic origin were exhausted, organizers pointed out how collectors turned once again to the Renaissance and the periods of Louis XIII, Louis XIV, Louis XV, Louis XVI, and finally to the First Empire to procure original pieces for their interiors, with items from the Restoration and Louis-Philippe next on the list.[21] Coming full-circle, the decorative objects from the recent past were inscribed in a longer, teleological history of revival styles.

Illuminated from above, this chronological display of objects was interrupted in the center by a vestibule endowed with the most monumental and highly decorated objects of the century, including a vase created for the 1889 universal exhibition by Huvé frères after a design by Paul Sédille[22] (Figure 6.4). This central vestibule

Figure 6.4 "Musée Centennal du Mobilier et de la Décoration."
Source: Maurice Le Corbeiller, *Musée Centennal des classes 66, 69, 70, 71, 97. Mobilier & Décoration à l'Exposition universelle international de 1900, à Paris. Rapport de la commission d'installation* (St.-Cloud: Bélin, [1902]).

stretched to the other side, the second part of the Museum, where organizers had planned a series of eight rooms along the same north/south axis, sparsely lit from the adjacent galleries through large bay windows. Electricity was banned from all these spaces for fear of potential fire, as well as due to a desire to maintain historical accuracy in a series of thematic period displays that could otherwise be accused of "sacrificing factual truths to the effects of theater and the diorama."[23] These "special rooms" were furnished and decorated following the period style of various eras from the country's recent history, and included a *Salon Louis XVI* (Louis XVI drawing room), a *Chambre à coucher, époque de la Constituante* (a bedroom from the time of the National Constituent Assembly), a *Pièce Directoire* (a Directory room), a *Salon Empire* (an Empire drawing room), the *Chambre de Talma* (the bedroom of famed *Comédie-Française* actor François Joseph Talma), a *Cabinet de travail Restauration* (a Restoration study), a *Chambre à coucher Louis-Philippe* (a Louis-Philippe bedroom), and a *Salon Napoléon III* (a Napoléon III drawing room). Brought to life with the help of the upholsterer-decorators Georges and Henri Rémon, who completed any missing elements from these made-up ensembles by creating imitation pieces reflective of the chosen period styles, the interiors created in three dimensions an impression of times long gone. They encouraged virtual immersion in interiors from the recent past, newly defined as a series following the nation's political regime changes.

Les Clous de l'Exposition: popular attractions at the fair

The period rooms of the *Musée Centennal* can thus be considered immersive environments similar to many of the popular attractions created for the 1900 universal exhibition, including Colibert's *Cour des Miracles*, which aimed to transport visitors to a different time or a different place.[24] Among the most successful creations in this vein were the displays put together by the famed fashion designer Félix within the *Palais du Costume* (Costume Palace). There, visitors found a history of women's fashion throughout the ages, with reconstructions of period attire displayed on mannequins in settings à la Musée Grévin, Paris' famed wax museum, founded in 1882.[25] The aim was to show "the woman, resurrected from century to century," critics explained, "living before the crowds" as directed by the tastes and customs of each era.[26] With period furniture created by the Parisian manufacturers Maison Jansen to match each historical setting, Félix was able to bring to life the matrons of a Roman atrium from Trajan's reign, Gallic women and interiors from the time of the Roman invasion, a Byzantine interior and a feudal interior from the Middle Ages inhabited by members of the fair sex, a Venetian Renaissance *tableau* showing an upper-class lady stepping out of her home and into a gondola, *aristocrates* from the reigns of Louis XIV and Louis XV in their private quarters, as well as scenes from the Revolution and Directory eras.[27] The fashions of the nineteenth century were also displayed, grouped in settings reminiscent of the First Empire through the Second and culminating with displays of contemporary ballroom gowns by Félix.

To celebrate trends in women's fashion during Napoléon's reign, for example, Félix chose to reconstruct a moment from the eve of the emperor's coronation, when Joséphine had modeled her attire in front of her husband (Figure 6.5). Consulting both visual and written sources from the period, Félix presented visitors with a meticulous reenactment of the moment, and included every element of the future empress' garb, including the jewelry—crown, tiara, and belt—created especially for her by

Figure 6.5 "La Veille du Sacre, 1804."
Source: Projet Félix, *Exposition Universelle de 1900: Palais du Costume – Le Costume de la Femme à Travers les Âges* (Paris: Imprimeries Lemercier, [1900]), 53.

the jeweler Marguerite.[28] The interior decoration of the room recalled the Egyptian and Pompeian fads of the time, while Joséphine's mirror was flanked by a pair of imperial eagles on their orbs, a reminder of Napoléon's military conquests and claims to Roman imperial ancestry. His back towards viewers, the wax embodiment of the future emperor permitted visitors to imagine that they, too, took part in the event. The illusion of virtual immersion in a different time and place was thus complete.

Scholars have demonstrated that by the 1890s the fairgrounds themselves, pavilions and immersive displays included, became the largest attractions of the world's fairs, engaging crowds and generating profits.[29] The 1893 World's Columbian Exhibition in Chicago not only became a "neoclassical dreamland" with an artificial lake and monumental sculpture but it also included a so-called "Midway Plaisance," an area fully dedicated to popular entertainment.[30] Relegated to private entrepreneurs, amusement venues garnered harsh competition, and dozens of proposals were submitted to the organizing committee for the 1900 Exposition Universelle alone. Although the French government had charged rent for private pavilions since 1889, space was limited and not all proposals could be accommodated.[31] In 1900, in addition to the *Palais du costume*, the *Vieux Paris* (Old Paris) and the *Village Suisse* (Swiss Village) were among the private attractions sanctioned by the organizers. Many others, including a *Pompéi vivante* (Living Pompeii) with a fake volcano, and a project titled *Une demi-lieue sous les Terres* (Half a league under the surface of the Earth), were rejected by the jury.

Une demi-lieue sous les Terres involved the construction of an inverted Eiffel Tower, measuring the same height as the original, albeit situated underground. With attractions on each level, much like Gustave Eiffel's above ground construction, its main role was to settle the contemporary debate about the existence of a fire in the center of the earth, an idea inspired in part by Jules Verne's novel *Journey to the Center of the Earth* (1864).[32] The fake volcano, on the other hand, as the press scandalously announced, "almost came into eruption in Paris, in the very center of Paris!" The city, indeed, came close to having its own crater, boiling matter, smoke, flames, and flowing lava, all directed towards the public's entertainment pleasure.[33] The *Pompéi vivante*, finally, would reconstruct life in the ancient city, with its customs and contemporary festivities, before the eruption of Vesuvius and would include replicas of famed temples, private houses, the gladiators' arena, and the city's forum. Supported by famous personalities at the time, such as Jean-Léon Gérôme, Émile Étienne Guimet, Eugène Muntz, Victorien Sardou, Jules Claretie, and Antonin Proust, the attraction would have been yet another example of a total work of art at the level of popular spectacle, with scenic painted canvases replacing three-dimensional reconstructions where space did not permit constructing the entire monument.[34]

From those attractions (at the time called *clous* or nails for their role in anchoring the exposition) that were accepted, however, the *Village Suisse* received most attention in the media, undoubtedly due to its innovative yet accurate reconstructions of existing regions from the Swiss Alps. Mountains, valleys, lakes, pastures, forests, and waterfalls were all included, as well as a village and town populated with real inhabitants from various Swiss regions. Critics could not withhold their excitement: "Mountains in Paris! Mountains as beautiful, as unexpected, as steep, as wild as the Swiss mountains, an extraordinary thing that deserves being seen from close-by," they exclaimed.[35] The attraction was designed by Charles Henneberg and Jules Allemand and occupied an area along Avenue Suffren and Avenue de la Motte-Piquet, concealing the quartier de Grenelle from public view almost entirely and transforming the Parisian horizon into an alpine landscape.[36] Close to three hundred inhabitants from Swiss villages dressed in costumes characteristic of their regions participated in the display, which included facsimiles of a Fribourg dairy well-known for its production of Gruyère cheese, the Mumpf guesthouse where the popular actress Rachel had been born, Rousseau's birth home from Geneva, Bern's famous towers, and the Bourg-Saint-Pierre inn where Napoléon had eaten while crossing the Saint-Bernard (Figure 6.6).[37] Men and women were accompanied by their favorite domestic animals, contributing together to an illusion that allowed Parisians to imagine themselves "transported out of France and suddenly dropped in the middle of the Swiss Alps."[38]

Constructed displays of this kind, especially in the context of displaced colonial villages, were not introduced at the 1900 world's fair. Rather, the practice of displaying natives from geographically removed regions had already been established at the Paris 1889 Exposition Universelle.[39] There, one could see a Senegalese village right next to a Congolese village and a New Caledonian village, among others.[40] Small-scale reconstructions of existing buildings from places "away from home" were also customary by this time. Henry Houssaye had explained as early as 1878 how world's fairs enhanced travel through displays such as the *Rue des Nations* (Street of Nations), arguing: "the universal exhibitions have significantly simplified the journey around the world. At the Champ de Mars, one scarcely needs half a day to visit the entire inhabited world."[41] Displays such as those created by Sweden and Norway at the 1878 Exposition Universelle had already offered the illusion of populated interiors in national styles using life-size mannequins.[42] Inspired by the wax figures introduced in France by Philippe Curtius and Madame Tussaud, these *tableaux vivants* offered

Figure 6.6 "Le Village Suisse. – Chalet d'Effretikon – Maisons de Berne."
Source: Ludovic Bachet, ed., *Le Panorama: Exposition Universelle 1900* (Paris: Librairie d'art, 1900), n.p.

new possibilities for imagining space and time. As an article in *Le Monde Illustrée* explained,

> the idea of representing interior scenes through life-size figures is extremely clever. This is the best way to truly interest the public and to fix in his mind the costumes and habits of different nations – more successful than the most truthful painting could ever be.[43]

Even the imaginary rendition of an entire street, if not an entire village or town, generated popular appeal when the *Rue du Caire* (Cairo Street) became one of the most famous attractions at the 1889 Paris world's fair.[44] Through the inclusion of native people, these spaces allowed for vicarious travel into a world removed from the everyday life experiences of exhibition visitors. Such constructions helped blur reality and representation in environments that engaged all the senses. But what distinguished the *Village Suisse* from these earlier examples was its designers' determination to remain faithful to archeological, geographic, and ethnographic truth at all costs and by all means possible.

For this purpose, as the readers of the *Journal de Genève* discovered, La Société du Village Suisse de Paris had amassed with the help of both French and Swiss investors no fewer than three million francs by 1898.[45] These funds had allowed Henneberg and Allemand to take molds from the rocks of the Swiss Alps, which they took back to their studio to create exact replicas for their own imitation mountains. One hundred workers intensively labored to duplicate these casts until enough were formed to cover the wooden skeletal framework designed for the fair, which itself had cost more than a million francs to produce. Real vegetation was transplanted from Swiss mountain regions and grown indoors until it could be relocated and acclimatized in Paris.[46] Finally, because the skeletal framework of the imitation mountains was constructed of wood, the cost of fireproofing and creation of an artificial wall to surround the one-kilometer long compound to prevent the threat of fire were added, as requested by the Paris Préfecture.[47]

Equal attention was paid to the elevation of the various buildings chosen for the Swiss Village. As *L'Eclair* explained in 1899, "Each house that springs up is both an art-work and an archeological reconstruction at the same time; and each will be supplied with authentic furniture and inhabited by Swiss locals in their national costumes." Further, all the objects displayed therein would be authentic, as Swiss peasants had insisted that the designers visit their private homes to get a sense of the antiquities they possessed. In the case of Napoléon's Bourg-St-Pierre inn, the building's owners had offered the very furnishings belonging to the room where the emperor had dined for display at the exhibition.[48] For the price of one franc,[49] one could thus discover a mock-up of Switzerland in Paris, so authentic that travel there became pointless: the Fribourg dairy made Gruyère on site, the Auberge du Treib offered Swiss wine and beer, and the Chalet de la Crèmerie proposed a cup of milk or Kohler chocolate to weary "explorers."[50] And, as one sat at the same table where Napoléon had dined, lis-tening to Swiss singers from the nearby Chalet de la Crèmerie or witnessing national dances in the main square against the exact recreation of Swiss rocks and real Swiss plants grown under the Parisian sky, the illusion of virtual travel to a different place— and time—was was complete. The Swiss Village was a total environment where all senses were engaged with the purpose of creating as accurate of an impression of a different reality as possible.

Art Nouveau and French patriotism on display at the fair

French officials were sympathetic to this "total" approach to entertainment and de-sign. Their excitement towards accurate reconstructions of geographically remote regions or historically removed eras emerges not just in widely popular displays such as Colibert's *Paris en 1400*, the *Village Suisse*, or Robida's *Vieux Paris*, a "miniature city containing reproductions of famous Parisian buildings 'mutilated' or destroyed from the Middle Ages to the eighteenth century"[51] that was given pride of place on the exposition's grounds, but also in the period rooms specifically cre-ated for exhibition within the French *beaux-arts* section. Indeed, in addition to the *Musées centennaux* mentioned earlier, French organizers supported the creation of a "retrospective" exhibition dedicated to French art of the past century. This would include both an "*oeuvre d'art*" section, including paintings and sculptures that served "no other purpose other than being artistic," and an "*objet d'art*" section, includ-ing "utilitarian" objects no matter their intended use.[52] To enhance the accuracy of the displays, and responding to contemporary trends that saw the *objets* and the *oeuvres d'art* increasingly intertwined in a variety of public exhibitions, *L'Exposition rétrospective* was to include "rest areas" (*salons de repos*) interspersed among the various exhibition rooms, which "gave for each characteristic period from these ninety years [past] the physiognomy of an artistic ensemble by combining paintings, sculptures, furniture, tapestries and even trinkets" (*bibelots*).[53]

The nascent vogue for museum period rooms had undoubtedly inspired exhibition organizers, Paris' Musée Carnavalet serving as a local example. There, a group of rooms once found elsewhere had been reconstructed and installed in 1878. Chief among them, for example, was Richelieu's former 1653 *Grand Cabinet* designed by Louis Le Vau, Charles Le Brun and Gerard van Opstal.[54] The organizers of the *salons de repos* of the 1900 *Rétrospective*, however, seemed as concerned with the preservation of past artistic achievements as with the imaginative recreation of a lost historical period aimed at showcasing the French genius in the realm of both the fine and the decorative arts. Indeed, this concern came at a time when the country was struggling with economic and national crises following scandals such as the failure of the Panama Canal project and the Boulanger and Dreyfus affairs.[55] The earlier defeat in the Franco-Prussian Wars, as well as the loss of Alsace and Lorraine to the German Kaiserreich, further bled Frenchmen's self-confidence, who, by 1897, feared another "German conquest," albeit in the economic realm. Historian Richard Mandell has even noted how the French "stole" the 1900 universal exhibition from the Germans, who had planned it years in advance. The prominent French journal *Le Figaro* wrote about how Germans "thought" the exhibition "would be a providential opportunity to demonstrate that the nineteenth was a German century, just as the eighteenth had been a French one."[56] France was not going to allow that to happen.

Fears of degeneration were equally rampant, and figures such as the aristocrat, the Jew, the homosexual, and even the new generation of intellectuals following Decadent or Symbolist precepts were accused of bringing France down the path to perdition.[57] The country came together under the banners of anti-Semitism and nationalism, with societies such as the 1882 *Ligue des patriotes* (Patriots' League) determined to train boys in physical education so that they could reconquer the territories lost to the Germans.[58] In this context of insecurity and fear of international competition, as art historian Nancy Troy has shown, France would reject the artistic achievements of foreign nations in favor of its own traditions.[59] The case of art dealer Siegfried Bing, whose *La Maison de l'Art Nouveau* – famed for giving an entire movement and style its name – had supported the work of foreign artists since 1895 is a case in point. As Troy explains, Bing's undertaking was condemned not only as "incoherent" but also as "a threat to the entrenched tradition of the decorative arts that the French identified with the 'Louis styles' of the *ancien régime*"[60] and with historicism more generally.

Yet the French still had to cope with Art Nouveau, increasingly defined as a rejection of the past and an espousal of a new decorative language immersed in the non-historic—"whiplash lines, vegetable curves, female hair, peacocks, sea weeds, lily pads, and swans"—especially at a fair where the German, Austrian, and Hungarian exhibits displayed whole rooms in the new style.[61] Bing, too, had opened a privately run pavilion dedicated to the new art, where he showed the work of artists Georges de Feure, Edouard Colonna, and Eugène Gaillard.[62] Critics were torn: while some rejected Art Nouveau in a grueling fashion, others attempted to make sense of it, even to define it as the culmination of a century-long history of *French* decorative arts, as we have seen with the *Musée Centennal du Mobilier et de la Décoration*.[63] Writing for the *Feuilleton du Journal des débats*, André Hallays, for example, proclaimed himself disgusted by the modern style, thanking the exhibition organizers for bringing its "burlesque" to light. A "monstrous thing," he continued, "the same furniture in the same style for all people. Nothing better to prove the emptiness of this pretense renaissance: the furniture in each house must vary according to climate and customs."[64] While he criticized Art Nouveau in general, however, Hallays praised the work of the French sculptor, cabinet-maker, and decorator Alexandre Charpentier for the Grands Magasins du Louvre furniture displays at the exhibition. Charpentier's contribution attracted similar reactions from other critics at the time. Writing for *Littérature et Beaux-Arts*, Gabriel Mourey singled the artist out as one where "racial traditions and the most vivid concern for modernism come together in perfect equilibrium." However, the Louvre dining-room came short of fully pleasing this writer, who found Charpentier erring through "an excess of truth in its natural ornaments, fruits, flowers, branches, etc."[65] As art historian Rossella Froissart Pezone explains, the difficulty that Charpentier encountered came from his desire to "create a new, 'modern' style while responding to the demands of a fearful industry and clientele, strongly attached to 'national traditions.'"[66]

Despite these mixed reactions, however, among the innovations proposed by both Charpentier for the Grands Magasins du Louvre department store and the artists associated with Bing's enterprise, was the presentation of decorative arts objects in room ensembles that also created the illusion of a complete artistic environment. Like the period rooms of the Musée Carnavalet, the period displays of nineteenth-century styles of the *Musée Centennal* and the *salons de repos* created for the *Rétrospective* in the French beaux-arts section, these rooms adopted a "total" approach to design, while suggesting a parallel concern with entertaining the crowds. The public was transported not into a world of the past, the picturesque, or the exotic but, as I have shown elsewhere, into a world of natural fantasy, where the boundaries between the organic and the inanimate were increasingly blurred.[67]

Le Musée Centennal: historicism, immersion, and nationalist propaganda

The competition between the old and the new, veiled in nationalistic terms, became the hallmark of the 1900 Exposition Universelle. Raymond Koechlin, the vice president of the Union Centrale des Arts Décoratifs, clarified this relationship when describing the society's own pavilion designed by the interior designer, ceramist, collector and art dealer Georges Hoentschel, stating that:

> the doctrines of the Union Centrale are hardly revolutionary ... if it has made an effort to bring contemporary art out of its rut, and it fears foolish eccentricity as much as it abhors servile copying, it knows that nothing lasts which is not connected by some thread to the past.[68]

This is very much what the organizers of the *Musée Centennal du Mobilier et de la Décoration* must also have felt when they struggled to define the country's century-long contribution to the field of decorative arts through a sequence of themed interiors organized around their recent past. However, instead of associating innovation with a revival of eighteenth-century craft traditions as Hoentschel had done for the pavilion of the Union Centrale or Bing through Georges de Feure's contribution to the *Pavillon de l'Art Nouveau*, they reorganized the entire history of nineteenth-century decorative arts in a way that facilitated virtual immersion in the domestic environments of their more immediate forefathers. The old, traditional labels associated with the "Louis styles" and earlier were now gone; in their stead came a series of new labels that nevertheless offered a sense of coherence and direct continuity, consonant with the assertion of a much-needed national confidence in the French decorative arts genius.

Indeed, in addition to the chronological displays of decorative art objects discussed earlier, the Centennial Museum displayed a variety of furniture pieces and decorative items organized in a series of eight national period rooms. Dedicated to the Louis XVI, the Constituent Assembly, the Directory, the Empire, the Restoration, the Louis-Philippe, and the Napoléon III eras, the rooms attempted to give an accurate impression of domestic life during these different political regimes. Much like with *Le Vieux Paris* or the *Village Suisse*, the public's reaction to these spaces was informed by the rooms' ability to facilitate virtual travel to a different time or place than the "here" and "now" of turn-of-the-century reality. Writing for *Le Journal* of November 7, 1900, for example, Jules Claretie confidently asserted that the rooms were "alive." "One could say that on these carpets will not appear mere shadows but will instead walk the living," he continued. Experiencing them would invoke a different kind of Pompeii, one without ruins, but where "all the furniture in place, the inhabitants would return. From Louis XVI to Louis-Napoléon, via Louis-Philippe, it is truly and chronologically a visit to times long-gone. It is a Memory Palace."[69]

The *Musée Centennal*, for Claretie, thus equaled "un Château du Souvenir" (a Memory Palace), or "a visit to times long-gone."[70] This sense of immersion into a past made present through the combination of objects in an aesthetically unified environment was accentuated by the authenticity of the items used in the formulation of the various décors. Whenever possible, the organizers attempted to acquire pieces original to the period recreated for each of the rooms on display. No fewer than one hundred and forty-six private collectors joined in the endeavor, their collectibles complementing items borrowed from the collections of the Comédie-Française, the Mobilier National, the Musée de la Manufacture des Gobelins, the Palais de Fontainebleau, and the Union Centrale des Arts Décoratifs.[71] The accuracy of the domestic reconstructions invoked was further assured by the incorporation of imitation period pieces whenever originals were lacking. Georges and Henri Rémon, the Museum's decorators, were charged with finishing off the decorations – *boiseries*, upholstery, floors, carpets, and wallpaper included – based on contemporary sources and according to the spatial constraints of the exhibition's layout.[72]

The first of these rooms, the Salon Louis XVI, displayed a set of Aubusson-upholstered furniture in gilt wood that seemed the perfect complement to a set of *boiseries* borrowed from the collection of the Musée des Arts décoratifs.[73] The newly matched objects were accompanied by a variety of decorative items from the period, which together strove to create the impression of a real drawing room from the time of the last eighteenth-century monarch. The room from the time of the Constituent Assembly was organized around a wooden blue bed, upholstered in cotton with red figures printed against a white background that showed dancers on the location of the former Bastille. Gray wood paneling from the collection of Cézard Dutocq completed the display, to which the decorators also added wall hangings matching in design the bed's upholstery. The Museum's organizers thus obtained a modest bedroom "similar to one that a member of the Assembly might have inhabited in 1792."[74] They were less fortunate with the Directory room, where the décor had to be composed from scratch to include a set of disparate furniture pieces and decorative items authentic to the period (Figure 6.7). Working with archival documents from the time, Georges and

Figure 6.7 "Pièce Diréctoire."
Source: Maurice Le Corbeiller, *Musée Centennal des classes 66, 69, 70, 71, 97. Mobilier & Décoration à l'Exposition universelle international de 1900, à Paris. Rapport de la commission d'installation* (St.-Cloud: Bélin, [1902]), Pl. XIX.

Henri Rémon created a set of painted panels in the semi-Louis XVI, semi-Pompeian taste of the moment, which they set against a wooden octagonal space decorated with mirrors and columns characteristic of these styles. A Savonerie carpet from the collections of the Mobilier National and a fireplace lent by Huvé completed the display. While critics such as Jules Claretie referenced the soirées of Mme. Récamier and the gracious forms of her armchairs, visitors' imagination animated the environment. In Claretie's words,

> on these armchairs of ancient forms, Mme. Récamier might have rested her semi-nude shapes. The beauties of the time would have sat there to hear Garat cooing and lisping. Perhaps a young general, already 'emperor,' recently returned from Italy, stepped on these same carpets in his victorious soldier shoes.[75]

Any discrepancies between furniture and decoration were kept to a minimum with the help of viewers' imagination, whose sense of the past was rekindled through "*la restitution de tout un milieu*," or the grouping of original objects and accurate reconstructions of objects from the same time period into a complete environment.[76]

Other rooms involved less work. The Empire drawing room was a case in point, with collectors eager to contribute entire furnishing ensembles for display (Figure 6.8). Wall paneling salvaged from buildings in the course of demolition was added to authentic furniture, bringing an extra touch of accuracy to the recreation of a salon from the days of Napoléon I.[77] So was the bedroom of the famed *Comédie-Française* actor François Joseph Talma, whose furnishings designed by Charles Percier had been purchased *en masse* upon Talma's death in 1826 (Figure 6.9). Georges and Henri Rémon were able to recreate for these original pieces a setting identical to their initial

Figure 6.8 "Salon Empire."
Source: Maurice Le Corbeiller, *Musée Centennal des classes 66, 69, 70, 71, 97. Mobilier & Décoration à l'Exposition universelle international de 1900, à Paris. Rapport de la commission d'installation* (St.-Cloud: Bélin, [1902]), Pl. XX.

Figure 6.9 "Chambre à Coucher de Talma."
Source: Maurice Le Corbeiller, *Musée Centennal des classes 66, 69, 70, 71, 97. Mobilier & Décoration à l'Exposition universelle international de 1900, à Paris. Rapport de la commission d'installation* (St.-Cloud: Bélin, [1902]), Pl. XXII.

Figure 6.10 "Salon Napoléon III."
Source: Maurice Le Corbeiller, *Musée Centennal des classes 66, 69, 70, 71, 97. Mobilier & Décoration à l'Exposition universelle international de 1900, à Paris. Rapport de la commission d'installation* (St.-Cloud: Bélin, [1902]), Pl. XXV.

one by copying the shape and décor of the very room which had housed them. This space was still in existence in a *hôtel particulier* in Paris, whose owner had given decorators permission to visit the room and take notes in situ.[78]

But it was the Napoléon III drawing room that attracted the most attention in the press among the various exhibits, evoking as it did the atmosphere of the most recent past (Figure 6.10). To many visitors, the room sparked memories from young adulthood. A guest commented how he "felt thirty years younger, about to enter grandmother's drawing room" to receive his New Year's gifts.[79] Of "screaming truthfulness,"[80] the rooms benefited from visitors' recollections, assisting organizers in their task of recreating the impression of earlier times and, quite often, lost spaces. Jules Claretie, once again, clarified the relationship between present and past in his evocation of a *Musée Centennal* room that reminded him of the former French president Marie François Sadi Carnot:

> There is, in fact, among these rooms, one where it seems that the Organizer of Victory, his eye on the border that he defends ... is reborn, reappears in front of his writing table, the map of France open on his desk.[81]

Planning to revisit the world's fair room at the same time that a monument dedicated to Carnot would be inaugurated in the city of Lyon, the critic found himself "close to the sons of the assassinated grand citizen, at least in thought."[82] Past and present were thus brought together under the auspices of the Museum as visitors' memories animated the period rooms.

Historian Leora Auslander has observed how political regimes had used style and taste "to represent and construct their powers" throughout the eighteenth and most of the nineteenth centuries.[83] However, with the advent of what she calls "the bourgeois stylistic regime" after 1880 and the state's loss of control over furniture production in the midst of free enterprise and an expanding market, Auslander argues that French furnishings and home decoration became divorced from the world of politics and power. The example of the *Musée Centennal* proves how political meanings in the aesthetic realm persisted into the 1900s, showing how nationalist agendas informed not only exhibit choices at international world's fairs but also people's choices in how to decorate their abodes. Collectors such as Henry-René d'Allemagne, for example, who had contributed objects to the Louis-Philippe period room at the exhibition, would go on to create interiors in the styles newly underscored by the *Musée Centennal*. D'Allemagne's *hôtel particulier* would boast a "Salon Directoire," proof that the rooms at the *Musée Centennal* had found their followers.[84]

"An accumulation of spectacles," the environments created by the installation committees of the 1900 French product sections 66 (Fixed Decoration), 69 (Furniture), 70 (Carpets), 71 (Mobile Decoration), and 97 (Bronzes) at the *Musée Centennal* presented new period styles as a sequence of theatrical and immersive displays in a format resembling other themed amusements at the exhibition. These politically themed environments elicited comparable reactions from the viewers, who seemed more than eager to virtually travel to a recent past. While the *Style Troisième République* per se did not feature among the eight period room displays, "contemporary art" *was* represented through works by Gallé, Majorelle, and Chéret, grouped together in the very last space of the *Musée Centennal*'s "museum" section. A display cabinet with floral inlay by Gallé, a wooden, double chest in somber hues by Majorelle, and three pastels by Chéret – a pastoral scene and two sketches for Baron Joseph Vitta's salon decoration – were thus the culmination of the chronological sequence of works from the nineteenth century. As the catalog concluded, "from Riesner and Clodion to Gallé, Majorelle and Jules Chéret, we have thus completed our work, rightfully proud of at least our points of departure and arrival."[85] The works of artists such as Louis Majorelle, Jules Chéret, or Émile Gallé were cast not only as the culmination of a century-long progression of styles but also as embodiments of a new Third Republic aesthetic, a present-day, ongoing historical and politically defined "theme" for which contemporary Frenchmen were the proud actors. Together, the chronological displays and eight period rooms of the *Musée Centennal* at the 1900 Exposition Universelle, while adopting a language seeped in theatricality and entertainment, reinforced a flagging perception among the French of their contemporary ongoing leadership in decorative arts production.

Notes

The author would like to thank Séverine Montigny and Marine Thézé of the Bibliothèque Historique de la Ville de Paris, Florence Quignard-Debuisson of the Collection Debuisson, and Yves Badetz of the Musée d'Orsay for supporting the research endeavors that made this chapter possible.

1 J. de Tourette, "L'Exposition de 1900: Paris en 1400. – La Cour des Miracles," *La Petite Gironde, Bordeaux* (23 Jan. 1899).
2 Anonymous, "Paris en 1400," *Le Petit Journal* (10 May 1899).
3 de Tourette, "L'Exposition de 1900."
4 Elizabeth Emery and Laura Morowitz, *Consuming the Past: The Medieval Revival in Fin-de-Siècle France* (Aldershot, England and Burlington, VT: Ashgate, 2003), 8.
5 The chapter thus follows the writings of Guy Debord, who has argued that "the entire life of societies in which modern conditions of production reign announces itself as an

immense accumulation of spectacles," to argue that both attractions and national displays at the 1900 world's fair employed similar organizational patterns to different ends. See Guy Debord, *Society of the Spectacle* (Detroit: Co-Op, A Black & Red Translation, 1970).

6 For an account of how the eighteenth century has inspired nineteenth-century French art and private life see Debora Silverman, *Art Nouveau in Fin-de-Siècle France: Politics, Psychology, and Style* (Berkeley: University of California Press, 1989), Meredith Martin, "Remembrance of Things Past: Robert de Montesquiou, Emile Gallé and Rococo Revival during the Fin-de-Siècle," and Andrew McClellan, "Vive l'amateur:" The Goncourt House Revisited," in Katie Scott and Melissa Hyde, eds., *The Rococo Echo: Art, Theory and Historiography from Cochin to Coppola* (Oxford: Voltaire Foundation, 2014). See Anca I. Lasc, "A Museum of Souvenirs: Adolphe Thiers, Collector of the Nineteenth Century," *Journal of the History of Collections* 28, no. 1 (Mar. 2016), 57–71 and Anca I. Lasc, "Interior Decorating in the Age of Historicism: Popular Advice Manuals and the Pattern Books of Édouard Bajot," *Journal of Design History* 26, no. 1 (Feb. 2013), 1–24 for an account of the Renaissance influence on nineteenth-century French interiors.

7 Maurice Le Corbeiller, *Musée Centennal des classes 66, 69, 70, 71, 97. Mobilier & Décoration à l'Exposition universelle international de 1900, à Paris. Rapport de la commission d'installation* (St.-Cloud: Bélin, [1902]), 13.

8 Auguste Marguillier, "Bibliographie des ouvrages publiés en France et à l'étranger sur les beaux-arts et la curiosité pendant le deuxième semestre de l'année 1902," *Gazette des beaux-arts* 28 (1902), 527.

9 Le Corbeiller, *Musée Centennal*, 13.

10 Ibid.

11 Richard D. Mandell, *Paris 1900: The Great World's Fair* (Toronto, ON: University of Toronto Press, 1967), 64.

12 Le Corbeiller, *Musée Centennal*, 13.

13 Ibid., 15.

14 Ibid., 18–19.

15 Ibid., 21.

16 Ibid., 21–22.

17 Ibid., 29.

18 Ibid., 30.

19 Ibid., 31.

20 While such terminology as "Style Louis-Philippe" or "Style Empire" or even "Style Second Empire" had been used before the 1900 exhibition, to my knowledge, there never existed a chronological sequence of stylistic revivals based on nineteenth-century political regimes that was publicly on display. Nor have I ever come across such an extensive – almost exhaustive – list of nineteenth-century revival styles in the written accounts of the period.

21 Ibid., 34.

22 Ibid., 17.

23 Ibid., 14.

24 Other kinds of public immersive environments available to Parisians around the same time as the 1900 Exposition Universelle included the wax museum, the panorama, the panopticon, the department store window as well as early cinema. See Anne Friedberg, *Window Shopping: Cinema and the Postmodern* (Berkeley, Los Angeles and London: University of California Press, 1994).

25 Anonymous, "Les Clous de 1900: Les derniers projets examinés par la commission," *L'Eclair* (9 Aug. 1897).

26 Anonymous, "Le Palais du Costume," *Le Temps* (17 Dec. 1897).

27 Projet Félix, *Exposition Universelle de 1900: Palais du Costume – Le Costume de la femme à travers les âges. Catalogue* (Paris: Imprimeries Lermercier, 1900).

28 Ibid., 52. The catalog cites Imbert de Saint-Amand's *La Cour de l'impératrice Joséphine* (Paris: Dentu, 1883) as a source for this information. A further exhibit of costumes, carpets, and embroideries from the third to the thirteenth centuries uncovered during A. Gayet's archeological excavations at Akhmin, Dronkah, Deir-el-Dyk, and Damiette in Egypt between 1898 and 1899 further attest to the desire for accuracy in the *Palais du Costume*'s representations. See Félix, *Exposition Universelle de 1900*, 60 and Thiébault-Sisson, "Promenades à l'Exposition: Dix siècles d'histoire du vêtement," *Le Temps* (22 Sep. 1900).

29 Paul Greenhalgh, *Ephemeral Vistas: A History of the Expositions Universelles, Great Exhibitions and World's Fairs, 1851–1939* (Manchester and New York: Manchester University Press, 1988), 46. According to Greenhalgh, the 1867 Exposition Universelle was the first to "use entertainment as a crowd-puller," 42.

30 Greenhalgh, *Ephemeral Vistas*, 43–46.

31 Ibid., 34.

32 Anonymous, "L'Actualité: Le Projet Paschal Grousset et le Feu Intérieur," *L'Éclair* (28 Mar. 1895).

33 Anonymous, "Un Volcan à Paris," *Le Temps* (17 Oct. 1897).

34 Anonymous, "Archéologie: Pompéi morte ou vivante?" *Journal de Genève* (21 Aug. 1899).

35 Anonymous, "L'Actualité: Le Village Suisse à l'Exposition de 1900," *L'Eclair* (15 Dec. 1899).

36 Anonymous, "Au Village suisse," *Le Temps* (12 Dec. 1899).

37 Thomas Grimm, "Exposition de 1900: Le Village Suisse," *Le Petit Journal* (7 Aug. 1899); Anonymous, "L'Actualité: Le Village Suisse à l'Exposition de 1900."

38 Grimm, "Exposition de 1900."

39 Robert W. Rydell, *All the World's a Fair: Visions of Empire at American International Expositions, 1876–1916* (Chicago: University of Chicago Press, 1987), 7. Also see Greenhalgh, 85–86.

40 Greenhalgh, 88.

41 Henry Houssaye, "Voyage autour du monde a l'exposition universelle," *Revue des deux mondes* (15 July 1878), 365.

42 Swedish and Norwegian exhibits had already included wax displays at the Philadelphia Centennial exhibition in 1876. It also appears that some of the *tableaux vivants* represented there were "repeated" at the 1878 exhibition in Paris. See Bruno Giberti, *Designing the Centennial: A History of the 1876 International Exhibition in Philadelphia* (Lexington: University Press of Kentucky, 2002), 139–145.

43 Anonymous, *Le Monde Illustrée* (5 Oct. 1878).

44 Delort de Gléon, *La Rue du Caire: L'Architecture Arabe des Khalifes d'Égypte a l'Exposition Universelle de Paris en 1889* (Paris: Librairie Plon, 1889). Despite efforts to achieve an authentic rendition of Cairo's real appearance, the 1889 Parisian Rue du Caire was not an exact restitution of any one street from Egypt, as confessed by the French architect responsible for this attraction, Delort de Gléon himself.

45 Anonymous, "Village suisse à Paris," *Journal de Genève* (4 Dec. 1898).

46 Anonymous, "L'Actualité: Le Village Suisse à l'Exposition de 1900."

47 Ibid.

48 Ibid.

49 Paris Exposition 1900, *Guide Pratique du visiteur de Paris et de l'exposition* (Paris: Hachette & Cie, 1900).

50 Ibid.

51 Elizabeth Emery, "Albert Robida, Medieval Publicist," in Richard Utz and Elizabeth Emery eds., *Cahier Calin. Essays in Honor of William Calin* (Kalamazoo, MI: Studies in Medievalism, 2011), 52.

52 Anonymous, "Exposition rétrospective et musées centennaux," *Revue Popularie des Beaux-Arts* (1898).

53 Ibid.

54 John Harris, *Moving Rooms: The Trade in Architectural Salvages* (New Haven and London: Yale University Press, 2007), 135. For more information on the development of period rooms, see Bruno Pons, *French Period Rooms, 1650–1800: Rebuilt in England, France, and the Americas* (Dijon: Faton, 1995).

55 Mandell, *Paris 1900*, 26.

56 Ibid., 28–31.

57 Charles Sowerwine, *France since 1870: Culture, Society and the Making of the Republic* (New York: Palgrave Macmillan, 2009), 91.

58 Ibid., 57.

59 Nancy J. Troy, *Modernism and the Decorative Arts in France: Art Nouveau to Le Corbusier* (New Haven, CT: Yale University Press, 1991), 3.

60 Ibid.

61 Mandell, *Paris 1900*, 74–75.

62 Ibid., 75.

63 Eugène Gaillard, for example, upon seeing the Werkbund displays at the exhibition, had reputedly stated: "Moi, je sors de là enragé, hurlant vive la liberté ! vive l'indépendance ! vive la personnalité et vive le combat pour l'art en ordre dispersé ! – tout se rejoindra, tout s'unifiera spontanément et beaucoup mieux grâce à notre seule unité de race et de tendances." See Eugène Gaillard, *Nos arts appliqués modernes. Le mobilier au Salon d'Automne de 1910. Impressions et opinions* (Paris: E. Floury, 1910), 13.

64 André Hallays, "En Flânant," *Feuilleton du Journal des débats* (12 Oct. 1900).

65 Gabriel Mourey, "L'Art décoratif à l'Exposition universelle," *Littérature et Beaux-Arts*, 809.

66 Rossella Froissart, "Les Arts décoratifs au service de la nation, 1880–1918," *Arts & Societies* (Séminaire du 17 juin 2005), www.artsetsocietes.org/f/f-froissart.html (visited 29 Feb. 2016).

67 Anca I. Lasc, "*Le Juste Milieu*: Alexandre Sandier, Theming, and Eclecticism in French Interiors of the Nineteenth Century," in *Interiors: Design, Architecture, Culture* 2, no. 3 (Nov. 2011), 277–306. David Raizman has also discussed the shift from displaying unique pieces of furniture to more "total" interiors in the context of universal exhibitions. See David Raizman, "Giuseppe Ferrari's Carved Cabinet for the 1876 Centennial Exhibition: Presentation Furniture in the Cultural Context of World's Fairs," *West 86th Street* 20, no. 1 (2013), 62–91. He further singles out the existence of literary presentation furniture at these world's fairs, including Gerrard Robinson and Thomas Tweddy's "Robinson Crusoe" sideboard of 1857, which also left behind historic and exotic themes. Patricia Lara-Betancourt further elaborates on the shift from unique furniture pieces to complete room displays in the context of the British department store. See Patricia Lara-Betancourt, "Displaying Dreams: Model Interiors in British Department Stores, 1890–1914," in Anca I. Lasc, Patricia Lara-Betancourt and Margaret Maile Petty, eds., *Architectures of Display: Department Stores and Modern Retail* (London and New York: Routledge, 2017).

68 Koechlin cited in Amy Ogata, "The Union Centrale des Arts Décoratifs Pavilion, Art Nouveau, and the *Cabinet d'Amateur* at the Fin de Siècle," in Daniëlle Kisluk-Grosheide, Deborah L. Krohn, and Ulrich Leben, eds., *Salvaging the Past: Georges Hoentschel and French Decorative Arts at the Metropolitan Museum of Art 1907–2013* (New York, New Haven and London: The Bard Graduate Center, The Metropolitan Museum of Art, and Yale University Press, 2013), 198. The notion "of an Art Nouveau born out of the fertile *French* rather than foreign soil," as Amy Ogata explains, "was thus born." Also see Silverman, *Art Nouveau* and Troy, *Modernism and the Decorative Arts*. According to these scholars, the French eighteenth-century past resonated the most with the turn-of-the-twentieth-century decorators, collectors, and designers.

69 Jules Claretie, "Les Chambres," *Le Journal* (7 Nov. 1900). The original French citation reads: "Elles sont vivantes et on dirait que, sur ces tapis, vont non pas apparaître des ombres, mais marcher des vivants. [...] A les étudier on feuillette, dirait-on, des chapitres d'histoire, on pénètre dans un autre Pompéi, mais sans ruines et où, tous les meubles étant en place, les habitants vont revenir. De Louis XVI à Louis-Napoléon, en passant par Louis-Philippe, c'est vraiment et chronologiquement une visite aux temps passés. C'est le Château du Souvenir."

70 Ibid.

71 Le Corbeiller, *Le Musée Centennal*, 7–9.

72 Ibid., 37. Originals, however, were to be clearly separated from imitations or copies by their identification as such on carefully designed wall labels centrally displayed within the rooms.

73 Ibid., 39.

74 Ibid., 41.

75 Claretie, "Les Chambres." The original quote reads: "Sur ces fauteuils, de tournures antiques, Mme Récamier a pu reposer ses formes demi-nues. Les belles merveilleuses se sont assises là pour écouter roucouler et zézayer Garat. Peut-être sur ce tapis le jeune général, retour d'Italie, a-t-il posé ses talons de soldat victorieux, déjà 'imperator.'"

76 Ibid.

77 Le Corbeiller, *Le Musée Centennal*, 43.

78 Ibid., 45.

79 Ibid., 51.

80 Claretie, "Les Chambres."

81 Ibid. The original quote reads: "Pour moi, je les revoyais, dimanche, précisément à l'heure solennelle où M. le Président de la République inaugurait à Lyon le monument du noble Président Carnot, et il me semblait que, par là, je me trouvais rapproché, au moins par la pensée, des fils du grand citoyen assassiné. Il y a, en effet, dans ces chambres, une chambre où il me semble que, devant sa table de travail, avec sa carte de France ouverte sur son bureau, - l'œil sur la frontière qu'il protège ... – renaisse, réapparaisse l'Organisateur de la Victoire." Indeed, Carnot had been a leader in the resistance movement under the 1870 government of national defense.

82 Carnot had been stabbed by an anarchist on June 24, 1894, and died on June 25 of the same year.

83 Leora Auslander, *Taste and Power: Furnishing Modern France* (Berkeley: University of California Press, 1996), 2.

84 Henry-René d'Allemagne, *La Maison d'un vieux collectionneur* (Paris: Librairie Gründ, [1948]). I thank Forence Quignard-Debuisson, of the Debuisson collection, for bringing this book to my attention.

85 Le Corbeiller, *Le Musée Centennal*, 36.

7 "Our country has never been as popular as it is now!"[1]

Finland at the 1900 Exposition Universelle

Bart Pushaw

Introduction

"Never has the civilized world cast its gaze upon our country as they do now," the painter Albert Edelfelt urgently wrote to his colleagues in Finland in 1898. "Finland's culture has been rightly praised. That is why the expectations for our participation at the world's fair of 1900 are so great."[2] Edelfelt's pressing request circulated among a group of artists who understood what was at stake in 1900. By this time it was well known that world's fairs were the premiere venue for displaying a distinctive national identity on an international stage. Here nationalism and the arts became intertwined, conveying messages about peoples and places with whom the intended audience rarely came into contact. For a small nation like Finland, participation in these fairs was pivotal since it offered unparalleled opportunity for international exposure. Edelfelt thus mobilized Finnish artists, architects, and designers into an aesthetic army marching in the front lines so their country could become an equal player in the international arena of the Exposition Universelle.

To declare Finland as an equal power in 1900 was to assert Finnish independence, at least culturally, from the Russian Empire, a risky claim in an era when the area was still a Grand Autonomous Duchy of Russia, the western hinterlands of a sprawling and powerful realm. Given Finland's subordinate status within Russia, it is astonishing that the Finns managed to present their own pavilion separate from the Russian pavilion, let alone on the Rue des Nations. Permission to create a separate Finnish pavilion might suggest harmonious relations between Russia and its autonomous borderland, but in this age of ostentatious nationalisms, Finno-Russian relations had never been more troubled.

Finnish nationalism emerged as a distinct phenomenon from Russian nationalism because the area of Finland had only been under Russian rule since 1809, after it was ceded from a weakened Swedish Empire following Russian victories in the Finnish War. The previous six centuries of Swedish hegemony over Finland permeated everyday life in the region well into the nineteenth century, most notably in the endurance of Lutheranism and Swedish as the language of culture and the elite. Inspired by Johan Herder's views about the inherent value of smaller cultures, Finland's Swedish-speaking intelligentsia sparked a movement in the early nineteenth century to assert a distinctly Finnish cultural identity, one based in local geography, traditions, and, most importantly, the elevation of the Finnish language, a Finno-Ugric tongue completely unrelated to neighboring Swedish or Russian and then spoken only among the rural peasantry. This nascent national movement inspired the poets Johan Ludvig Runeberg and Zacharius Topelius, who extolled the beauty of Finland in verse, while folklorist Elias Lönnrot collected and compiled Finnish folk songs into an epic—the *Kalevala*, first published in 1835—narrating heroic tales of a primordial Finland, thereby bestowing the native peasantry with a powerful ancient mythology and history. These pillars of Finnish identity—language, religion, a nascent literature, and mythological history—were at odds with expanding Russian imperialism in the late nineteenth century, which built its powerful expression upon its cultural and linguistic Slavic identity and Eastern Orthodox faith. As one Finn noted in 1890, "the existence of this little people, happy within itself beside the Russians, wounds Russian sensibilities."[3]

When Edelfelt wrote to his colleagues about international attention turned towards his country, he referred to an international image of Finland—most accessible through the music of Jean Sibelius (1861–1957)—and international awareness about draconian Russification policies coming from St. Petersburg.[4] Thus the Finnish pavilion not only sought to articulate a clear vision of Finnish identity, but also forcefully proclaim Finland as distinct, if not completely separate, from Russia. The pavilion itself, "small and unassuming" among surrounding grandiose displays, evinced a clear vision of the Finnish people for international viewers, leading one to admire the Finns' "quality of vigor, solidity, robust health, a hint of willfulness balanced by a positive attitude."[5] In an article dedicated to the Finnish pavilion in the German periodical *Dekorative Kunst*, one reviewer declared the pavilion was "the aesthetic success of the entire Boulevard of Nations."[6] Contemporary international critics agreed the Finns had created a distinctly Finnish visual identity complementing their cultural (if not also political) autonomy.

The works of Finnish participants appeared overwhelmingly "national," the pavilion being filled with images of ancient warriors from Finnish mythology or lyrical winter landscapes. So complete was the impression of unity that one painting depicting a scene of contemporary industry stood out as an incongruous note. Beda Stjernschantz's *The Glassblowers* presented viewers with an aestheticized representation of the contemporary working class, portraying technological innovation that was visually in accordance with Arts and Crafts ideals. Stjernschantz's painting warranted critical attention specifically because, as one critic described, "the national note completely disappears, to be replaced by the technique and the range of vision of the cosmopolitan schools."[7]

Throughout twentieth-century criticism and scholarship, the Finnish Pavilion at the 1900 Exposition Universelle has been understood as undeniably Finnish, ushering in a distinct "national style." Some scholars even consider the pavilion part and parcel of the first "golden age of Finnish design," prefiguring a teleological development towards the ascent of twentieth-century Finnish design firms such as Marimekko.[8] This emphasis on national solidarity obscures the complex power relations and multiethnic strife in turn of the century Finland, an issue manifest in Beda Stjernschantz's enigmatic *The Glassblowers* and other works in the pavilion, notably Emil Halonen's relief sculptures. Alongside an analysis of Stjernschantz's little known painting and Halonen's sculptures, this essay reconstructs the Finns' path to unexpected success in Paris in 1900, elucidating how Finnish architects, artists, and designers effectively cultivated a singularly national visual vocabulary for a foreign audience. By emphasizing the artifice of a cohesive Finnish national identity at the fin-de-siècle, however, this essay also strives to reveal what is at stake when the ideals of the Aesthetic movement and the nationalism of world's fairs mask multiethnic conflict, inviting a reconsideration of our assumptions regarding national unity at these international venues.

Ensuring participation in Paris

In an age when strong pan-Slavic, pro-Russian, and anti-separatist sentiments emanated from St. Petersburg to the borderlands, the fact that Finland even had a pavilion at the 1900 Exposition Universelle was a remarkable feat in and of itself. While Finland had a presence at earlier fairs, it was marginal and subsumed under Russian sections. Kerstin Smeds has argued that we cannot determine why Finland gained permission for a pavilion since primary sources merely indicate how it occurred. In fact, Smeds contends that "by autumn 1897 there was no real reason *not* to give the Finns the potential to build a pavilion"[9] because Finno-Russian relations at the time were relatively neutral. Finns eagerly awaited an official invitation from Paris, as was customary for world's fairs, but it never arrived. Consequently the Finnish- and Swedish-language press launched an intense campaign advocating for Finnish participation in 1900.[10] Ultimately the work of three people secured a distinct Finnish space for the fair of 1900: politician Leo Mechelin, artist Albert Edelfelt, and engineer Robert Runeberg (Figure 7.1).

Leo Mechelin, a senator from Finland and later champion of liberal democracy, skillfully negotiated permission from the tsar for Finnish participation on the grounds that other autonomous regions of the Empire, namely the Caucasus and Siberia, would have a presence at the fair.[11] While Mechelin handled the political side of the pavilion between Russia and France, Albert Edelfelt, named Finnish Art Commissioner for the fair, was seminal for cultural matters.[12] Edelfelt was the first Finnish artist to establish a well-respected reputation among Parisian art circles in the 1880s

Figure 7.1 The artists Albert Edelfelt, Albert Gebhard, Axel Gallén (Akseli Gallen-Kallela) and Emil Wikström in Paris, Spring 1900.

Source: Reproduced from Bertel Hintzen, *Albert Edelfelt*, del II. 1888–1905 (Söderström & Co förlags AB, Helsinki, 1954), 150. © Finnish National Board of Antiquities, Helsinki.

and the first Finn to have a work in the Musée du Luxembourg. By 1899, Edelfelt's connection with the Parisian art world was never stronger, and the artist utilized his vast network of contacts to ensure that Finland would have its own pavilion. Equally significant to Edelfelt was the engineer Robert Runeberg, who spoke Russian well and thus negotiated between the demands of St. Petersburg and the desires of Helsinki. While some Finnish nationalists were wary of Runeberg's intimate connections with the Russian government—his Finnish patriotism was less blatant than Edelfelt's or Mechelin's—his contacts were pivotal to ensuring bilateral cooperation.[13]

Permission was granted on two conditions: first, the pavilion would be located at a "second-rate" location between Luxembourg and Bulgaria at the back of the Quai d'Orsay. This location, undesirable as it may have been, actually placed the Finnish pavilion at quite a distance from the Russian building. As such, the second requirement stated that pavilion must bear the inscription "Le Pavilion Finlandais, Section Russe," leaving no room for interpretation of Finland's political status. At this point, these stipulations did not concern the Finns. Rather, they were relieved that their pavilion was a certainty at all.

"Our Small, Very Small Finnish Church"

Already in 1898 Finnish organizers were anxious about ways to distinguish their pavilion from others. One correspondent in the Swedish-language paper *Hufvudstadsbladet* declared,

> One learns that in Paris each country's pavilion must be prepared in the style and materials characteristic for each land. We lack our own style. The Finnish wooden style, which has sailed under our flag before, is an artistically made product constructed more or less by thoughtful architects. But the worst is that such a construction will be seen as Russian in style since 'Finnish ornamentation' in Russian building styles is so great and significant. As far as we are concerned, we are against a pavilion made of wood.[14]

The Finns were rightly concerned, for the Russians ultimately built a pavilion entirely of wood. Helsinki, Finland's imperial capital, boasted elegant neoclassical architecture, but since the city had expanded only after it gained capital status under Russian rule in 1809, the buildings lacked "Finnish" qualities. Wooden architecture, on the other hand, was the most typical across Finland and most characteristic of its people. But as this *Hufvudstadsbladet* writer suggested, wooden architecture was certainly not unique to Finland, and could easily draw up similarities with Russian culture for international viewers. With peasant wooden structures and neoclassical style clearly out of the running, commissioners were at an impasse regarding what material could constitute a distinctively Finnish style.

Finnish architects ultimately turned to stone, a clever choice in a land where rocky granite earth is abundant. From Reykjavík to Riga, stone became a primary medium in the construction of national styles throughout Northern Europe.[15] As a native material, stone communicated resilience, strength, and even indigenous identity. The use of stone in new architecture also demonstrated the resourcefulness of new local stone industries and the triumph of new modern technologies.[16] While Finns learned these technologies abroad, their successful implementation became a testament to the power of a modern Finnish industry. The winning design for the pavilion applied this vogue for stone in an imaginative rendering of a rural stone church by the young architects Herman Gesellius, Armas Lindgren, and Eliel Saarinen (Figure 7.2).

It was no accident that Finland's pavilion resembled a church. By 1900 religion was one of the clearest differences between Finns and Russians. Russification campaigns beginning in the 1880s promoted the adoption of Eastern Orthodox Christianity among the Lutheran populations of what is today Finland, Estonia, and Latvia. But for the Finns (as well as the Estonians and Latvians), Lutheranism was not only a pillar of their identity, but also a marker of their connection to continental European culture and history. Rather than the central plan typical of Orthodox churches, the pavilion was laid out like a basilica with a bell tower at the crossing. In its monochromatic gray and white color scheme, the pavilion also departed from the exuberance of the colored domes of Russian churches and evoked comparison with Christian architecture of the High Middle Ages.[17]

Reproductions of the pavilion in contemporary journals often focused on the building's decorative flourishes, suggesting the building was more ornate than the basilicas it supposedly emulated. To be sure, Gesellius, Lindgren, and Saarinen decorated the exterior of their imagined Finnish church with discrete floral and faunal motifs, including soapstone bear heads encircling the main entrance, squirrels around the smaller entrance, large pinecones, and frogs. Sculptor Emil Wikström even created four bears standing guard on the roof. Finland's church-like pavilion was thus more medievalizing than it was medieval, more modern pastiche than accurately historical. Art historians Michelle Facos, Thor Mednick, and Janet Rauscher have argued that the style of the pavilion was so eclectic that it would have been completely unfamiliar to a Finn.[18] When the initial plan was revealed in

Figure 7.2 The Finnish Pavilion at the 1900 Exposition Universelle, Paris. Photograph, 275 × 208 mm.

Source: © Finnish National Board of Antiquities, Helsinki.

Finland, wary viewers thought the tower would give off "Asiatic or Hindu impressions."[19] Such architectural eclecticism was actually representative of the way in which pavilion officials distilled a diverse culture into a single representation, inevitably simplifying, essentializing, and exoticizing an unknown people to gain international acclaim and respect, raising questions about agency in (self-)representation and reception, to which we will return below.

For a small nation such as Finland to create a clear and cohesive vision for an international audience, it was important to simplify rather than elaborate. The co-existence of too many styles would muddy an already murky image of the distant Northern land. Wanting to ensure clarity, the painter Akseli Gallen-Kallela urged Saarinen, the pavilion's main architect, "simplify, simplify, and throw those sweet arabesques and mawkish [forms] copied from *The Studio* to hell!"[20] In the end, the pavilion's Art Nouveau flourishes were restrained, revealing an austere, mostly unadorned façade. The building's eclecticism gave way to architectural simplicity, which impressed contemporary critics. In *Dekorative Kunst*, a reviewer favorably compared the Finnish pavilion with Denmark's small exposition nearby, noting both "testify in common to a simple and honest purpose not to seem more than they are, not to promise more than they can fulfill."[21] If Finland's pavilion was modest, it was understandably so given the small amount of space it was allotted— some scholars even contend it was the smallest pavilion at the fair of 1900.[22] But Edelfelt's reflection on the pavilion as he viewed it from a distance turned its rugged simplicity into a virtue:

> There stands our small, very small Finnish church—all its features marked with nobility, all its details full of style—and tells about a poor people that has suffered and sung itself into this state of simple beauty that one misses everywhere in this carnival orgy.[23]

The outer appearance of Finland's pavilion thus communicated strength and resilience, a concept Finns call *sisu*, and it entranced the fair's attendees, luring them inside to see what else Finland had to offer.

"His Nation's Soul in Color and Line"

Immediately upon entering the pavilion, the soaring height of the bell tower would have guided the viewer's gaze upwards, where a mythical painted world suddenly unfolded (Figure 7.3).

Viewers encountered a wise sage and his band of rugged soldiers, a field of venomous vipers, and the dimming fires of a pagan ritual. This was the world of the *Kalevala* epic as told in Akseli Gallen-Kallela's series of frescoes. For some viewers, Gallen-Kallela's frescoes appeared to be "apparently rude and almost archaic" yet also "original, truly learned, and sophisticated."[24] Painted directly onto the walls of the pavilion, the frescoes were part of the pavilion's decorative program rather than a contribution to Finnish "fine arts." This distinction determined whether or not artworks would be included in "Finland's home" or within the Russian delegation in the fair's Palace of Fine Arts. Architect Jac Ahrenberg voiced his concern for this distinction, proclaiming:

Figure 7.3 Interior of the Finnish Pavilion and its construction team at the 1900 Exposition
         Universelle, Paris. Photograph, 208 × 272 mm.
Source: © Finnish National Board of Antiquities, Helsinki.

> it is better that Finnish artworks are completely absent rather than being mixed
> in with paintings by other nationalities [...] since merging the art and expression
> of life of two peoples, so different in temperament, race, religion, and upbringing
> as the Finns and the Russians would give a completely false image of each other's
> ideals in art.[25]

Despite such debates, examples of Finnish painting in the Palace of Fine Arts were
officially displayed alongside, but not among, Russian artworks.[26]

Most of the Finnish paintings were not displayed at the pavilion, with the exception
of a series of fourteen panel paintings by Finland's leading artists portraying the country's geography, ethnography, and everyday life. This series of paintings was placed in
the Finnish pavilion above the sections designated for the industries they represented,
reinforcing the pavilion's didactic nationalistic program. Juho Rissanen's *Winter
Fishers*, for instance—a depiction of two hearty red-cheeked fishermen clad in heavy
pelt and fur coats on a frozen lake patiently awaiting a catch—was above the Fisheries
Union display. The activity in Rissanen's painting is not industrial, reflecting instead
a traditional pursuit and way of life in Europe's Northern climes. During the long,
harsh winters, ice fishing was a means to keep food on the table and vital to life in the
countryside. By depicting rural peasant fishermen rather than urban industrial harbor
fishermen—a distinction visible in the figures' costume and location—Rissanen's scene

could be considered "ethnographic." The commission and inclusion of Rissanen's painting emphasized the importance of winter in Finnish life, a theme already present in other panel paintings, such as Väinö Blomstedt's silent winter landscape or Pekka Halonen's scene of a hunter working his way uphill on skis in a snowy pine forest.[27]

Gallen-Kallela's four *Kalevala* frescoes ultimately attracted the most attention and proved to be the lasting image of the pavilion. The *Kalevala*, the collection of Finnish folksongs compiled into narrative form in 1835 by the doctor and scholar Elias Lönnrot, intended to elevate native Finnish mythology to a pan-European level. Lönnrot based the *Kalevala* primarily on folk songs gathered from Karelia, a region in eastern Finland where nineteenth-century Finnish nationalists had ostensibly located the origins of a pure, unadulterated form of traditional Finnish folk life, admiring the primordial relationship between people and the rugged land.[28]

For the pavilion's interior, Gallen-Kallela took artistic liberty with scenes inspired from the *Kalevala*, "The Forging of the Sampo," "The Defense of the Sampo," "Ilmarinen Plowing the Viper-Field," and "Heathendom and Christendom." In *Defense of the Sampo*, the old sage Väinämöinen and his soldiers fiercely protect the *sampo*, a magic mill that ensured the eternal prosperity of the Finnish nation and its people against the impending attack of the harpy-witch Louhi, the "Mother of Darkness." The eagle-like figure of the witch and her placement in the composition was a decisive measure by Gallen-Kallela. Just as the Russian threat came from the east to the Finns, Louhi swoops in on the soldiers from the right. Anti-Russian sentiment was also implied in *Ilmarinen Plowing the Viper-Field*, where the epic's mythical blacksmith tills a field teeming with venomous vipers, agitated and ready to attack. The vipers were colored red, blue, or white, colors of the flag of the Russian Empire. In the final version, one viper even had a golden crown, an undeniable allusion to the Romanov monarchy. One reviewer in Finland commented how "the battle between Finnish and Russian culture [...] has been cleverly hidden, and could be interpreted merely as illustrations of the rune songs of the Kalevala."[29]

Indeed, despite references to the violence and barbarism of the Russian Empire in Gallen-Kallela's frescoes, international critics either did not comprehend them or did not want to allude to them, focusing instead on the artist's innovation. One commenter in *Dekorative Kunst* mentioned the frescoes revealed that Gallen-Kallela was

> A Finn through and through, he has reborn his nation's soul in color and line, given it life in his art. Compared with most contemporary work in this area, his pictures come to us like true songs of heroes after a period of feeble decadence, like ocean gales and mountain storms after the dusty city air and hothouse perfumes of our all too artificial gardens.[30]

Scholars typically mark 1900 as the moment Gallen-Kallela became a celebrity figure and Finland's national artist *par excellence*, exemplifying both the artist's own meteoric rise to international fame as well as the growing prestige of the visual arts in Finland as a whole.[31] Such accounts often verge on the hagiographical, as if, in the words of art historian Janne Gallen-Kallela-Sirén (the artist's great-grandson), "the Kalevalian heroes that Gallén frescoed on the ceiling of the Finnish pavilion had cast a magical spell upon the audience critics below."[32] According to most scholarship on the pavilion, the vision of Finland was so cohesive and distinctively national that the success of 1900 heralded a new Finnish style. But such cohesion assumes uniformity in what the Finnish national project was, and that the exhibitors in Paris worked towards the same goal.

In 1904, art historian Julius Meier-Graefe would allude to Gallen-Kallela's frescoes in his seminal *Entwicklungsgeschichte der modernen Kunst,* where he connected the artist's works directly to expressions of Finnish character:

> because we do not have time to read the *Kalevala*...art becomes a cry of subjugated identity; the grim immobility of the imagery has an impact reminiscent of battle-scenes fought out among primitive peoples. The menace conveyed is no empty gesture, but signals a people in uproar, full of passionate inspiration, reaching metaphysical heights and turning into something terrible.[33]

The inclusion of Gallen-Kallela in Meier-Grafe's history of modern art in 1904 is a testament to how quickly the artist was able to put Finland on the European cultural map. Also evident in Meier-Graefe's analysis is a connection between art and race. By using the term "primitive peoples," not only does the author imply that Finns are inclined towards inevitable violence ("something terrible"), but he also invokes a real, brutalizing discourse about power, racial hierarchies, and notions of the savage made manifest in the spectacle of live human displays at world's fairs lasting well into the twentieth century (see below). Race was still a key issue in turn of the century Finland, and as such it bears looking into the surprising contradictions surrounding contemporary racial theory and the pavilion's projected image of national and ethnic solidarity.

The ambiguity of race at the Finnish pavilion and fine arts display

Upon more critical examination of the artworks on display at the pavilion and the Fine Arts building, contradictions regarding Finnish national solidarity abound in the submissions of Finnish artists, especially around ethnicity. Emil Halonen's much-lauded wooden relief sculptures, which "give such brief but telling characteristics of life in our region"[34] according to one Finn, included a representation of a Sámi reindeer herder (Figure 7.4).

Figure 7.4 Emil Halonen, wooden relief sculptures in the Finnish Pavilion, with Sámi and reindeer in the center.

Source: Reproduced from *Dekorative Kunst, illustrierte Zeitschrift für angewandte Kunst*, vol. 6 (Munich, 1900), 461. © Bayerische Staatsbibliothek, München, urn:nbn:de:bvb:12-bsb00087500-5.

The Sámi, indigenous peoples who live mostly around the Arctic Circle in the area of Sápmi (known pejoratively in Nordic languages as Lappland, reaching across the northernmost areas of Norway, Sweden, Finland, and Russia), were oppressed and ostracized for centuries at the behest of dominant powers (in many respects they remain so to this day). Often demonized as the dimwitted, primitive "Other" for their persistent nomadic culture and historical refusal to assimilate, the Sámi were also racially "othered" despite the fact they are, like the Finns and Estonians, ethnically Finno-Ugric. The Sámi were one of many peoples put on display at world's fairs, though they avoided this fate in 1900.[35]

In the colossal 1893 tome *Finland in the Nineteenth Century*, on display in the press room at the pavilion in multiple languages and clearly directed at a foreign audience, a passage describes Finland's indigenous population:

> They call themselves Sámi and think of themselves as relatives of the Finns. The Finns do not want to accept this relationship, which is rather distant. A Sámi is not the stepbrother of the Finns, or even a cousin. The Sámi is small, thin, supple, with black hair and brown eyes, sometimes inactive without thinking, sometimes impetuous, lively and curious as a child, tender hearted, direct, sensitive, easily disappointed and easily frightened, a child of nature who lacks the basis and depth within the character of the Finns.[36]

The author of *Finland in the Nineteenth Century* was at pains to divorce the native Sámi from their Finnish brethren, citing Sámi "ignorance" and "naiveté" as evidence for their divergence from the Finns. If the Sámi were unequivocally not Finnish, why would Halonen create a reindeer herder for the Finnish pavilion, and why would Finnish critics still consider Halonen's works so characteristic of their country? Within the series, the Sámi man was near to a representation of naked Finns beating themselves with birch branches in a sauna, a tradition Finns consider distinctively Finnish. But for a foreign audience unfamiliar with Finnish customs, what was any less "impetuous" about this bathing ritual?

Chiseled and strengthened by their resilience in harsh Northern climes, Nordic artists considered their upbringing in and around rugged nature central not only to their identity but also their artistic expression.[37] Halonen's *Sámi and Reindeer* features a similar coarseness in its appearance. Carved from roughly hewn native pine, Halonen's scenes evoked a certain bio-mystical primitivism in their materiality and production. By foregrounding an indigenous person as subject matter, Halonen's *Sámi and Reindeer* tapped into the European obsession with displaying foreign "types," revealing that Finland, too, offered its own kind of exoticism. As portrayed in the pavilion, the Sámi became a foil against which to evaluate the civilized, cosmopolitan nature of the Finns, distinguishing peoples otherwise muddled together in contemporary racial theory.

In his highly influential 1855 *Essai sur l'inégalité des races humaines*, one of the first pseudo-scientific studies to assert the importance of a racialized hierarchy of power, Joseph Arthur de Gobineau ranked the Finns as part of the inferior "yellow" races along with the Mongols.[38] As late as 1878, the "question of the Finnish race" was still a hot topic of debate, luring foreign scientists to study Finnic craniology.[39] Given the unstable status of the Finnish "race" in the nineteenth century, some Swedish-speaking Finns asserted their difference as "Svecomans," as opposed to the Finnish-speaking "Fennomans," bolstered by the work of Gobineau and others, which placed Germanic peoples at the top of the global racial hierarchy. Although by 1900 many Swedish-speaking Finns proudly adopted Finnish names and promoted the Finnish national project, a small but loud faction of Svecomans forcefully proclaimed a distinction between Swedes and Finns, since any connections with the Sámi were unthinkable.

The Svecoman elite lambasted the interior of the 1900 Finnish pavilion, declaring it "too Fennomane."[40] The pavilion's exhibition of the Finnish Friends of Handicraft (Fin. *Suomen Käsityön Ystävat*, Swe. *Finska Handarbetets Vänner*), which had been actively cultivating a national Finnish style in textiles since 1879, included a work with a "Mordvinian motif," drawn from the distant Mordvins, another Finno-Ugric people who historically have lived near the center of the Volga River.[41] Including a Mordvinian pattern certainly emphasized the Finno-Ugric roots of Finland and distanced the country from its more recent connections with Sweden. The loudest critics of the "Finnishness" of the Finnish pavilion were also Swedish-speaking industrialists of an older generation, whose investment in Finnish industry provided the financial backing for the pavilion in the first place.

The majority of artists, architects, and designers in Finland came from this Swedish-speaking upper class. For historians of Nordic art, this fact is so common that it hardly receives any attention, yet it played a crucial role in the cultural politics of the era. It is impossible to generalize the relationship between Swedish-speaking artists and the Finnish national project; each artist determined it on an individual level. Some fully promoted Finnish nationalism and Fennicized their names, such as Axel Gallén who adopted the name Akseli Gallen-Kallela. For Beda Stjernschantz, her status as a Swedish-speaker led her to create remarkable Symbolist works, tinged with a strange National Romantic sensibility of what Finnish identity could be.[42] Stjernschantz's experiences with ethnic and linguistic divides in Finland eventually brought her to an isolated Swedish-speaking population on the island of Vormsi off the coast of Estonia in 1895, but her questioning of Finnishness occurred even earlier in a more subtle and compelling way in *The Glassblowers*, her 1894 painting selected for display in Paris at the 1900 Exposition Universelle (Figure 7.5).

Dirt and diversity in Beda Stjernschantz's *The Glassblowers*

In Finland, authentic nationalistic expression was tied not only to subject matter, but also to artistic process. Those artists deemed most patriotic often worked in solitary studios nestled among pristine forests of soaring pines and firs, reflecting a connection to their native environment and inspiring their creative endeavors. Such rhetoric around the solitary artist was often gendered as masculine, since intimate interaction with the harsh realities of the Nordic landscape was understood to corrupt the demure interiority of upper-class women striving to become professional artists. Mythologized through the works of male artists and articulated as unequivocally national, the forested region around Karelia would become an important site for the enterprising Beda Stjernschantz and her painting *The Glassblowers*.[43] Yet rather than the rugged wilderness so identified with Finnish art of the period, her painting presents us with an evening interior view of a factory, focused on seven male figures engaged in various phases of blowing glass to create bottles.

Figure 7.5 Beda Stjernschantz, *The Glassblowers*, (1894), oil on canvas, 175.5 × 179.5 cm.
Source: K. H Rendulin museo © K. H. Renlundin museo, Keski-Pohjamaan maakuntamuseo, Kokkola, Finland.

In Karelia, Stjernschantz stayed with family friend and medical doctor Karl Magnus Gadd, whose practice focused on administering to the workers toiling in the nearby glass factory in the town of Pitkäranta. In a letter to her sister, Stjernschantz revealed she had actually seen the doctor perform a necropsy on a dead factory worker, noting, "[the doctor] took the lungs, which were too enlarged and thoroughly permeated with soot."[44] The glass factory was so filthy that the artist exclaimed that, "mostly I walk around like a filthy swine, since the factory is so dirty and smoky that nothing, not even colors, stay clean."[45] While Stjernschantz laments the absence of cleanliness, her painting presents us with pristine workers, seemingly immune to the smoke from the fire behind them. An array of brown tones creates a monochromatic sense of unity in the background, but the clear lilacs and pinks of the dress worn by a woman in the left-hand corner, and especially the interplay of pastel oranges and pinks reflecting on the shirt of the man in the foreground, reveal none of the miasmal smoke to which Stjernschantz alluded in her letters. As the artist had seen with her own eyes, the dirt, smoke, and grime of the factory was disastrous for the health of its workers, yet the workers of *The Glassblowers* do not toil, but may even take joy in their craft. The artist presents harmonious industry, where soot and grime "soil" but little of the canvas. How can we explain this apparent contradiction between Stjernschantz's visceral experience of the factory and its aestheticized representation on canvas? Does her painting merely celebrate the happy, fulfilled Karelian craftsman working in the remote Finnish heartland?

The aesthetic cleanliness of Stjernschantz's figures is even stranger when juxtaposed with Eero Järnefelt's 1893 *Under the Yoke (Burning the Brushwood)*, which reveals soot-covered peasants enduring the brutal realities of rural existence. In representing an interior scene, Stjernschantz strategically avoided Karelia's rugged nature, which was coded as masculine. Yet Stjernschantz also wrote in letters that she worked with her paints and canvas on *The Glassblowers* inside the factory, most often between midnight and 3:00 a.m., to capture the full effect of the fire's brilliance. Stjernschantz was literally covered in soot as she feverishly recorded the environment around her. By flouting gender conventions in *what* can be painted and *where*, the artist ennobled her workers with a pristine beauty in the most unexpected of places.

But how did this painting function six years later in the international context of Finnish participation in the 1900 Exposition Universelle? Since her works so rarely explicitly engaged Finnish nationalism, Stjernschantz was an unlikely contender for inclusion at the fair. One of her most important and well-known works, *Everywhere A Voice Invites Us...*, depicts Estonian peasant children in a bucolic monochrome Symbolist landscape, not Finnish children.[46] Paintings of industry were extraordinarily rare among Nordic artists of the period. Later twentieth-century art historians characterize turn of the century Nordic art as "anti-urban" modernism, distinctive for its images of wild nature, whether in sweeping panoramas or as spaces of internal Symbolist contemplation.[47] As one of few paintings depicting industry, Stjernschantz's artwork clearly merited inclusion, especially as viewers connected technology with progress.

Such a reading, however, oversimplifies how Stjernschantz's painting functioned in 1900, enticing us to reconsider the success of the Finnish pavilion. An economic downturn in the Russian Empire in the late 1890s forced the factory in Stjernschantz's painting to close in 1899; as of 1900, when the painting reached an international audience, the aestheticized rural workers were ostensibly unemployed. Russian-language pamphlets about the factory from 1896 reveal that the only languages spoken at the factory were Russian, and occasionally German with outsiders; never Swedish, and certainly never Finnish since the workers were Orthodox, not Lutheran.[48] These workers in Pitkäranta were thus Russian, not Finnish—a crucial distinction given that the entire purpose of the Finnish pavilion and arts display was to distinguish Finland culturally from Russia.

Karelia, that perceived center of ancient Finnish culture that inspired so many Finnish artists in their search for the "authentic" and unequivocally "national," was also a center of intermixed Finnish and Russian cultural contact. Finns were acutely aware of the ethnic divide among Karelians and their feelings about the Russians were complex. One Finnish artist lamented their Russian "smell," yet appreciated their close-knit familial ties.[49] This artist was interested in Russian Karelians in an anthropological "folk" sense, while also characterizing them as disgusting and dirty.

Stjernschantz represented clean and idealized Russian glassblowers; in an era of anti-Russian sentiment among patriotic Finns, her painting can be seen as a subtle critique of the homogenizing forces of Finnish nationalism, a project that had limited her participation due to her sex. If so, the symbolism was not noted by the pavilion's commissioners, who likely saw *The Glassblowers* as no more than a convenient addition, an image of modernized industry otherwise little represented in the pavilion. Stjernschantz's deliberate idealization of rural Russian artisans reveals the tenebrous nature of the national vision of the Finnish pavilion and the exclusionary forces of a homogenizing nationalism.

Finland's international efficacy

In December of 1900, a short article in the paper *Åland* declared, "Our art exhibition has been a great success – we are hearing it from everywhere nowadays – as is the pavilion. The question is whether or not the entire pavilion will be honored to the French state, if the city of Paris pledges its upkeep."[50] Many Frenchmen were extremely enthusiastic about the Finnish pavilion, and Gustave Soulier, editor-in-chief of *Art & Décoration*, published an extensive celebratory feature article dedicated solely to the Finnish pavilion.[51] Despite best efforts, however, the pavilion, including Gallen-Kallela's frescoes, was destroyed, though many of its paintings were later bequeathed to Helsinki's state gallery.

Although the pavilion proved temporary, its unexpected success had a lasting impact, especially for aspiring Estonian and Latvian artists who similarly lived in a multilingual and ethnically diverse society under Romanov rule and strove to create a distinctly national art.[52] Many Estonian artists would travel to Helsinki to emulate the methods of Gallen-Kallela and his compatriots, hoping to achieve similar international admiration.[53] Janis Rozentāls's 1910 monumental frieze of ancient Latvian mythological figures adorning the Riga Latvian Society Building has been interpreted as a direct response to Gallen-Kallela's *Kalevala* frescoes.[54] This almost cult-like status the Finnish pavilion enjoyed among Baltic artists was predicated not only on the validation of an authoritative international jury, but also on the assumption of national solidarity.

An American review of the Exposition Universelle attests to why the Finns were so elated with their accomplishments in 1900:

> To the American mind Finland seems a place so far away to be almost beyond the pale of civilization, and yet the men from that land appear very like our own intelligent citizens of highly favored and beloved America. And, in truth, we are all very much alike.[55]

Finland's display of its mythological history, panoramic landscapes, educated populace, and thriving industries communicated the idea that Finland was, in fact, a real place just as civilized and cultured as the imperial powers the world over, populated with hardworking, self-sufficient people. The presentation in Paris successfully filled a lacuna in global knowledge about Finland, but it also created an exclusionary vision of Finnish identity.

The inclusion of Beda Stjernschantz's *The Glassblowers* among the Finnish entries seemingly promoted Finnish industry, but the actualities of the artwork's production and the artist's own background evinced a more nuanced and unstable image of Finnishness. Similarly, Emil Halonen's relief sculpture of a Sámi reindeer herder conflicted with the idea that Finnish culture had little in common with its indigenous population as argued in the pavilion's promotional materials. Even the fact that Finland's disgruntled Svecoman faction was in an uproar about the pavilion further proves that the Finland presented at the Exposition Universelle was not universally accepted by nor representative of the country's entire populace. Albert Edelfelt's celebratory words about Finland's unprecedented popularity certainly rang true in 1900, but such an achievement was the result of a temporary, constructed, and fictitious spectacle of national solidarity, not of the complicated intersection of cultural diversity and multiethnic exchange which characterized lived Finnish reality at the time.

Notes

1 Albert Edelfelt to Akseli Gallen-Kallela, 23 June 1899, quoted in Janne Gallen-Kallela-Sirén, "Axel Gallén and the Constructed Nation: Art and Nationalism in Young Finland, 1880–1900" (Ph.D. diss., New York University, 2001), 570.

2 Albert Edelfelt to Akseli Gallen-Kallela, 1898, Quoted in Kerstin Smeds, *Helsingfors-Paris: Finlands utveckling till nation på världsutställningarna* (Helsinki: Finska Historiska Samfundet, 1996), 279.

3 Quoted in George C. Schoolfield, *Helsinki of the Czars: Finland's Capital, 1809–1918* (Columbia, SC: Camden House, 1996), 199.

4 For an excellent summary of the cultural impact of Russification in Finland in the late 1890s, see Glenda Dawn Goss, *Sibelius: The Composer's Life and the Awakening of Finland* (Chicago: University of Chicago Press, 2009), 239–253.

5 "'Finnland' auf der Pariser Weltausstellung," in *Dekorative Kunst* 12 (1900): 457.

6 "Finnland," *Dekorative Kunst*, 457.

7 William Walton, André Saglio, and Victor Champier, *Chefs d'Œvre of the Exposition Universelle* (Philadelphia, PA: Georgie Barrie and Son, 1900), 45.

8 John Boulton Smith, *The Golden Age of Finnish Art: Art Nouveau and the National Spirit* (Helsinki: Otava, 1985), Marianne Aav, "The First Golden Age of Finnish Design," in *Now the Light Comes from the North: Art Nouveau in Finland*, eds. Ingeborg Becker and Sigrid Melchior (Berlin: Bröhan Museum, 2002), 89–94.

9 Smeds, *Helsingfors-Paris*, 337.

10 "Meddelanden från Industristyrelsen i Finland," XXVI–XXX, (Helsinki: 1899), 58; *Hufvudstadsbladet*, October 5, October 9, and December 2 1897.

11 *Suomen Teollisuuslehti*, March 1, 1898.

12 Marika Hausen, "Finlands paviljong på världsutställningen i Paris," in *Finkst sekelskifte. En konstbok från Nationalmuseum*, ed. Eva Nordenson (Stockholm: Nationalmuseum, 1971), 48.

13 Smeds, *Helsingfors-Paris*, 337–342.

14 "Den finska paviljongen vid världutställningen," *Hufvudstadsbladet*, June 26, 1898.

15 Sixten Ringbom, *Stone, Style, and Truth: The Vogue for Natural Stone in Nordic Architecture, 1880–1910* (Helsinki: Suomen Muinaismuistoyhdistys, 1987), Ritva Wäre, *Rakkenettu suomalaisuus: Nationalismi viime vuosisadan vaihteen arkkitehtuurissa ja sitä koskevissa kirjoituksissa* (Helsinki: Suomen Muinaismuistoyhdistys, 1991), Barbara Miller Lane, *National Romanticism and Modern Architecture in Germany and the Scandinavian Countries* (Cambridge and New York: Cambridge University Press, 2000), Pekka Korvenmaa, ed., "Signals from the Periphery: National Romanticism in Finland & Central Europe," *Centropa*, 2 (January 2002), Charlotte Ashby, "The Pohjola Building: Reconciling Contradictions in Finnish Architecture Around 1900," in *Nationalism and Architecture*, eds. Raymond Querk, Darren Deane, and Sarah Butler (Farnham and Burlington, VT: Ashgate Publishing Press, 2012), 135–146.

16 Ashby, "The Pohjola Building," 142.

17 Janne Gallen-Kallela-Sirén, "Axel Gallén and the Constructed Nation: Art and Nationalism in Young Finland, 1880–1900," (PhD diss., New York University, 2001), 574–575; Pekka Korvenmaa, "Who are we? Where do we come from? Where are we going?," in *Now the Light Comes from the North: Art Nouveau in Finland*, eds. Ingeborg Becker and Sigrid Melchior (Berlin: Bröhan Museum, 2002) 72–73.

18 Michelle Facos, Thor J. Mednick, and Janet S. Rauscher, "National Identity in Nordic Art: Perceptions from Within and Without in 1889 and 1900," in *Centropa*, 3 (September 2008), 215.

19 *Hufvudstadsbladet*, August 4, 1898.

20 Gallen-Kallela-Sirén, "Axel Gallén," 50.

21 Quoted in Ingeborg Becker, "The Wilderness and the City Lights – Northern Polarities, or Finland's Artists between National Romanticism and the International Scene. The Case of Akseli Gallen-Kallela," in *Now the Light Comes from the North: Art Nouveau in Finland*, eds. Ingeborg Becker and Sigrid Melchior (Berlin: Bröhan Museum, 2002), 14.

22 Facos, Mednick, and Rauscher, "National Identity," 217.

23 Gallen-Kallela-Sirén, "Axel Gallén," 591–592.

24 Walton, Saglio, and Champier, *Chefs d'Œvre*, 41.

25 Jac Ahrenberg, "Expositionen af finska konstverk, afsedda att utställas i Paris 1900," in *Finsk tidskrift för vitterhert, vetenskap, konst och politik* (Helsinki: Finsk tidskrifts tryckeri aktiebolag, 1900), 69.

26 Most contemporary reviews of this section of the Palace of Fine Arts treated the displays of Russia, Poland, and Finland as three distinct sections within a larger display of the Romanov Empire.

27 Over half of the paintings displayed at the Finnish exhibits outside of the pavilion at the 1900 Exposition Universelle were winter scenes. Riitta Konttinen, *Sammon takojat: Nuoren-Suomen taiteilijat ja suomalaisuuden kuvat* (Helsinki: Otava, 2001), 234–235.

28 For a general overview of Kalevala imagery, see Riitta Ojanperä, ed., *The Kalevala in Images: 160 Years of Finnish Art inspired by the Kalevala* (Helsinki: Ateneum Art Museum, Finnish National Gallery, 2009). For a thorough analysis and insightful critique of Gallen-Kallela's Kalevala imagery, see Gallen-Kallela-Sirén, "Axel Gallén."

29 "Finska paviljongen," *Borgå Nya Tidning*, May 4, 1900.

30 "Finnland," *Dekorative Kunst*, 463.

31 The exhibition catalogue *Akseli Gallen-Kallela: Eurooppalainen mestari* (Helsinki: Helsingin taidemuseo, 2011) suggests such a phenomenon.

32 Gallen-Kallela-Sirén, "Axel Gallén," 565–566.

33 Quoted in Becker, "The Wilderness and the City Lights," 20.

34 "Finland på världsutställnigen," *Hangö*, April 14, 1900.

35 Cathrine Baglo, "På ville veger? Levende utstillinger av samer i Europa og Amerika," (PhD diss., Tromsø University, 2009).

36 Quoted in Tuija Hautola-Hirvioja, "The Image of the Sámi in Finnish Visual Arts before the Second World War," *Acta Borealia* 2 (2006): 101.

37 Michelle Facos, *Nationalism and the Nordic Imagination: Swedish Art of the 1890s* (Berkeley: University of California Press, 1998), 74–105.

38 Smeds, *Helsingfors-Paris*, 155.

39 Ibid., 166–170.

40 Ibid., 339.

41 Smeds, *Helsingfors-Paris*, 138. The activities of the Finnish Friends of Handicraft have already received excellent treatment in Charlotte Ashby, "Nation Building and Design: Finnish Textiles and the Works of the Friends of Finnish Handicrafts," *Journal of Design History* 23 (2010): 351–365.

42 Bart Pushaw, "Ruotsalainen Karjala? Beda Stjernschantzin teos *Kaikkialla ääni kaikuu…* (1895) – Ett svenskt Karelen? Beda Stjernschantz verk *Överallt en röst oss bjuder…* (1895)," in *Beda Stjernschantz. Ristikkoportin takana – Bakom gallergriden*, ed. Itha O'Neill (Helsinki: Suomalaisen Kirjallisuuden Seura, 2014), 180–196.

43 Ville Lukkarinen and Annika Waenerberg, *Suomi-kuvasta mielenmaisemaan. Kansallismaisemat 1800- ja 1900-luvun vaihteen maalaustaiteessa* (Helsinki: Suomalaisen Kirjallisuuden Seura, 2004).

44 Itha O'Neill, "Ristikkoportin takana – Bakom gallergrinden," in *Beda Stjernschantz* (Helsinki: Suomalaisen Kirjallisuuden Seura, 2014), 35.

45 O'Neill, "Ristikkoportin takana," 35.

46 Pushaw, "Ruotsalainen Karjala."

47 Kirk Varnedoe, ed., *Northern Light: Realism and Symbolism in Scandinavian Art, 1880–1910* (New York: Brooklyn Museum, 1982).

48 The pamphlet states: "In order to improve the moral well-being of the inhabitants, [an Orthodox] priest was invited and schools were opened, where teaching takes place in Russian and German so that children learn in their native language." G. Grendahl, *Питкаранта: Краткое описание питкарантского месторождения,рудников и заводов* (St. Petersburg: Tipo-Litografiya A. E. Vineke, 1896), 43.

49 Gallen-Kallela-Sirén, "Axel Gallén," 231.

50 *Åland*, December 1, 1900.

51 Gustave Soulier, "Le Pavilion de Finlande à l'Exposition universelle," *Art & Décoration* 8 (1900): 1–11.

52 Korvenmaa, "Signals from the Periphery."
53 Karin Hallas-Murula, *Soome-Eesti. Sajand arhitektuurisuhteid* (Tallinn: Eesti Arhitektuurimuuseum, 2005), 21–79; Bart Pushaw, "Art and Multiculturalism in Estonia and Latvia, circa 1900," in *A Companion to Nineteenth-Century Art. From Revolution to World War*, ed. Michelle Facos (Boston: Wiley-Blackwell, 2018).
54 Jeremy Howard, "Latvian National Romanticism and *Art Nouveau*: Origins and Synthesis," in *Romantisms un neoromantisms Latvijas mākslā*, ed. Elita Grosmane (Rīga: Izdevniecība AGB, 1998), 144–147.
55 *Paris Exposition Reproduced from Official Photographs…* (New York, Akron, OH, and Chicago: The R. S. Peale Company, 1900), not paginated.

8 "A revelation of grace and pride"

Cultural memory and international aspiration in early twentieth-century Hungarian design

Rebecca Houze

For the 1904 Louisiana Purchase Exposition in St. Louis, Missouri, architect Pál Horti represented Hungary with a striking installation. He arranged a suite of interiors in a gated, wooden structure that was reminiscent of a rural Hungarian dwelling and village church. Some of the rooms featured traditional Hungarian crafts, while others showcased modern designs. Just two years later sculptor Géza Maróti decorated the Hungarian rooms at the 1906 Milan International Exhibition in an Art Nouveau style, which critics interpreted as a modern, and particularly elegant, expression of Hungary's "Byzantine" past. The National Hungarian Applied Arts Association supported both exhibits in conjunction with the Hungarian Ministry of Education and Religion and Hungarian National Museum and School of Applied Art, institutions that fostered and promoted Hungarian decorative art for an international market. Though many of the same modern artists were represented in both St. Louis and Milan, their work was quite differently framed in each venue. Horti emphasized vernacular culture as the source for modern Hungarian design in his exhibit in St. Louis, whereas Maróti highlighted the fashionable, urbane result of Hungarian design education and innovation, which was in keeping with contemporary Austrian, German, French, Italian, and Scandinavian trends. Through the first years of the twentieth century Hungarian artists and architects oscillated between vernacular and international visions of modernity, each of which had their supporters. This chapter examines Hungary's participation at four international exhibitions of decorative art in Paris (1900), Turin (1902), St. Louis (1904), and Milan (1906). These events promoted new design concepts to critics, artists, and consumers, paving the way for innovative collaborations between designers and manufacturers across national borders. Moreover, for Pál Horti and Géza Maróti the Hungarian presence at world's fairs opened doors to new cross-cultural friendships and design commissions in the Americas, which demonstrate the role international exhibitions played in constructing an often fluid national identity.

Hungary at that time was the eastern half of Austria-Hungary, the political entity comprising two semi-independent historical countries, joined by the 1867 Compromise. The Dual Monarchy raised the status of Hungary from one of the many regions of the former Austrian Empire to a self-governing kingdom, a political structure that endured only until the end of the First World War in 1918. The two countries were not economic equals within the unwieldy alliance, but rather equivalent political and bureaucratic institutions, which both remained relatively discontented. Whereas Austria believed that it had lost both territory and power in the arrangement, Hungary struggled to overcome its historical subjection to Austria's ruling Habsburgs, and to assert its economic and industrial independence as a modern Western European nation.

Both Austrian and Hungarian manufacturers favored Italian Renaissance revival styles in the third quarter of the nineteenth century. After the 1867 Compromise, however, Hungarian designers and design critics increasingly advocated styles that were visually distinct from those of their Austrian competitors, and which alluded to particularly resonant periods of Hungarian history. Hungarian designers attempted to find a visual language that was contemporary, yet which also visibly evoked ancient patterns and motifs associated with the mythical and sparsely documented Eastern origins of the Magyar peoples.

To express these complex aspirations in an architectural style, or to represent them with a few objects in an exhibition abroad, was no easy task. The installations in St. Louis and in Milan both drew upon artistic concepts with which Hungarian architects and artists had experimented a few years earlier at the 1900 Exposition Universelle in Paris, and at the 1902 First International Exhibition of Decorative Art in Turin. At those events, designers impressed the public with unusual, creative installations that synthesized, with varying degrees of success, Hungary's nineteenth-century folkloric revival with the chic cosmopolitanism of Art Nouveau and Secession styles which had become fashionable across the industrialized world in the first years of the twentieth century.

Hungary struggled to present itself as both modern and culturally unique to world audiences at international exhibitions between 1867 and 1911, often exploiting and repurposing common cultural stereotypes towards new ends.[1] At the Paris 1867 Universal Exposition, for example, visitors enjoyed a colorful "gypsy band" that performed at a picturesque Hungarian tavern (*csárda*), linking notions of Hungarian identity to the sensuous pleasures of food, wine, music, and dance, enhanced by colorful costumes, and exotic rhythms. At the 1873 Vienna World's Fair, new relationships between industrial, urban Budapest and Hungary's more remote villages were highlighted in the "Ethnographic Village." Furnished wooden farmhouses typical of particular regions or characteristic of certain ethnic groups (Hungarian, Saxon, Romanian, Szekler, and Serbian) delighted visitors with their attractive interiors.[2]

The image of the Hungarian restaurant with its spicy paprika dishes, wine, and "gypsy band," complicates the pastoral idyll of the pretty Hungarian farmhouse with charming decoration. Urban entertainment contrasts with rural domesticity in those two different versions of Hungarian national identity. Similarly complicated are the more masculine stereotypes of the mounted Hungarian herdsman (*gulyás*) of the Great Plain (*Alföld*), and the dashing gentleman in gala dress. The Romantic image of the herdsman, with his embroidered sheepskin cloak (*suba*) and felt coat (*cifraszűr*), was a favorite motif for Hungarian writers and artists throughout the nineteenth century. The *cifraszűr* was politicized as a symbol of nationalistic resistance, and worn by revolutionaries, while the fancy gala dress made reference to the ornate uniforms of Hungarian hussars, fifteenth-century mounted warriors, who acquired reputation as swashbuckler adventurers.[3]

The 1900 Exposition Universelle, Paris

The 1900 Exposition Universelle in Paris provided an opportunity to reinterpret old stereotypes of Hungary and Hungarian culture that had circulated at national and international exhibitions for several decades. Hungary's national pavilion in Paris was well received by visitors who marveled at the astonishing displays of military and historic interiors. The pavilion was a smaller version of the building complex in Budapest's City Park, which had been designed by architect Ignác Alpár for the 1896 National Hungarian Millennial Exhibition.[4] The most remarkable of the historical buildings on which Alpár had based his design was the Vajdahunyad castle (in present-day Hunedoara, Romania), which had been the residence of Hungarian statesman and military leader, János Hunyadi (1406–1456) and his son, King Mátyás Hunyadi, also known as Matthias Corvin (1443–1490). The elder Hunyadi was celebrated for his defense of Hungary against Ottoman invasions in the fifteenth century with the help of hussars, while Matthias Corvin was remembered for his introduction of Italian Renaissance culture to the Hungarian court.

The interior decorations of the national pavilion in Paris included an impressive "hall of weapons," as well as mural paintings and other objects, which dramatized Hungary's thousand-year-long history. In his description of the pavilion, Jenő Radisics, director of the Budapest Museum of Applied Arts, recounted the legendary history of Árpád's arrival in the Carpathian basin.[5] Árpád, heroic founder of Hungary, was believed to have descended from the House of Attila, fifth century king of the Huns. The national pavilion thus reinforced Romantic images of Hungary's origins among heroic Central Asian horsemen, who had passed on, presumably, their formidable, passionate, warfaring, oriental Magyar spirit to their present-day descendants. At the same time, the rich historicist furnishings in the pavilion reminded visitors of Hungary's centuries of artistic and intellectual culture. The ethnic difference of the Magyar peoples from their Germanic and Slavic neighbors was projected in formal terms in Paris by the unavoidable comparison among the Hungarian, Bosnian, and Austrian national pavilions, situated side by side on the street of nations along the Quai d'Orsay. While Hungary's national pavilion recalled the country's long history, some of which preceded Habsburg rule, Austria's baroque building, designed by architect Ludwig Baumann, celebrated the height of Hapsburg imperial culture in the eighteenth century. Between them the diverse peoples of Bosnia-Herzegovina were highlighted in a pavilion that drew upon Islamic architecture, and which featured a mural of pan-Slavic heritage painted by the prolific Art Nouveau illustrator Alphonse Mucha.[6]

Across the fairgrounds, in one of the galleries along the Esplanade des Invalides, both Austria and Hungary were further represented with displays of decorative art. Suites of furnished rooms featuring the work of contemporary designers and manufacturers were grouped together by exhibiting nation. The linear arrangement of rooms encouraged visitors to visually compare the new interior design trends of different countries, which they encountered as a progression, one after another. Along the east side of the Esplanade des Invalides were exhibits representing Japan, Austria, and Hungary, followed by Denmark, Italy, Great Britain, the United States, Germany, Russia, and Belgium. France occupied the entire length of the facing building along the west side of the Esplanade des Invalides. As a result of its greater visibility and its appeal to critics at the fair, French decorative art became synonymous with the refinement and elegance of Art Nouveau, to which furnishing designs from other nations were compared.[7]

Architects Zoltán Bálint and Lajos Jámbor designed an unusual installation to represent Hungarian applied art. Several furnished interiors were connected by a series of decorative archways, heavy with colorful floral patterns in high relief (Figure 8.1). The display made reference to Budapest's evocative Hungarian National Museum

Figure 8.1 Hungarian installation of decorative arts at the 1900 Exposition Universelle, Paris.
Designed by Zoltán Bálint and Lajos Jámbor.
Source: Museum of Applied Arts, Budapest.

of Applied Art, designed by architect Ödön Lechner in 1896 in conjunction with Hungary's National Millennial Exhibition that year. Bálint's and Jámbor's painted plaster installation quoted the Budapest museum's decorative scheme, including its sensuous cladding in green and yellow glazed ceramic tiles, produced by Hungary's world-renowned Zsolnay manufactory. Their decorative motifs recalled those of embroidered costumes worn by mounted herdsmen in Hungary's Great Plain, as well as the colorful patterns of traditional textiles, which could still be found throughout the Hungarian part of the monarchy. At that time Lechner was Hungary's most influential artistic theorist. He believed that a modern "language of form" could and should emerge from the visual vocabulary of Hungary's traditional arts, the origin of which he and others traced to ancient India and Persia.[8]

Architects Ödön Faragó, Ede Toroczkai Wigand, and Pál Horti each contributed work for the Paris 1900 exhibition. Their furnishings and decorative art objects, including heavy wooden furniture with wrought iron fittings, metal pendants and brooches with elaborate gothic decoration, and ceramic vessels with oriental motifs, echoed the historicism of the national pavilion.[9] These artists, at the forefront of modern Hungarian design, would continue to attract attention at the fairs in Turin, St. Louis, and Milan over the next several years. Just as Lechner had incorporated motifs drawn from Hungarian folk art into the glazed ceramic ornament of the Museum of Applied Arts and other modern buildings in Budapest in the 1890s, Faragó, Wigand, and Horti were inspired by the Hungarian craft revival.[10] Both amateur and academically trained ethnographers and art historians at the time collected textiles, costumes, ceramics, and other examples of traditional folk art, which they feared would soon disappear as a result of rapid industrialization.[11] Their motivations were scholarly, artistic, political, and commercial. The interesting patterns of woven and embroidered textiles fueled the creative imagination while also serving as distinctive visual symbols of Hungarian identity, which could be exploited by the contemporary furnishing industry. At the same time, concern for the plight of the rural peasant farmer in the new economy led benevolent enthusiasts to support rural cottage industries, and to encourage the production of traditional crafts for an outside market.[12]

One of the Hungarian interiors on display in Paris was a dining room designed by Ödön Faragó (Figure 8.2). Its sturdy wooden table and chairs with leather strip upholstery incorporated both medieval and folk influences in their decoration. Two dish cabinets, a tall case clock, and a side table featured similar forms. Stylized floral patterns, painted on the wall, were mirrored in the design of the carpet. The embroidered linens and colorful ceramic dishes in the room reminded visitors of the cheerful interiors of Hungarian peasant farmhouses, such as those that had been so popular at the 1873 Vienna World's Fair.

Also featured among the Hungarian exhibits in Paris was a carpet designed by Pál Horti, which received a gold medal. Judith Koós writes that Horti's modern design, synthesized from traditional crafts, was among the first examples of Hungarian

Speisezimmer; Prof. E. Faragó, ausgeführt in den staatlichen gewerblichen Fachschulen.

Figure 8.2 Ödön Faragó, Dining room, exhibited at the 1900 Exposition Universelle, Paris, Hungarian section.

Source: George Malkowsky, ed., *Pariser Weltausstellung in Wort und Bild* (Berlin, 1900), p. 437. Courtesy of the Center for Research Libraries, Chicago.

Art Nouveau.[13] Horti used the term "Secession" to describe the new movement in Hungarian applied arts, in effect identifying the new direction with similar trends which had emerged in the 1890s in Vienna, Munich, and Belgium.[14] Horti, Faragó, and Wigand attempted to develop decorative elements with both local specificity and general international appeal. They adopted the linear, organic, vegetal forms, which at the time were associated with Belgian architect Henry van de Velde, Austrian architect and founding member of the Vienna Secession Josef Hoffmann, and the French designers Eugène Gaillard, Édouard Colonna, and Georges de Feure, whose elegant, urbane furnishings were exhibited in Paris by art dealer Siegfried Bing. But their strong emphasis on folk elements left the Hungarian designs hovering somewhere between authentic country village and modern fashionable city, without settling very comfortably in either place.

The Hungarian decorative arts in Paris generally received positive reviews in the popular French press and in the many illustrated catalogues and reports published for the fair, but they failed to attract the attention of some critics in the leading international decorative arts journals.[15] A few Austrian writers, however, who did take notice, were uncomfortable with the strong emphasis on folk art that they perceived in the Hungarian exhibits. W. Fred (pseudonym for Alfred Wechsler) reported for the Austrian Museum of Applied Art's journal *Kunst und Kunsthandwerk* that the Hungarian exhibition, which seemed to be pursuing a "national style" (a term he used with derogatory distaste), hardly even deserved mention. In his opinion, there was no evident effort in those designs to move forward and to pursue an innovative, contemporary language of form.[16] More likely, however, Wechsler had taken offense at what he perceived as a subversive strategy to reject and undermine Austrian dominance in the field of modern decorative furnishings with overt references to Hungarian folk art. By contrast, Gabriel Mourey, French correspondent to *The Studio*, admired the "freshness" and "novelty" of the experimental designs. Contradicting W. Fred (Wechsler) Mourey described the Hungarian furnishings as both "logical" and "honest" as they looked ahead toward the future.[17]

Turin 1902: first international exposition of modern decorative art

As a result of his accomplishments in Paris, Pál Horti was selected to design the installation of Hungarian decorative arts in Turin. For Elek Koronghi Lippich, head of the art department of the Hungarian Ministry of Education, Horti's approach was grounded in native artistic tradition, yet also spoke the international language of Art Nouveau. In 1898, two years before the Paris exhibition, Jenő Radisics hosted an exhibition of decorative arts with contributions by Louis Comfort Tiffany, Alphonse Mucha, Otto Eckmann, Émile Gallé, Frida Hansen, and many other Western European and North American artists at the Museum of Applied Arts in Budapest. The work of leading Hungarian artists and design firms, such as János Vaszary, József Rippl-Rónai, the Zsolnay manufactory, and Pál Horti, was likewise included within the international context of the exhibition.

Horti's installation in Turin reflected the Budapest museum's new orientation towards international arts and crafts that had evolved from the theories of William Morris, John Ruskin, and Walter Crane, and their followers in the second half of the nineteenth century. His exhibit featured an elaborate entrance composed of two rectangular gateways, which were embellished with stylized decorative motifs that resembled the floral patterns in traditional Hungarian embroideries (Figure 8.3). The rounded, curvilinear ornaments combined with the geometric structure of the whole and generated a visual tension, which visitors found eye-catching and modern. Italian art critic Vittorio Pica admired the Hungarian decorative arts, which, he wrote, had been understandably exhibited separately from the Viennese displays in order to draw attention to their distinct stylistic characteristics. He appreciated the way in which the Hungarian exhibit blended historical motifs from the Orient, the Middle Ages, and the Renaissance period of King Matthias Corvin with recent European trends, demonstrating that originality could indeed be found within the legacy of the past.[18]

Horti's stylized arabesques adorning the entrance to the Hungarian section of decorative art in Turin can be imagined as embroidered ornaments derived from folk costume. They also resemble the eyes of a peacock feather, and were similar to his graphic ornaments for the cover of *Magyar Iparművészet*, the newly rebranded journal of the Society for Applied Art and Hungarian National Museum of Applied Art, beginning in 1897 (Figure 8.4). In contrast to the ornamental archway designed by Jámbor and Bálint for the Hungarian installations at the Paris 1900 Exposition Universelle, or to Faragó's floral motifs adorning the walls of his dining room there, Horti's decorative patterns in Turin were abstracted and flattened. They were reminiscent of Alphonse Mucha's many stylish, orientalist Art Nouveau illustrations of the period. One might also notice a lineage in the work of nineteenth-century Aesthetic Movement artists such as Aubrey Beardsley, James McNeill Whistler, or the many designers for London's Liberty & Co., who employed the peacock feather as decorative motif in shades of turquoise, violet, and gold.[19] Whereas the oriental inflections of Faragó's floral patterns in Paris pointed to Hungary's supposed ancient Central

Figure 8.3 Hungarian installation of decorative arts at the Turin 1902 First International Exhibition of Modern Decorative Art. Designed by Pál Horti.
Source: Photographed by Antal Weinwurm. Museum of Applied Arts, Budapest.

Asian origins, preserved in the floral motifs of traditional Hungarian folk art, Horti's orientalism in Turin was contemporary, cosmopolitan, and fashionable.

Austrian curator Fritz Minkus, director of Vienna's Imperial Lace School (*k. k. Central-Spitzencurs*), was surprised by and pleased with the Hungarian installation in Turin. He saved the best for last in his lengthy review of the works shown by participating nations, concluding his report with a description of mosaic decorations by Miksa Róth and Róbert Scholz in blue and violet, which complemented Horti's architectural framework.[20] Minkus found especially appealing a dining room arranged by Ede Toroczkai Wigand, with furnishings of green-stained wood and decorative iron fittings, which was modeled on peasant furniture; it featured the endearing wall hanging, *Little Girl with Kitten*, designed by Hungarian artist János Vaszary. Sarolta Kovalszky produced the tapestry at the Németelemér weaving workshops in Torontál County (Vojvodina, in present day Serbia), which she established to study and preserve traditional kilim carpet patterns in the form of scherrebek (skaerbaek) tapestries, an imported Danish weaving technique that had become especially popular with Art Nouveau and Jugendstil designers in Germany and Scandinavia.[21]

Most significant for Minkus was that the Hungarian display in Turin stylistically departed from that of the Hungarian decorative arts exhibited just two years earlier in Paris. Minkus, like W. Fred (Wechsler), had been critical of the nationalistic overtones of the Hungarian exhibits there, which he recalled as a "truly ugly" display. Perhaps he had perceived Jámbor's and Bálint's decorative scheme of 1900 as more strident

Figure 8.4 Cover of *Magyar Iparművészet* 3 (1900). Illustrated by Pál Horti.
Source: Museum of Applied Arts, Budapest.

and confrontational than Horti's incorporation in Turin of subdued "national" motifs, used with "discretion" and "good taste," which were "blended with modern ideas."[22] And perhaps Wigand's dining room set in the style of peasant furniture was presented with a lighter touch. It is also possible that Wigand's domestic interior with simple, "healthy" furnishings, recalled for Minkus the sort of peasant farmhouse interiors that had so pleased visitors at the 1873 Vienna World's Fair a generation earlier. Minkus may have spent more time looking at the Hungarian design in Turin in part because the better-known Vienna Secession designers Josef Hoffmann and Koloman Moser, who were busy with multiple exhibitions abroad, had been unable to contribute any work to the Turin show, a turn of events that W. Fred (Wechsler) found regrettable.[23]

St. Louis 1904 Louisiana Purchase exposition

Hungarian exhibits at the 1904 Louisiana Purchase Exposition were visible in several places on the fairgrounds in St. Louis, including in the Agriculture, Transportation, and Mines and Metallurgy buildings, in the Art Palace, and in the handsome appearance of Hungarian Commissioner-General Györgyi Szogyeny, who wore "the most picturesque dress" in the opening day procession. In a passage that recalls the romanticized, masculine image of the oriental Magyar in gala costume, one visitor noted:

> The magnificent uniform was a relic of the old Hunnish days and a fine example of almost "barbaric splendor," albeit the Hungarians have long since been recognized as among the most virile and progressive of European peoples. With all their intelligence and developed traits of several centuries of European culture, they still retain some of that savage spirit which, as oriental invaders, once made them the dread of the West, hence the persistency with which they have retained their racial unity in the dual empire.[24]

Despite that fairgoer's obvious pleasure in the "barbaric splendor" of the "virile," "savage" Magyar, the Hungarian exhibit that received the most attention from the general public was the installation of decorative arts and traditional crafts in the Manufactures Building.[25] The display celebrated, by contrast, an image of Hungary as wholesome and pious, with many references to the healthy domestic abode, women's handicrafts, the peasant, and the child.

Elek Lippich again selected Pál Horti to organize the exhibit in St. Louis. This time, however, Horti used a different approach. Rather than repeating the successful Art Nouveau synthesis of abstract forms and delicate folk motifs that had pleased critics, including the Austrian Fritz Minkus, in Turin just two years earlier, Horti chose a vernacular theme. For the installation in St. Louis he built two wooden towers joined by a wide gate (Figure 8.5). The distinctive shape of the square wooden tower with steeply pitched roof was typical of church architecture throughout Transylvania, including in Kalotaszeg, an area of Hungarian speaking villages to the west of Koloszvár (Cluj-Napoca, in present day Romania), many of which had been fortified with walls and observation towers as defense against foreign invaders in the eighteenth century. It was also reminiscent of the rectangular towers of Transylvanian Gothic castles, including the fifteenth-century Vajdahunyad Castle, one of the buildings upon which the Hungarian national pavilion at the 1900 Paris Exposition ultimately had been based.

Figure 8.5 Hungarian installation of decorative arts at the St. Louis 1904 Louisiana Purchase Exposition. Designed by Pál Horti. View of wooden towers.
Source: Museum of Applied Arts, Budapest.

Horti's installation may refer more specifically to one of the picturesque Calvinist Reform churches, such as that in Körösfő (Isvoru Crişului, in present day Romania), with its tall, shingled steeple echoed by four smaller turrets of the same type at each corner of the spire. The churches had special meaning for Hungarian modern artists who traveled to Kalotaszeg to collect art and to sketch architecture and folk costumes in the early twentieth century. For Aladár Körösfői Kriesch, Sándor Nagy, Laura Kriesch-Nagy, Mariska Undi, and the other artists and architects affiliated with the Gödöllő art colony (see below, 158), the rural villages in Kalotaszeg were ideal communities, in which art and life were integrated. Moreover, as Katalin Keserű has argued, the Protestant Church was the institution that enabled a secular vernacular culture to flourish in Hungary.[26] One such church is also featured in Kriesch's mural, "Kalotaszeg Church Procession," which was mounted inside the gated exhibition space (Figure 8.6). The distinctive wooden tower had also been featured atop the Hungarian Forestry Administration building at the 1873 Vienna World's Fair, where it was situated among the Hungarian farmhouses in the Ethnographic Village. Horti replicated the iconic tower in St. Louis, drawing upon its Romantic associations with the Transylvanian forest, with wood as natural resource, and with the rustic farmhouse that had been symbol of Hungarian national identity for at least a quarter century.[27]

Horti's gate was an elaborate version of the traditional Székely gate found among farmsteads in parts of Transylvania populated by Hungarian-speaking Székely peoples. The particular form of the gate was so similar to "Perso-Turkish models," wrote Aladár Körösfői Kriesch some years later, "that we may justifiably regard them as the last survivals of the wooden architecture of the primitive Hungarians."[28] One such gate had been built for the traditional Székely farmhouse in the Ethnographic Village at the 1873 World's Fair in Vienna.[29] Another was prominently featured as one of

Figure 8.6 Hungarian installation of decorative arts at the St. Louis 1904 Louisiana Purchase Exposition. Designed by Pál Horti. View of mural by Aladár Körösfői Kriesch.
Source: Museum of Applied Arts, Budapest.

the grand entrances to the fairgrounds at the 1896 National Millennial Exhibition in Budapest.[30] The wooden Székely gates, frequently decorated with carved or painted ornament, typically featured two doors: one large enough for a horse; the other for people to pass through. The traditional Székely gate also incorporated a covered dove-cote into its structure, as did Horti's version in St. Louis.

At the turn of the twentieth century, Transylvanian Kalotaszeg and the Székely region (*Székelyföld*), both in Romania today, occupied a special place in the imagination of Hungarian artists and ethnographers. They believed the people living in these regions still practiced ancient Hungarian customs, and represented the authentic Magyar spirit which had been preserved in mountain villages that remained relatively untouched by industrialization. The iconic forms of the Kalotaszeg church, the Székely gate, and the colorful embroidered folk costumes featured in Kriesch's mural resonated as symbols of Hungarian identity as much as they were appreciated for their aesthetic qualities.[31] Visitors to Horti's installation in the Manufactures Pavilion in St. Louis passed through the Székely gate into a courtyard bounded on one side by a second, aristocratic gate of elaborate wrought ironwork, linking the rural and aristocratic histories of Hungary in the minds of visitors as had been attempted four years earlier with the diverse exhibitions at the 1900 Paris Universal Exposition. It was the rural theme that dominated the installation, however. Embroideries and woven rugs (*háziipar*) displayed around the perimeter of the gated area gave the exhibition the atmosphere of a village market stall (Figure 8.7).

Within the gated area visitors could view several rooms that exhibited Hungarian decorative arts, including Zsolnay ceramics, fine jewelry and enamelwork, mosaics designed by artist Miksa Róth, rugs made by Sarotla Kovalszky, and embroideries produced by women associated with Hungarian home industries associations. Mariska Undi designed a child's nursery for the fair with toys and furnishings, Ede Toroczkai

Figure 8.7 Hungarian installation of decorative arts at the St. Louis 1904 Louisiana Purchase
Exposition. Designed by Pál Horti. View of folk embroideries.
Source: Museum of Applied Arts, Budapest.

Wigand arranged a lady's sitting room, and Horti designed a number of pieces of
furniture in walnut, including a "quaint" dining room, likely similar to the ones that
he had published in *Magyar Iparművészet* that year.[32]

Undi and Wigand, like Kriesch, were members of the Gödöllő art colony, founded
outside Budapest in 1901, and modeled on the British Arts and Crafts principles of
William Morris and his followers.[33] One of the group's goals was to preserve and revive
traditional Hungarian crafts (*háziipar*), such as embroidery and weaving, by learning
the old techniques and using them for the foundation of a modern, Hungarian decorative art rooted in the authenticity of tradition. Undi, Kriesch, Wigand, and Kovalsky
(whose tapestry looms were subsequently moved to Gödöllő) attracted attention in
St. Louis; their work had been financially supported and promoted by official applied
arts institutions in Hungary: the Museum and School of Applied Art, the Hungarian
Applied Arts Society, and the Ministry of Education. Horti's celebration of decorative
wooden architecture, and the Gödöllő artists' folk art revival, likely resonated in
the United States, where the craft aesthetic was particularly popular, and where the
collection and preservation of Native American arts was well underway.[34] This might
also help explain why Horti decided to stay in the United States to develop his own
approach to furniture design, with its emphases on wooden construction and use of
indigenous folk motifs. American designers had absorbed the ideas of British Arts
and Crafts movement through Gustav Stickley's magazine, *The Craftsman*, which
promoted William Morris' teachings to an American audience.[35]

In his long report on the 1904 St. Louis fair for *Magyar Iparművészet*, Horti
lamented Hungary's decision not to erect a national pavilion.[36] Indeed, its scattered

exhibits, with the most interesting examples of applied art in the Manufactures Building, rather than in a national pavilion or in the Palace of Fine Arts, meant that it was overlooked by international correspondents for *Kunst und Kunsthandwerk*, *Deutsche Kunst und Dekoration*, *Dekorative Kunst*, and *The Studio*, who neglected to mention Hungary at all in their accounts. The official reports published in the United States, on the other hand, each praised the installation in the Manufactures Building. The Hungarian art critic Károly Lyka published a bilingual catalog of the Hungarian arts on display, including both fine and applied art, but the book seems to have gone unnoticed by the foreign presses, even though Lyka had also just introduced the work of three Hungarian modern artists associated with the Gödöllő colony, Kriesch, Wigand, and painter Sándor Nagy, in the widely read German language journal *Dekorative Kunst* shortly before the exhibition.[37]

By contrast, and perhaps more significantly, Horti's designs of walnut dining room furniture and his architectural framework for the St. Louis installation, especially its wooden towers, attracted the attention of American furniture makers with whom he continued to work following the exhibition. For the next two years Horti remained in the United States to study indigenous Native American arts and American manufacture.[38] By early 1905 he was working in New York, where he wrote that he had designed furniture, carpets, and piano cases in order to help fund his travels, which were primarily supported by Kálmán Györgyi, Secretary-General of the Hungarian Applied Art Association.[39] Horti spent the year traveling through North America, during which time he designed furniture for Charles P. Limbert in Grand Rapids, Michigan, and the Shop of the Crafters, established in 1904 by Oskar Onken in Cincinnati, Ohio, including a "Mission style" piano decorated with "Indian motifs" (Figure 8.8).[40]

Horti assessed examples of American furniture design at the fair in St. Louis, and reported that many lacked the modernity and innovation he noticed among the German decorative arts, the interiors in the Austrian pavilion, and even the Japanese exhibits. Perhaps Horti hinted at a potential market for his own work, imagining the contributions he could make to American design by working with Charles Limbert and Oskar Onken; he had likely already made those contacts by the time his review was published in *Magyar Iparművészet*. In contrast to his criticism of contemporary American furniture design, Horti greatly admired the new American ceramics. He especially praised the quality and craftsmanship of the Rookwood Potteries, some of which featured incised ornament derived from the stylized patterns of Native American ceramics.[41] Such designs would have stood out to Horti and the Gödöllő artists who were similarly interested in reviving the patterns and techniques of traditional Hungarian folk art.

Horti must have seen in the Fine Art Palace in St. Louis the diverse collection of Native American artifacts, including decorative Pomo baskets, Navajo blankets, and Pueblo ceramics. Photographer and art writer George Wharton James and ceramicist Ernest Batchelder contributed pieces from the American Southwest.[42] The fair's Art Director, Halsey C. Ives, had envisioned an inclusive art exhibition of both fine and decorative art, and displayed the Native American objects at the prompting of Frederic Allen Whiting, chief of the St. Louis fair's Applied Art Division. Whiting, a prominent editor and Secretary-Treasurer of the Boston Society of Arts and Crafts, believed that "some of the best crafts work done in the country is done among the Indians. I cannot see why the Indian should not be considered as an artist craftsman as well as any other worker."[43] Furthermore, as Elizabeth Hutchinson notes, visitors

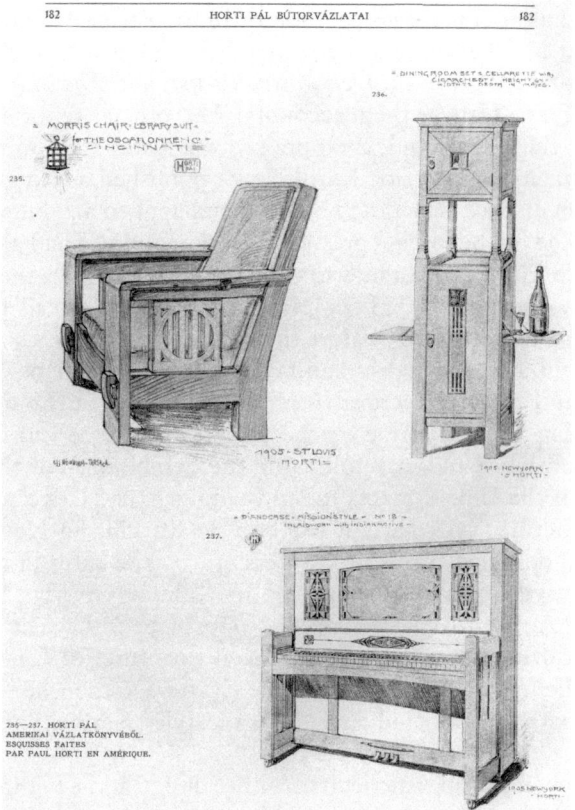

Figure 8.8 "Mission style" furniture including piano with "Indian motifs." Designed by Pál
    Horti for Shop of the Crafters, 1907.
Source: Illustrated in *Magyar Iparművészet*. Courtesy of the National Széchényi Library, Budapest.

to St. Louis greatly admired Native American craftspeople, who made and sold their
wares in the Indian Village organized by the fair's Anthropology Department.[44]

It is tempting to consider that Horti might have chosen to construct the wooden
church towers and Székely gate for the St. Louis fair intentionally to catch the eye
of American furniture makers, who were so interested in wood craftsmanship. It is
more likely, however, that Horti's subsequent relationships with Limbert and Onken
were simply a fortuitous result of the cross-cultural contacts made possible by the
fair. What is certain, however, is that both Hungarian and American designers were
drawn to indigenous sources and motifs. Horti's study trip, his sketchbook, and his
avid interest in Native North American and Aztec culture mirrored the ethnographic
adventures of his fellow artists in Hungary who similarly researched and collected examples of traditional material culture from rural villages. Ede Toroczkai Wigand and
Aladár Körösfői Kriesch added the place names "Toroczkó" (Torockó/Rimetea, in
present day Romania) and "Körösfői" to their own names to signify their (imagined
if not actual) ancestral connections to the Hungarian land. The exhibition of Native
American dwellings, costumes, cultural traditions, and arts and crafts in the United

States paralleled the display of traditional village architecture, woodcarving, textiles, and other folk art in Central Europe throughout the 1890s. This is particularly evident in the case of the "Kwakiutl Village" and "Cliff Dweller" (Pueblo) exhibits at the 1893 World's Columbian Exposition in Chicago; the Moravian Village at the 1895 Czechoslavic Ethnographic Exhibition in Prague; and the Ethnographic Village at the 1896 National Millennial Exhibition in Budapest. Other examples include the 1894 Exhibition in Lviv (today in Ukraine) and the American Indian Congress at the 1898 Trans-Mississippi Exhibition in Omaha, Nebraska.[45]

Following his sojourn in the United States and Canada, in 1906 Horti embarked on a voyage around the world, traveling to Mexico, South America, Japan, and India. According to Judith Kóos and Zsolt Somogyi, the artist contracted both malaria and yellow fever, which hindered his travels and his enthusiastic studies of world artistic culture. He eventually succumbed to the illnesses and passed away in Bombay at the age of 42. One cannot help but wonder what the Hungarian designer might have accomplished had his career not been cut short, or how his own work might have evolved over the years had he been able to continue his research.

1906 Milan International Exposition

Hungarian artists and designers returned to Italy in 1906 for the Milan International Exposition. The event celebrated the opening of the new Simplon tunnel from Switzerland to Italy through the Alps, which made possible a continuous railway from Paris to Milan. "Work" was the theme of the exhibition, which highlighted transportation and new technology, while also devoting attention to the display of manufactured objects, and other entertainments, which had been standard at world's fairs for more than half a century.[46] Many of the decorative arts, including the Austrian exhibits, were shown in national pavilions. The Hungarian display, however, was located in the grand Pavilion of Decorative Art, where it occupied the rounded, wedge-shaped corner of an impressive building designed by Italian architect Sebastiano Giuseppe Locati. The details of the exhibit were described and illustrated by Kálmán Györgyi, Secretary-General of the Hungarian Applied Arts Association, the group responsible for organizing the Hungarian exhibitions of decorative arts abroad.[47] Géza Maróti arranged the ambitious installation with Ödön Faragó. Together they decorated more than 30 rooms, corridors, and galleries in which visitors could survey a variety of traditional Hungarian crafts and modern interior designs.

According to several accounts, the highlights of the exhibition were Maróti's dramatic entrance hall and spacious gallery, framed by a series of semicircular archways ornamented with a low-relief pattern of wheat sheaves, and featuring an enormous sculpture; an allegorical representation of "Genius," a thirteen-foot tall female figure with raised arms holding two laurel wreaths above her head (Figure 8.9).[48] Piroska Ács writes that Maróti developed a unique silver coating for the walls, which he used with a layer of yellow shellac and a gray patina to mimic the look of old churches.[49] The lower third of the walls were clad in rectangular Zsolnay ceramic tiles, as Ács describes, in "bluish gold" eosin glaze. In the center of the room visitors enjoyed a fountain sculpture of ceramic wild ducks with a metallic "rainbow sparkle."[50]

Beyond the jewel-like entrance hall visitors could wander through several gallery spaces. The central Court of Doves (*Sala dei Colombi*) also designed by Maróti, featured lace, jewelry, and silver objects in glass cases surrounded by small alcoves

Figure 8.9 Hungarian installation of decorative arts at the Milan 1906 International Exhibition. Designed by Géza Maróti.

Source: Museum of Applied Arts, Budapest.

and connecting rooms, in which could be seen innovative interiors arranged by Pál Horti, as well as by Ede Toroczkai Wigand, Sándor Nagy, and Aladár Körösfői Kriesch, the artists belonging to the Gödöllő artists' colony, whose work had been so admired in St. Louis. At the far end of the exhibition was a large room, in which were displayed embroidered textiles and other traditional crafts (*háziipar*) in a lively setting. Ödön Faragó decorated the walls in a colorful painted style reminiscent of his attractive dining room installation at the Paris 1900 Exhibition Universelle six years earlier. Faragó also designed, in a similar manner, the exhibitions of work produced at the various state vocational schools, as well as interiors by Budapest furnishing manufacturers Gábor Steinbach and Endre Thék. Visitors in the rooms of commercial furnishings were obliged to cross a gallery of Zsolnay ceramic vases and other decorative objects, in order to reach the exhibition of crafts. The route through the exhibition likely underscored Ödön Lechner's belief in the relationship of traditional crafts (*háziipar*) to modern applied arts (*iparművészet*).

Italian architect and architectural critic Alfredo Melani reviewed the event for *The Studio*. Among the works of all nations represented he found the Hungarian exhibits to be of greatest interest, writing, "Hungary ... has formally asserted herself at Milan, and the display is one of lofty expression—aesthetically, politically, and nationally: In a word, it is a revelation of grace and pride."[51] The Hungarians, Melani wrote, had communicated a modern ambition that emerged from "the genius of their race," and which could be seen in all branches of the decorative arts from sculpture and furniture to jewelry, textiles, and ceramics.[52] Perhaps he unconsciously associated the "genius" of the Hungarian race with Maróti's allegorical sculpture in the entrance hall. For example, Melani noted a parallel between the folk-inspired designs of Faragó and the Gödöllő artists, with their colorful floral motifs, and the work of Hungarian landscape painters "who chant the hymn of the mountains, the valleys and the meadows of a homeland."[53] Implicit in Melani's review is the sense that Hungarian rural authenticity had been transformed alchemically into urban fashionability.

Melani especially admired the examples of needle lace by Árpád Dékáni, Ede Wigand's dining room furniture executed by József Mócsay, and the three colorful rooms of an "artist's abode" arranged by Sándor Nagy, which displayed carpets made by Laura Kriesch-Nagy (the sister of Aladár Körösfői Kriesch and wife of Sándor Nagy). The carpets were similar in spirit, he wrote, to the woven tapestries designed by Aladár Körösfői Kriesch, and executed by Leo Belmonte, which depicted scenes of Transylvanian peasants and their horses. These were displayed to enchanting effect in one of the two rooms representing the Hungarian Ministry of Education, the organization that had supported many of the modern decorative artists.

The naïve style of Hungarian folk-inspired furnishings was perhaps best represented in Milan by designer Vilmos Veszely's delightful nursery and wooden children's toys, which were singled out by both Melani and Györgyi in their reviews. The colorful child-like spirit of the rooms designed by Veszely, as well as by Faragó and the Gödöllő artists—much like Mariska Undi's playful children's nursery in St. Louis, and like János Vaszary's tapestry, *Little Girl with Kitten*, from the Turin exhibition—represented one aspect of what Melani considered "modern." He noticed in them an independent spirit that was not merely technical or expressive, but rather a manifestation of something more complex. "There is genuine soul in these efforts," he wrote, "...we discover in modern Hungarian art rather a psychological than a physical form, something intimate and profound which penetrates deep into the mind."[54]

It is also clear from Melani's account, however, that he was attracted to the sensuous, luxurious colors and exotic, iridescent effects of Maróti's rooms and halls, through which visitors entered the Hungarian installation in Milan, and which formed their first impression of Hungarian design. He appreciated the "Turco-Byzantine" character, as he described it, of those rooms, which he believed was a richer source of inspiration than the overly referenced Italian Renaissance. In other words, he seemed to perceive an approach in the Milan installation that differed from the exhibitions of Hungarian history in 1896 and again in 1900, in which various artistic periods, including that of King Matthias Corvin, were celebrated in the building and interiors of the Vajdahunyad castle. The Milan exhibition, by contrast, was a more evolved and refined version of the folk-art inspired approach that Lechner and his followers had pioneered at those events. The Hungarian artists, Melani wrote, "have embroidered the flowers of their art, in order once more to affirm the ineffaceable characteristics of their race."[55]

Maróti's preference for abstract geometric forms, seen for example in the rectangular wall tiles in the Court of Doves, was also apparent in some of his modern furniture designs, especially the cabinet with a zigzag pattern of chevrons that was exhibited in the music room in Milan (Figure 8.10). The bas-relief frieze of putti playing musical instruments was designed by artist Edé Telcs for the exterior of the new Royal National Hungarian Academy of Music (today the Franz Liszt Academy of Music) in Budapest, where Maróti's Genius sculptures were also later installed.[56] The strongly geometric forms in Maróti's work recall the shift to rectangular geometric patterns, for example in the Wiener Werkstätte furnishings designed by Josef Hoffmann and Koloman Moser beginning in 1903, but they also look ahead to the vibrant art deco

Figure 8.10 Géza Maroti, Music room, exhibited at the Milan 1906 International Exhibition.
Source: Museum of Applied Arts, Budapest.

vocabulary he would later explore in the Americas. In 1927 Maróti and his family moved to the United States at the prompting of their friend Eliel Saarinen, who invited Maróti to help with the decorative scheme of the new Cranbrook Academy of Art buildings in Bloomfield Hills, Michigan, a suburb of Detroit.[57] Maróti remained in the United States until 1932, during which time he received more commissions for ornamental work with architect Albert Kahn on Detroit high-rise buildings, including the elegant 1928 art deco Fisher Building, for which he designed a lavish cycle of decorative mosaics.[58]

Maróti's friendship with Saarinen, which led to his North American commissions, was likely the result of cross-cultural exhibitions that reflected the national-international impulse of Art Nouveau at the turn of the twentieth century. The Finnish pavilion at the 1900 Paris Universal Exposition, designed by Gesellius, Lindgren, and Saarinen, was a poetic expression in wood of Finnish national myth and traditional crafts. It was widely acclaimed at the fair, and was a model for the kind of modern expression of national identity that Hungarian designers sought.[59] In 1907 the Hungarian National Museum of Applied Arts in Budapest held an exposition of modern Finnish art and design, which featured the work of Saarinen and painter Akseli Gallen-Kallela, champion of the Finnish folkloric revival.[60] Letters reveal that Maróti traveled to Finland in 1907 and again in 1920, when he stayed with Saarinen and his family at Hvittrsäk, a villa on the shores of Lake Vitträsk that served as an art colony for many years. Saarinen in turn visited Maróti in Hungary at least twice, in 1908 and 1911, when he stayed at the artist's family home in Zebegényben.[61]

Although it is likely that Maróti and Saarinen knew each other before Maróti's success in Milan, that event paved the way for Maróti's subsequent career. It was there that his work was noticed by Antonio Boari, architect of the Mexico City opera house, who had traveled to Italy on behalf of the Mexican government to find artists to help collaborate on the decorative scheme for the Palace of Fine Arts in Mexico City. In 1908 Boari commissioned Maróti to design a large glass window for the Mexican National Opera Theater (*Palacio de Bellas Artes*).[62] In subsequent years Maróti continued to work outside Hungary, designing a permanent exhibition hall representing Hungary for the Venice Biennale, and organizing another major exhibition in London. Most stylistically striking was the pavilion he designed for the 1911 exhibition in Turin, which featured a tower with pitched roof meant to evoke the shape of "Attila's tent." The style of the Turin pavilion parallels that of Hungarian architect Károly Kós and his circle in Budapest. Kós's memorable buildings inspired by Transylvanian vernacular architecture share some characteristics with Pál Horti's installations in St. Louis, though they belong to the next generation of artistic and architectural experiments with modern design.[63] Maróti's 1911 pavilion in Turin was quite different from the modern-gothic skyscrapers he would later decorate in the United States. Nevertheless its eclectic and wide-reaching embrace of exotic, historical forms animated its design much in the way that American Art Deco celebrated a world of fantasy, from Ancient Egypt to the American Southwest. It is not surprising, perhaps, that Maróti's last work, still unpublished, was an art historical study of the lost city of Atlantis.[64]

A phoenix rises from the ashes

On August 3, 1906, just five months after it opened, the Hungarian section of decorative arts in Milan burned to the ground. It is difficult to imagine how painful it must have been for the artists who had put forth such creative effort on behalf of Hungary to assert itself once and for all as a modern, progressive, stylish nation to see their works disappear amidst the rubble of the ruined pavilion. Accounts published later noted the symbolic significance of the last object to succumb to the flames, Maróti's allegorical Genius sculpture from the exhibition's entrance hall. The Hungarian artists quickly put together a new installation, albeit less grand than the original. Fortunately Aladár Körösfői Kriesch's mural of Kalotaszeg peasants from the 1904 fair in St. Louis was still available. It was newly mounted on the wall in a decorative framework designed by Ödön Faragó. Hungary's enthusiastic construction of a second exhibit greatly impressed the international community, which praised the Hungarian spirit for its ability to "rise from the ashes."[65] Hungary's ability to persevere in the face of misfortune was as a metaphor for the irrepressible Magyar spirit that had successfully fought against Ottoman invaders and which resisted Hapsburg oppression.

Perhaps historians have placed too much emphasis on the ways in which various countries articulated narratives of national identity at world's fairs. Though this was clearly one important aim of such events, and one that was most frequently observed by art and design critics at the time, it is also possible that such an emphasis has overshadowed the intriguing implications of the many unusual and innovative displays that were mounted around the turn of the twentieth century. The Hungarian Ministry of Education, with the assistance of the Hungarian Society for Applied Art, controlled the narrative of Hungarian cultural identity at world's fairs by designing its "otherness," a strategy that may have served to marginalize Hungarian designers whose work received mixed reviews in Paris. By creatively reshaping the image of Hungary and Hungarian design at Turin, St. Louis, and Milan, however, and by looking deep within, but also beyond Hungary's borders for inspiration, Horti and Maróti took part in an international conversation that would result in lively and diverse expressions of modern design on both sides of the Atlantic in the first decades of the twentieth century. Pál Horti influenced his American colleagues, employers, and patrons, just as he himself was influenced by American crafts at the St. Louis fair. Géza Maróti achieved professional success as a result of his involvement in early twentieth-century international exhibitions. Though he ultimately returned to Hungary following his travels around the world, his glass mosaic in Mexico City, and some of his colorful decorations of North American skyscrapers can still be seen today.

The diversity of Hungarian installations on the world stage in less than one decade reveals that "national identity" was an elusive concept in the early twentieth century. It was shaped by political and economic aspiration as well as by an appreciation for urban style and the fashionability of modern spaces and objects. In Hungary, national identity was closely tied both to cultural memory and to the tensions brought to the fore by industrialization, which introduced new concepts of race and gender, and which changed traditional relationships between craft and industry, and between center and periphery.

Notes

The author thanks Zsolt Somogyi, design historian and archivist at the Hungarian National Museum of Applied Art in Budapest, for sharing with me his research on Pál Horti, and materials related to Hungarian exhibitions of decorative arts. Many thanks are due as well to the generous staff members at the St. Louis Museum of Art, the St. Louis Public Library, and the Missouri Historical Society in St. Louis. The Northern Illinois University Division of Research and Innovation Partnerships supported the research for this project. An early version of this paper was presented in 2004 at the Annual Meeting of the International Committee on Design History and Design Studies in Guadalajara, Mexico.

1 Terri Switzer, "Hungarian Self-Representation in an International Context: The Magyar Exhibited at International Expositions and World's Fairs," in Michelle Facos and Sharon Hirsch (eds), *Art, Culture and National Identity in Fin-de-Siècle Europe* (Cambridge: Cambridge University Press, 2003), 160–185; Miklós Székely, "From Figure to Pattern: The Changing Role of Folk Tradition in Hungarian Representations at Universal Exhibitions (1867–1911)," in Dagnosław Demski, Ildikó Sz. Kristóf, and Kamila Baraniecka-Olszewska (eds) *Competing Eyes: Visual Encounters with Alterity in Central and Eastern Europe* (Budapest: L'Harmattan, 2013), 190–212.

2 Austrian curator Jacob von Falke particularly admired the interior of the Székely house at the 1873 Vienna World's Fair. Jacob von Falke, "Das Kunstgewerbe," in Carl von Lutzöw (ed), *Kunst und Kunstgewerbe auf der Wiener Weltausstellung 1873* (Leipzig: Verlag von E.A. Seemann, 1875), 82.

3 Mónika Lackner, "Die erste volkskundliche Sammlung Ungarns: Zur Präsentation der ungarischen Volkskultur auf der Wiener Weltausstellung von 1873," in Éva Bajkay, Katalin F. Dózsa, and Marianna Hergovich (eds), *Zeit des Aufbruchs: Budapest und Wien zwischen Historismus und Avantgarde* (exh. cat. Vienna: Kunsthistorisches Museum, 2003), 103–21. See also István Györffy, *A cifraszűr* (The Cifraszűr) (Budapest: Nap Kiadó, 1930); Katalin Földi-Dozsa, "How the Hungarian National Costume Evolved," in Polly Cone (ed), *The Imperial Style: Fashions of the Habsburg Era* (exh. cat., New York: The Metropolitan Museum of Art, 1980), 75–88; Mönika Lackner and Péter Granasztói, *Cifraszűr/Hirtenmantel. Vom alltäglichen Kleidungsstück zum nationalen Symbol* (exh. cat., Kittsee, Ethnographisches Museum Schloss Kittsee/ Vienna: Österreichisches Museum für Volkskunde, 2003).

4 Miklós Székely, "The Resetting of the Main Historical Group From the Millennium Exhibition to the Paris Universal Exhibition of 1900," in Miklós Székely (ed), *Ephemeral Architecture in Central-Eastern Europe in the 19th and 20th Centuries* (Paris: L'Harmattan, 2015), 33–50.

5 Eugéne [Jenő] de Radisics, *Le Pavillon Historique de la Hongrie á l'Exposition universelle de Paris en 1900* (Paris: Librairie Centrale des Beaux-Arts, 1900). See also Lajos Németh, "Art, Nationalism, and the Fin de Siècle," in Gyöngyi Éri and Zsuzsa Jobbágyi (eds), *A Golden Age: Art and Society in Hungary 1896–1914* (Budapest: Corvina, 1997), 19–29; and János Gerle, "Hungarian Architecture from 1900 to 1918," in Dora Weibenson and József Sisa (eds), *The Architecture of Historic Hungary* (Cambridge, MA: MIT Press, 1998), 223–244.

6 Milan Hlaváčka, Jana Orlíková, and Peter Štembera (eds), *Alfons Mucha—Paříž 1900: Pavilion Bosny a Hercegoviny na světové výstavěl/Alphonse Mucha—Paris 1900: The Pavilion of Bosnia and Herzegovina at the World Exhibition* (Prague: Obecní dům, 2002).

7 Gabriel Mourey, "Round the Exhibition. Part I. The House of the 'Art Nouveau Bing'," *The Studio*, 20, no. 89 (August 1900): 164–180. See also Debora Silverman, *Art Nouveau in Fin-de-Siècle France: Politics, Psychology, and Style* (Berkeley: University of California Press, 1989), 284–314.

8 Anthony Alofsin describes this tendency in modern architecture as a "language of myth." Anthony Alofsin, *When Buildings Speak: Architecture as Language in the Habsburg Empire and Its Aftermath, 1867–1933* (Chicago, IL: University of Chicago Press, 2006), 127–176.

9 Kálmán Györgyi, "Az iparművészet a párisi kiállitáson," *Magyar Iparművészet*, 3, no. 5 (September 1900): 209–304.

10 Juliet Kinchin, "Hungary: Shaping a National Consciousness," in Wendy Kaplan (ed), *The Arts and Crafts Movement in Europe and America: Design for the Modern World* (Los Angeles, CA: Los Angeles County Museum of Art, 2004), 142–177.

11 A few noteworthy examples of these activities include the set of woven and embroidered textiles collected by Flóris Rómer and János Xántus for exhibition at the 1873 Vienna World's Fair; artist József Huszka's sketches of traditional Hungarian folk art in the 1880s; ethnographer János Jankó's field work in preparation for an ethnographic village of traditional vernacular architecture at the National Hungarian Millennial Exhibition in 1896; and art writer Dezső Malonyay's five-volume survey of traditional Hungarian folk art, *A magyar nep művészete* (1907–1922), with contributions by Aladár Körösfői Kriesch and other artists connected to the Gödöllő colony. For further see David Crowley, "The Uses of Peasant Design in Austria-Hungary in the Late Nineteenth and Early Twentieth Centuries," *Studies in the Decorative Arts*, 2, no. 2 (Spring 1995): 2–28; Katalin Keserű, "The Workshops of Gödöllő: Transformations of a Morrisian Theme," *Journal of Design History*, 1, no. 1 (1988): 1–23; Rebecca Houze, *Textiles, Fashion, and Design Reform in Austria-Hungary Before the First World War: Principles of Dress* (Farnham: Ashgate, 2015).

12 Among the most influential efforts to revive local crafts were the workshops established in Kalotaszeg by writer Etelka Gyarmathy; the Némtelemér weaving workshop in Torontál County, organized by Sarolta Kovalszky and Anatól Streitmann; the Halas lace workshop in Kiskunhalas, founded by artist Árpád Dékáni; and the home industries associations supported by Hungarian Countess Sarolta Zichy and Habsburg Archduchess Isabella. For further see Kálmán György, "A magyar hímző-háziiparról" (Hungarian Home Industry

Embroidery), *Magyar Iparművészet*, 3, no. 5 (1926): 51–53; Ferenc Németh, *A torontáli szönyeg. A szönyegszövés negyed százada Bánátben Streitmann Antaltól Kovalszky Saroltáig* (The Torontál Carpet: A Quarter Century of Weaving in Banat from Antál Streitmann and Sarolta Kovalszky) (Budapest: Forum Könyvkiadó, 1993). Juliet Kinchin suggests that the Hungarian Ministry of Education's decision to move Kovalszky and her looms to Gödöllő reveals that official design policy was more concerned with developing a market for modern decorative arts than with providing relief to impoverished regions in the countryside. Kinchin, "Hungary: Shaping a National Consciousness," p. 157.

13 Horti's carpet featured a stylized central quatrefoil motif in shades of blue and green with border of organic, undulating petal shapes in red and silver on a gold ground. Judith Koós, *Horti Pál élete és művészete 1865–1907* (Budapest: Akadémiai Kiadó, 1982), 79; Judith Koós, "A Hungarian Pioneer of Art Nouveau: Pál Horti," *Ars Decorativa*, 2 (1974): 173–189.

14 Koós, "A Hungarian Pioneer of Art Nouveau: Pál Horti," p. 174.

15 Rebecca Houze, "National Internationalism: Reactions to the Austrian and Hungarian Decorative Arts at the 1900 Paris Exposition Universelle," *Studies in the Decorative Arts*, 12, no. 1 (Fall-Winter 2004–2005): 55–97.

16 W. Fred, "Interieurs und Möbel auf der Pariser Weltausstellung," *Kunst und Kunsthandwerk*, 3 (1900): 331–352.

17 Gabriel Mourey, "Round the Exhibition. IV. Austrian Decorative Art," *The Studio*, 21, no. 92 (November 1900): 118–120.

18 Miklós Székely, "La critica italiana e ungherese sulle esposizioni universali in Italia fra il 1900 e il 1914," *Nuova Corvina* 21 (2009): 114–127.

19 Horti may have seen ceramic vases with prominent peacock feather designs, one by Rookwood Potteries, another by Tiffany, at the 1898 Spring International Exhibition in Budapest and the Paris 1900 Universal Exposition, respectively. They are both illustrated in *Magyar Iparművészet*. Mihalik, "Az iparművészeti múzeum tavaszi nemzetközi kiállítása," p. 268; Györgyi, "Az iparművészt a párisi kiállitáson," p. 262.

20 Fritz Minkus, "Die erste internationale Ausstellung für moderne dekorative Kunst in Turin," *Kunst und Kunsthandwerk* 5 (1902): 402–450.

21 The tapestry, which now resides in the Budapest Museum of Applied Art, is illustrated in Andrej Szczerski, "Central Europe," in Karen Livingstone and Linda Parry (eds), *International Arts and Crafts* (exh. cat., London: V&A Publications, 2005), 238–251. See also Sabine Wieber, "The Warp and the Weft: Tradition and Innovation in Skærbæk Tapestries, 1896–1903," *Journal of Design History*, 28, no. 4 (May 2015): 331–347.

22 Minkus, "Die erste international Ausstellung für moderne decorative Kunst in Turin," p. 449.

23 W. Fred, "The International Exhibition of Decorative Art at Turin: The Austrian Section," *The Studio*, 27, no. 116 (November 1902): 130–134.

24 John Wesley Hanson, *The Official History of the Fair: St. Louis, 1904. The Sights and Scenes of the Louisiana Purchase Exposition. A Complete Description of the Magnificent Palaces, Marvelous Treasures and Scenic Beauties of the Crowning Wonder of the Age* (St. Louis: John Wesley Hanson, 1904), 442. See also T. W. Park, "The Addresses by Prominent Personnages and the Ceremonies of the Opening Day Witnessed by the Largest Attendance of People in the History of World's Fairs," *World's Fair Bulletin*, 5, no. 8 (June 1904): 4 ff.

25 Mark Bennitt, Frank Parker Stockbridge, et al (eds), *History of the Louisiana Purchase Exposition: Comprising the history of the Louisiana Territory, the story of the Louisiana Purchase and a full account of the Great Exposition, embracing the participation of the sates and nations of the world, and other events of the St. Louis World's Fair of 1904* (St. Louis: Universal Exposition Publishing Company, 1905), 237–284.

26 Keserű, "The Workshops of Gödöllő," p. 5. See also George Starr, "Art and Architecture in the Hungarian Reformed Church," in Paul Corby Finney (ed), *Seeing Beyond the Word: Visual Arts and the Calvinist Tradition* (Grand Rapids, MI: William B. Eerdmans Publishing Company, 1999), 301–342.

27 Bjarne Stoklund, "How the Peasant House Became a National Symbol: A Chapter in the History of Museums and Nation-Building," *Ethnologia Europaea*, 29 (1999): 5–18. Many thanks to Miklós Székely for sharing with me his knowledge of the iconic Hungarian

fiatorony. He writes that the fifteenth-century Saint Stephen Tower in Nagybánya (Baia Mare, Romania), rebuilt in 1619, is one of the oldest examples. The wooden churches, widespread throughout Máramaros (Maramureș in present-day Romania) and Kalotaszeg (Țara Călatei in present-day Romania), which often feature a similar tower form, became an important symbol of the region; however, their origin is not well documented.

28 Aladár Körösfői Kriesch, "Hungarian Peasant Art," in Charles Holme (ed), *Peasant Art in Austria and Hungary* (London: The Studio, 1911): 31–46.

29 The pavilion was constructed to represent Hungary's timber industry, which was built in "a unique Hungarian style—that of the little country churches, which travelers so admire." (...un padiglione, il quale per la sua architettura, rappresenta da per sè stesso una originalitá di stile propio dell'Ungheria, inquantochè gli fu dato l'aspetto di una di quelle chiesuole di villaggio, tutte construite in legname, e che sono pel viaggiatore di si mirabile effetto.) "Le Foreste dell'Ungheria," in Edoardo Sonzogno (ed), *L'Esposizione universal di Vienna del 1873 illustrata* [Milan], vol. 1, no. 5: 39–40. The building is no. 8, "Pavillon der königliche ungarische Staatsforst-Verwaltung," in Zone II on the map of the fairgrounds in Jacob von Falke, *Die Kunstindustrie auf der Wiener Weltausstellung 1873* (Vienna: Gerold, 1873).

30 Rebecca Houze, "Home as Living Museum: Ethnographic Display and the 1896 Millennial Exhibition in Budapest," *Centropa*, 12, no. 2 (May 2012): 131–151. See illustration p. 140.

31 The Körösfő church is one of many Calvinist Reformed Churches in Transylvania. German merchants and settlers introduced the Calvinist reforms to Hungary in the sixteenth century, where many Hungarian-speaking peoples adopted them, such that the Reformed Church by the end of the nineteenth century was strongly identified as Hungarian. Furthermore, the (Hungarian) Reformed Church remained particularly associated with Transylvania, which, under Ottoman rule for many years, had not been as greatly influenced by the Habsburg Counter-Reformation. Because of this particular historical situation, George Starr writes, the Reformed Church "had long stood for Hungarian resistance to Austrian hegemony in political as well as religious spheres." Starr, "Art and Architecture in the Hungarian Reformed Church," p. 330.

32 Descriptions of the rooms in English may be found in the following sources: Skiff and Ives (eds), *Official Catalogue of Exhibitors: Universal Exposition St. Louis, U.S.A., 1904*, 241–244; David R. Francis, *The Universal Exposition of 1904* (St. Louis: Louisiana Purchase Exposition Company, 1913), vol. 1, 397; Bennitt, Stockbridge, et al, *History of the Louisiana Purchase Exposition*, 271–272; James W. Buel, *Louisiana and the Fair. An Exposition of the World, its People and their Achievements*, 10 vols (Saint Louis: World's Progress Publishing Co., 1904), vol. 9. Division CXXXVIII. Manufactures and Varied Industries, 3431–3433; "Hungary" and "Hungarian Section," *World's Fair Bulletin 5*, no. 10 (August 1904): 10, 36. Official reports of the Hungarian installations in St. Louis were published in *Magyar Iparművészet*. See Ferenc Szécsén, "A St. Louisi Világkiállítás," *Magyar Iparművészet* 7, no. 2 (March 1904): 89–92, in which are published illustrations of Horti's plans for the installation. Follow-up reports were published in *Magyar Iparművészet* 7, no. 4 (July 1904): 175–186, and *Magyar Iparművészet* 7, no. 5 (September 1904): 208–215.

33 Keserű, Katalin, "The Workshops of Gödöllő: Transformations of a Morrisian Theme," *Journal of Design History*, 1, no. 1 (1988): 1–23. See also Katalin Géller, Mária G. Merva, and Cecília Őriné Nagy (eds), *A gödöllő művésztelep 1901–1920/ The Artists' Colony of Gödöllő* (Gödöllő: Gödöllő Városi Múzeum, 2003).

34 The Cincinnati company, Rookwood Pottery, founded by ceramic painter Maria Longworth Nichols Storer, and Newcomb Pottery, featuring the work of women students at Tulane University's Newcomb College in New Orleans, were among the many American Arts and Crafts schools and manufacturers represented in the exhibition of Original Objects of Art Workmanship at the St. Louis fair. Frederick J. V. Skiff (ed), *Official Catalogue of Exhibitors. Universal Exposition St. Louis, U.S.A. 1904. Department B. Art* (St. Louis: Official Catalogue Company Inc., 1904), 75–90.

35 The first issue of *The Craftsman*, published in October 1901, was dedicated to the theme, "William Morris. Some Thoughts Upon His Life: Work and Influence." *The Craftsman* also published reviews of the German decorative arts and of the works exhibited by Tiffany &

Co. at the St. Louis fair. Gustav Stickley, "The German Exhibit at the Louisiana Purchase Exposition," *The Craftsman*, 6, no. 5 (August 1904): 489–506; "Tiffany and Company at the St. Louis Exhibition," *The Craftsman*, 7, no. 2 (November 1904): 169–183. See also Wendy Kaplan, "America: The Quest for Democratic Design," in Kaplan, (ed), *The Arts and Crafts Movement in Europe and America*, 246–283.

36 Pál Horti, "A St. louisi világkiállítás," *Magyar Iparművészet* 7, no. 6 (November 1904): 249–313.

37 Károly Lyka, *Official Illustrated Catalogue Fine Arts Exhibit, Hungary, St. Louis Exposition, 1904* (Budapest: V. Hornyánszky 1904); Karl Lyka, "Drei ungarische Künstler: Alexander Nagy, Aladár Kriesch, Eduard Wigand," *Dekorative Kunst* 14 (April–September 1904): 642–648.

38 In a December 8, 1904 letter to Hungarian art historian József Mihalik Horti explains that he will stay in America to learn more about "Indian handicrafts, Aztec art treasures, and equipment used in furniture factories." Zsolt Somogy, "Pál Horti's Late Works in the United States of America," *Ars Decorativa* 25 (2007): 105–122; Zsolt Somogyi, "An Adaptable Applied Artist: Pál Horti's American Furniture," *Ars Decorativa* 26 (2008): 131–143.

39 Horti's New York activities are documented in letter dated April 25, 1905 to Elek Koronghi Lippich, head of the Hungarian Ministry of Religion and Public Education's art department. Somogyi, "Pál Horti's Late Works," p. 106. See also Piroska Ács, "Kálmán Györgyi (1860–1930), Heart and Soul of the National Hungarian Applied Arts Association," *Ars Decorativa* 25 (2007): 135–146.

40 Somogy, "Pál Horti's Late Works," p. 110. See also Juliet Kinchin, "Modernity and Tradition in Hungarian Furniture, 1900–1938: Three Generations," *Journal of Decorative and Propaganda Arts*, vol. 24, Design, Culture, Identity: The Wolfsonian Collection (2002): 64–93. A desk and chair designed by Pál Horti for the Shop of the Crafters, c.1905, with Horti's signature peacock feather motif, are illustrated in Kinchin, "Hungary: Shaping a National Consciousness," p. 161.

41 Horti, "A St. louisi világkiállítás," p. 257.

42 Elizabeth Hutchinson, *The Indian Craze: Primitivism, Modernism, and Transculturation in American Art, 1890–1915* (Durham: Duke University Press, 2009), 125; Skiff and Ives (eds), *Official Catalogue of Exhibitors: Universal Exposition St. Louis, U.S.A., 1904*, 80–81, 90.

43 Hutchinson, *The Indian Craze*, 125.

44 Hutchinson, *The Indian Craze*, 127. See also Robert W. Rydell, *All the World's a Fair* (Chicago, IL: University of Chicago Press, 1984), 160–167.

45 See for example Marta Filipová, "Peasants on Display: The Czechoslavic Ethnographic Exhibition of 1895," *Journal of Design History* 24, no. 1 (2011): 15–36; Houze, "Home as Living Museum," 131–151; Josh Clough, "'Vanishing' Indians" Cultural Persistence on Display at the Omaha World's Fair of 1898," *Great Plains Quarterly* 25 (Spring 2005): 67–86; Paige Raibmon, *Authentic Indians: Episodes of Encounter from the Late Nineteenth-Century Northwest Coast* (Durham, NC: Duke University Press 2005).

46 Pietro Redondi (ed), *Città effimera: Arte, technological esotismo all'esposizione internazionle di Milano del 1906* (Milan: Edizione Gabriele Mazzotta, 2015).

47 Kálmán Györgyi, "Az iparművészet a Milánói kiállításon," *Magyar Iparművészet* 9, nos. 4–5 (July–September 1906): 162–219, 285–296. See also *Guida itenerario per visitare celermente l'esposizione: con due accuratissime e dettagliate piante: edifice, gallerie, padiglioni, ecc., elencati in ordine progressive ed alfabetico* (Milan: C. Abbiati & C., 1906).

48 János Gerle, "Maróti Géza építészet és szobrászat határán" (Géza Maróti at the border of architecture and sculpture), *Magyar Építőművészet*, 83, no. 6 (1983): 50–51; Elemér Czakó, *A Magyar iparművészet szereplése a milanoi nemzetközi világkiallitáson 1906-ban* (Hungarian Applied Arts Display at the Milan International Exhibition in 1906) (Budapest: Országos Magyar iparművészeti museum, 1907). The bronze sculpture was destroyed in the tragic November fire described below. When Hungary rebuilt its display,

Maróti recast the sculpture in duplicate. The new twin Genius figures stood on either side of the new entrance to the Hungarian section. After the fair the sculptures were mounted on the exterior of the new Royal National Hungarian Academy of Music (Franz Liszt Academy) in Budapest. The allegory of Genius as protective guardian spirit can be traced to classical antiquity. Maróti's version recalls other contemporaneous stylized female figures holding laurel leaves, including those framing the entrance to the Ernst Ludwig Haus in the Darmstadt art colony, designed by Vienna Secession building architect Joseph Maria Olbrich (1899–1901). Vienna's chief architect and fellow member of the Secession, Otto Wagner, incorporated a similar motif of winged "angels" on Vienna's postal savings bank (*Postsparkasse*) in 1905. Maróti was likely familiar with both examples to which he connected his own Hungarian Secession style. See Anthony Alofsin, *Frank Lloyd Wright: The Lost Years, 1910–1922. A Study of Influence* (Chicago, IL: University of Chicago Press, 1993), 38–39. Alofsin notes that Wright, who was influenced by Olbrich, was also drawn to the "winged victory" motif, but did not adopt the complete allegorical context that had been a feature of the Vienna Secession. The precise relationship between the Genius allegory, guardian angel, and winged victory as employed by these artists needs closer examination.

49 Piroska Ács, "The Exhibition Pavilions of Géza Maróti in Milan and Venice," in Éva Csenkey and Ágota Steinert (eds), *Hungarian Ceramics from the Zsolnay Manufactory* (exh. cat., Bard Graduate Center for Studies in the Decorative Arts, Design, and Culture/ New Haven, CT: Yale University Press, 2002), 211–216.

50 Ibid.

51 Melani, Alfredo. "Hungarian art at the Milan Exhibition," *The Studio*, 39, no. 162 (September 1906): 300–309.

52 Ibid.

53 Ibid.

54 Melani, "Hungarian art at the Milan Exhibition," p. 304.

55 Melani, "Hungarian art at the Milan Exhibition," p. 306.

56 The Music Academy was founded by Hungarian composer Franz (Ferenc) Liszt in 1875. In 1907 the academy moved from its original location, a Neo-Renaissance building on the Andrássy út, into a new building, designed by architects Kálmán Giergl and Flóris Korb, which featured an Art Nouveau decorative scheme inside with painted mural and frescos by Aladár Körösfői -Kriesch, and mosaics and stained glass windows by Miksa Róth. The exterior façade was embellished with the Genius sculptures by Maróti, friezes depicting the history of music by Telcs, and a large statue of Franz Liszt by sculptor Alojos Stróbl. Györgyi, "Az iparművészet a Milánói kiállításon," p. 189.

57 Mark Coir, "The Cranbrook Factor," in Eeva-Liisa Pelkonen and Donald Albrecht (eds), *Eero Saarinen: Shaping the Future* (New Haven, CT: Yale University Press, 2006), 29–44; Robert Judson Clark and Andrea P. A. Belloli, *Design in America: The Cranbrook Vision, 1925–1950* (exh. cat., Detroit Institute of Art, 1984), 278–279.

58 János Sturcz, "'Vederemo—It's Just Been Won!' Géza Maróti's Competition Design for the Rockefeller Center," *Ars Decorativa* 11 (1991): 11. Saarinen came initially to Chicago in 1922 to submit a design for the *Tribune* tower competition. Though his design was awarded only second prize the event secured for Saarinen a teaching position at the University of Michigan and later architectural commissions in the Detroit area. Maróti worked not only in Detroit and New York, but also in Chicago on the 1930 Foreman State Bank Building in Chicago, designed by architects Graham, Anderson, Probst and White. For a complete list of Maróti's American commissions, see Ács, "The Exhibition Pavilions of Géza Maróti in Milan and Venice," p. 216.

59 On the Finnish pavilion at the 1900 Paris Exposition, see Bart Pushaw, "Finland at the 1900 Exposition Universelle," Chapter 7 in the present volume.

60 "A Finnek," *Magyar Iparművészet*, 11, no. 1 (1908): 1–26; Torsten Stjernschantz, "A Finn Művészet és iparművészet," *Magyar Iparművészet*, 11, no. 1 (1908): 27–43.

61 Janos Gerle, "Maróti Géza építészet és szobrászat határán," *Magyar Építőművészet* 6 (1983): 50–51.

62 Sára Ivánfy-Balough and Imre Jakabffy, "Géza R. Maróti," *Ars Decorativa* 4 (1976): 127–147.

63 János Gerle, "Hungarian Architecture from 1900 to 1918," pp. 236–242. The Hungarian pavilion at the 1906 Jubilee Exhibition in Bucharest, designed by architects Géza Aladár Kármán and Gyula Ullmann, was reminiscent of Zoltán Bálint's and Lajos Jámbor's lively and controversial installations at the 1900 Paris Universal Exposition, but they also looked ahead to the fanciful buildings that Kós designed for the Budapest zoo between 1908 and 1910. See Shona Kallestrup, "Romanian 'National Style' and the 1906 Bucharest Jubilee Exhibition," *Journal of Design History* 15, no. 3 (2002): 147–162.
64 Ács, "The Exhibition Pavilions of Géza Maróti in Milan and Venice," p. 216.
65 Proska Ács (ed), *"Mi vagyunk Atlantisz": Vederemo!: Mároti Géza, 1875–1941* (Budapest: Iparművészeti Múzeum, 2003).

9 When the local is the global

Case studies in early twentieth-century Chinese exposition projects

Susan R. Fernsebner

This chapter explores the ways in which an early twentieth-century contingent of Chinese state and commercial elites, serving as exposition managers, presented a Chinese nation at a series of national and international expositions held between 1904 and 1915, and negotiated among diverse communities with interests in how "China" was construed at these events. These elites' mobilization of texts as well as objects gave expression to the fissures of a colonial and capitalist modernity—a modernity in which they sought an equal claim.

The exposition managers' activities simultaneously occurred with a political restructuring of their home state and provide valuable insights regarding Chinese elites' own nationalist concerns. At the turn of the twentieth century, a Manchu imperial court ruled China as part of a multi-ethnic empire under the Qing dynasty (1644–1911). In the subsequent decade, both the Qing court and urban, reformist elites would seek to reinvent the state amid the introduction of a New Policies program (1901–1911). Swept aside in the 1911 Revolution, the Qing Empire would be replaced by a Republic of China the following year. Yet China's exposition projects flourished throughout this political transition, a time in which the state was being radically restructured and the role of the nation's citizen-subjects newly formulated.[1]

This chapter follows a sequence of Chinese exposition projects over approximately a decade at the start of the twentieth century, beginning with Chinese participation in the St. Louis Louisiana Purchase International Exposition of 1904, a project overseen by the Qing dynasty's Maritime Customs Service. Here we find a moment of crisis as well as transition. Chinese visitors to the fair in St. Louis joined public rituals devoted to specific notions of "civilization," the natural and social sciences, and an explicit hierarchy of nations. At the same time, China's exposition participants, including official attendants and prominent attendees, also criticized the nation's representation at this event. They targeted both the displays and the agents responsible for their presentation, especially European and American officials serving the imperial house. A study of the 1904 fair and two other cases that follow, namely China's first national exposition in 1910, the Nanyang Exposition, and China's participation in the San Francisco Panama Pacific International Exposition of 1915, offers a view of the activism of Chinese elites and their articulations of a nationalist project.

Several important shifts occurred over this decade. One is the rising ambition of Chinese exposition organizers to use the fairs as a means to promote economic nationalism—and particularly an export-oriented, industrial economy—amid global competition and colonialist threats. China's exposition managers believed the fairs would further this aim by inspiring new kinds of participation and discipline within a domestic audience that they aimed to mobilize for this new economy. At the same time, an increasingly complex domestic, transnational, and émigré audience also brought their own critical voices to bear on these displays. Indeed, as Chinese subjects asserted agency at world's fairs, we also see that this subjectivity itself was not monolithic. We find a state with a transnational constituency—within and beyond the country's borders, real and imagined—and definitions of the nation itself that were also mosaics of contested meaning.

Colonialist critiques: St. Louis 1904

Presentations of Chinese displays at world's fairs in the nineteenth century were largely managed by foreign diplomats, merchants, and missionaries. China's world-famous porcelain works of Jingdezhen, for example, was presented at the Crystal Palace in 1851 in an exhibit by British Consul Rutherford Alcock.[2] Echoing a tradition of the curiosity cabinet, Alcock's display presented these goods as simultaneously foreign and familiar, exotic and fetishized. Contemporary European rhetoric held that Chinese decorative arts were not innovative, revealing instead, as Matthew Digby Wyatt intoned, China's "national character for imitation." Yet paradoxically, Europeans also claimed Chinese arts, setting them on a timeline in the development of European styles. Alcock's display also offered, it seemed, "the prototypes," of objects regarded as "essentially European and modern in their origin."[3] Invocations of "imitation" and "prototype" within an exhibition catalogue simultaneously marked hierarchies of space and time within a global order. Chinese critics meanwhile sought their own voices amid these negotiations of display.

China's involvement in its own national displays began in 1873, as European and American employees of the Qing Dynasty's Imperial Customs Service served as managers of Chinese presentations at the Vienna World's Fair. The Customs Service continued to manage displays at expositions throughout the nineteenth century, up until the international exposition at Liège in 1905. An institutional product of China's nineteenth-century "unequal" treaties with Britain, France, and the United States, the Customs Service began in 1854 and was famously led by the former British vice-consul Robert Hart (serving 1863–1910).[4] Hart's management style was exacting and minute, and his oversight of exposition projects fared similarly—with instructions that literally determined the color of the buttons worn by exposition officials at the fairs.[5]

Chinese goods carried layers of meaning during this time, particularly at the turn of the century. Amid imperialist armies' looting in Beijing after the Boxer Uprising of 1900, as James Hevia recounts, foreign soldiers and civilians collected plunder for their own auctions. Silks, furs, furniture, porcelain, and bronzes—both antiques and fakes—and other "curios" were hawked and bid upon by residents of legations and treaty ports.[6] These looted objects circulated globally, marking colonialist claims on a Chinese patrimony. Just a few years later, as European and American agents of the Qing Customs Service mobilized Chinese displays for the 1904 exhibition in St. Louis, their staging invoked Orientalist frames of meaning and also spawned new voices of critique.[7]

American organizers viewed the St. Louis Exposition as a "manifesto of racial and material progress," as Robert Rydell has detailed.[8] The event's anthropological exhibitions and conferences advanced the evolutionary philosophies of Charles Darwin and Herbert Spencer amid hierarchical discourses of nations and civilizations.[9]

Timothy Mitchell argues that the turn of the century exposition was an arena of races, objects, and representations set forth for a colonial subject to negotiate. At the same time, that subject did not simply observe.[10] Chinese participants also joined the 1904 St. Louis Exposition and directly critiqued the displays presented in the name of their nation by the Chinese Customs Service, including its European and American agents. In addition to a Chinese pavilion presented among the international halls, and a "Chinese Village" on the Pike (run by an independent concessionaire, Lee Toy), Chinese goods were displayed at two exhibit halls at the fair, the Liberal Arts Hall (the main site for Chinese displays) and the Education Hall. China offered only limited displays in other theme halls at St. Louis. While enthusiastic participation on the part of Chinese entrepreneurs resulted in a rich collection of goods being sent, China's exposition organizers failed to reserve adequate space. Compounding this issue, poor and crowded displays, weak accompanying texts, and even shifting display locations over the course of the event all sabotaged China's showing (Figure 9.1).[11]

In the months following the St. Louis exhibition, two Chinese attendees, Chen Qi and Chen Huide, presented a book-length commentary on the displays they had visited. As did most American and European commentators, they saw expositions as a ritual of competition among nations in linked realms of civilization and industrial progress. These authors feared many Chinese overlooked the value of these events in helping to improve society—particularly in transforming an agricultural nation into an industrial one. The position of the Chinese nation and its representation in a global arena was also threatened, a problem Chen Qi and Chen Huide framed almost poetically in an invocation of cultural icons. "How can we take our gods and present them to posterity as nothing but old, decaying puppets," they questioned, "good only for contributions to museums as objects of display?"[12] The St. Louis exposition coincided

Figure 9.1 Chinese Exhibit, Palace of Liberal Arts. St. Louis.
Source: Reproduced from *The Greatest of Expositions Completely Illustrated* (St. Louis: Official photographic Company, 1904), 46. Photograph courtesy of the Smithsonian Libraries and Dibner Library of the History of Science and Technology.

with an increasingly vocal debate over Chinese national representation and, not coincidentally, a turning point in China's own exposition management. At the same time that Chinese exposition participants were increasingly assertive amid the exposition's competitive symbolic hierarchies, they were also taking steps to claim authority over the management of national exhibition itself.

Chen Qi and Chen Huide attended the St. Louis Exposition as representatives of local provinces. Both were well-connected and part of a rising professional class in China at the turn of the century. Chen Qi (1878–1925) was a young Chinese military officer who had been dispatched by Zhao Ersun, the governor of Hunan Province, to accompany Hunan's exposition goods to St. Louis. At the event's conclusion, Chen Qi joined a high commission of Qing officials to survey political, socioeconomic, and military affairs in the United States and Europe before returning to China.[13] Chen's fellow critic, Chen Huide (1881–1976), attended the exposition as a member of Hubei Province's delegation. He remained in the United States after the event to earn a degree from the Wharton School at the University of Pennsylvania and returned to a prominent career in business in China.[14] Trained in new professional schools at the turn of the century, both men were steeped in concerns of national development. Their volume on the fair would be endorsed with the calligraphic inscription of Zhang Jian (1853–1926), a noted reformer and industrialist. Meanwhile, their critique echoed imperatives articulated by a broader class of Chinese elites including Confucian officials, reformist intellectuals, merchant-entrepreneurs, and educators invested in a discourse tied to a global "commercial war."[15]

In their book, Chen Qi and Chen Huide described in detail the different halls and attractions at the exposition while noting logistical errors and poor organization that led to a weak or entirely absent representation of the Chinese nation at the St. Louis Exposition. The authors specifically targeted China's expo officials for failures of proper exhibition technique and management. For instance, they asserted, displays that provinces had prepared for the Agriculture Hall or for the Mines and Metallurgy Hall were simply left to a "chaotic" presentation in the Palace of Varied Industries due to a lack of advanced preparation, weakening China's profile in comparison with other nations. They also noted that many provinces had in fact assembled extensive mineral displays for the Mines and Metallurgy Hall, but their collections were not accompanied by the necessary explanatory materials (e.g., charts, diagrams, or statistics). Displays in the Liberal Arts Hall had similar problems (Figure 9.2).

The lack of Chinese experts on site to deliver lectures and share expertise remained a disappointment. The authors lamented that few visited the displays and that those who did seemed to scoff at their disorder. "Alas," they complained, "our nation's endeavors do not meet the measure of her people."[16] The authors also sadly observed the paucity of China's national symbols at the fair. While the flags of other nations flew at "over five hundred sites" throughout the exposition fairgrounds, Chen and Chen informed readers that China's national flag was displayed at merely four sites at the exposition "but it is disgraced in [many] places beyond these."[17]

Specific objects with particularly loaded implications were the focus for additional Chinese critiques of displays in St. Louis. Chinese diplomatic Consul Xia Xiefu, making his own visit to the exposition, discovered with dismay a display of "women's little bound-foot shoes and opium implements."[18] Agitated at this disgrace, he immediately ordered its removal. Yet the Chinese Customs Service official Francis Carl grew angry in discovering that the display had vanished. He promptly scolded the

Figure 9.2 "Curious Chinese Devices in Manufactures." China's displays at the Palace of Liberal Arts included models of "a mill for the manufacture of bean-oil and bean-cake," as well as "sedan-chairs and chair coolies; [a] salt-factory, wind-mill of lateens sails and vermicelli factory in the order named."

Source: Mark Bennitt, *History of the Louisiana Purchase Exposition* (St. Louis: Universal Exposition Publishing Company, 1905), 291. Photograph courtesy of the Smithsonian Libraries and Dibner Library of the History of Science and Technology.

unfortunate subordinate who had removed it and had the display restored. Only a renewed protest by Xia brought about the permanent removal of the bound-foot shoes and opium pipes. In their account of this scandal, Chen Qi and Chen Huide note evidence of Carl's investment in the display in the fact that his own wife had brought "three-hundred pairs of Chinese women's little-footed shoes" as merchandise to peddle at the fairgrounds.[19] Aspersions or reality (or both), the suggestion indicates several dynamics at play. One is that the shame incurred by the equation of these objects—opium pipes and shoes for bound feet—with a Chinese nation was made possible by dominant discourses and exhibitionary practices associated with an Orientalism already common at the exposition. Another dynamic is that this display was also enmeshed in a commercialism that overlapped with the same global hierarchies embraced by the event's participants. Chinese elites were staking their own claims amid critiques of the 1904 St. Louis Exposition. They sought greater control of representation, and would achieve it, at least on one important level, by reclaiming control of exposition management from the Western agents of the Customs Service. As a second goal they also aimed for a new command of market competition.

A national exposition as a means to global "livelihood"

Chen Qi was a major figure in China's turn of the century exposition events who joined a coalition of commercial elites and state sponsors in managing domestic and international projects. Just as they sought better control over the modes of national representation abroad, Chinese officials would seek to mobilize similar technologies and spectacles in domestic events, directly advocating for China's first national fair.

Zhang Jian, who had endorsed Chen Qi and Chen Huide's report on the St. Louis exposition, was a leading supporter of exposition projects, as was the high-ranking Qing official Duanfang (1861–1911), among others. By the end of the same decade, Duanfang and Chen Qi would lead China's first national exposition project held under Qing auspices: the Nanyang "Encouraging Industry" Exposition to be held in Nanjing in 1910.[20]

Duanfang, the Liang-Jiang governor-general and former co-leader of the Qing state's 1905 investigative mission abroad (in which Chen Qi had also participated) presented a petition in 1908 to the Qing imperial court calling for a national exposition. He invoked the success of European and American expositions as well as those of Japan, arguing that the value of these events lay in their presentation of the tools, products, and methods of diverse industries for collective examination and the encouragement of success through competition.[21] Gaining imperial assent, China's first national exposition was held as proposed during the summer of 1910 in the old Southern Capital, Nanjing, with the stated goal to "promote industry and enlighten the people."[22] Chen Qi was appointed Managing Director while Xiang Ruikun, a graduate of the Industry and Commerce Department at Japan's Meiji University, was appointed Managing Vice-Director (Figure 9.3). The event was a joint undertaking of state and private sponsors.

In staging this exposition, Managing Director Chen Qi and his fellow organizers sought to stage a new kind of event for a mass audience in China, one that would help to shape a disciplined population to serve its nation. Organizers framed the exposition as a studied inventory of the nation's goods, both for domestic consumption amid foreign competition and export promotion, while placing a growing emphasis on science and industrial production. Spurred by the concerns of colonialism and

Figure 9.3 Vice-Director Xiang Ruikun (Left) and Director Chen Qi (Right) of the Nanyang "Encouraging Industry" Exposition of 1910.

Source: *Nanyang quanye hui jinian ce* (Shanghai: Jicheng tushu gongsi, 1910.)

international competition, exposition officials aimed to use a national exposition to mobilize a people in the name of economic nationalism.

Chen Qi promoted the exposition through a series of speeches asserting the power of exhibition as a means of economic development and a national indoctrination of China's people. Speaking to audiences of officials, industrialists, merchants, journalists, and academics, Chen invoked both market capitalism and colonialist threats. He noted that the exposition had come amid "an observation of each nation's competition within China, and at the same time out of an observation of China's own competition with other nations."[23] As he explained, that competition was no longer simply the battle of armies, but now a competition of "livelihood" (*shengji*). A broad term that exposition organizers frequently deployed, "livelihood" referred to the satisfaction of people's material necessities and also was linked to a nation's need to advance economic production itself.

Livelihood was also tied to colonialist competition globally. The very first exposition circular, for example, recounted the incursions of imperialism in Asia. Readers were reminded of the United States' colonization of the Philippines; Russia's construction of the trans-Siberian railroad; and Japan's war with Russia amid its encroachment in Korea and Manchuria. The author of this circular, moreover, argued that a more fundamental change was taking place. A nation-versus-nation competition was not the only challenge that gripped the world, but also the competition of "individual versus individual" within those nations. The solution offered by Chen was to embrace the exposition grounds as a site for this competition. Here, public display and shared comparisons of technology, agriculture, industry, and commerce would advance a national project.[24] China's exposition organizers sought to promote a revitalized economy in the face of global competition. They imagined the exposition grounds as an arena for raising the quality of national commodities to compete in domestic and global markets, and, at the same time, for encouraging a disciplined and patriotic labor force at home.

Exposition organizers thus imagined the national event simultaneously serving the purposes of economic development and social indoctrination. Just as Chen Qi envisioned China's first national fair as one that would lay a "national foundation" for the "progress of agriculture, mining, industry, and commerce," he also declared that "progressive nations utilize new people [*xin min*]" to bring about national prosperity.[25] Organizers encouraged a broad audience for the exposition, with students, soldiers, and youth granted discount tickets. Discounts were also provided for servants and "poor people," and the attendance of peasant groups and "overseas Chinese" was celebrated.[26] Exposition managers intended China's first national exposition as both classroom and national spectacle (Figure 9.4).

One province, Zhili, has left a detailed record of its own engagement in the exposition project. Zhili held a preparatory exhibition in December, 1909, in Tianjin. A published account of a visitor's tour noted that one could view porcelain, lacquer ware, paper, seals, embroidery, silks, and a variety of woven and dyed textiles. Readers learned that "all kinds of brass, pewter, wood and iron wares, all made by this very province's workers" were also on ready display, along with medicines of all sorts and items of fine art.[27] Tianjin's exposition officials extended invitations to representatives of foreign consulates, an opportunity to promote trade, while reports offered a careful record of the items and display categories to which each of the consulates devoted the most attention.[28] Zhili also sent its own contingent and goods to the national exposition in 1910, as did the other provinces of the empire.

Figure 9.4 Officials, soldiers, and, on the far right, female attendants gathered under the flags of diverse nations in a new kind of public ceremony under the auspices of the Qing dynasty for opening day ceremonies at the Nanyang Exposition.
Source: *Nanyang quanye hui jinian ce* (Shanghai: Jicheng tushu gongsi, 1910.)

The Nanyang national exposition featured themed halls including a Machinery Hall in addition to the broader category of Industrial Arts, as well as an Armaments Hall alongside the Agriculture Hall, Fine Arts, Education, and others. Another hall exclusively featured items contributed by the overseas Chinese community, with displays advertised as revealing both commercial strength and a link to an ancestral nation. Wealthy Chinese from Southeast Asia financed this pavilion. In addition to the commercial goods presented at the hall, the presentations of objects made by native peoples in Southeast Asia also piqued the interest of observers. The sound of a snake-skin drum was said to "beckon visitors to the displays of Batavian objects," including different types of weapons, among other things "never before seen by Mainland people." Displays of local Batavian curiosities were juxtaposed with Chinese commercial displays. The impressive array of commercial goods for export offered by overseas Chinese at this hall stood as testament to the "hundreds of years that our nation's people have been engaged in business and trade in Southeast Asia."[29] These displays at the national exposition also spoke to the influential investment, both commercial and political, that a Chinese community living in Southeast Asia held in China itself.[30] While exposition managers celebrated the overseas Chinese community at the fair, other displays of imported goods invoked concerns of encroaching foreign powers. At two halls at the national event, visitors to the fair could also survey the export goods of foreign nations that maintained an active and intrusive presence within China, including Britain, Germany, Japan, and the United States.[31]

Directors Chen Qi and Xiang Ruikun hoped to mobilize a "new people" through the exposition to serve a globally competitive Chinese economy. Zhili province offered one visual record of the objects and displays assembled for audiences at both its preparatory exhibition in 1909 and the national event in 1910 that echoes this agenda.[32] At the preparatory fair in Tianjin, displays advertised the strengths of new industrial manufactures; just as others, such as an exhibit of textiles manufactured at a training center for reformed criminals, revealed ways in which both the product and the force that created it were being improved (Figures 9.5 and 9.6). Manufactured goods similarly dominated displays that overlapped with themes of education, science, and industry at the national fair in 1910. There, under the category of Education, visitors encountered shelves stacked with bottles of medicines labeled and arranged by the Tianjin Army Medical Academy's Pharmaceutical Division. Nearby, electric devices, tools for work in chemistry, as well as the study of gases and air all awaited perusal. Beyond these stood models of ships, carts, and transport vessels, and also a model of the human body with bones, organs, arteries and veins, all inviting examination. At the national fair in 1910 products from the industries of Zhili province also accompanied the education displays. Carefully arranged soap products stood as testimony to the growing chemical industries in the region and also directly mirrored a display of the same product—with the words "Dragon Best Soap" emblazoned on labels in both English and Chinese—in the hall's pharmacy section (Figure 9.7). Other manufacturers advertised their brands less prominently but were still present on labels and cards in the textile and clothing display, which included hats, shoes, and wardrobe items. Agricultural displays offered wheat and grain products, with beverages and biscuits also prominently featured. A significant number of the displays indicated the sites of these goods' creation, including local training schools, such as that for a new-style thread spinning machine and for a prominent arrangement of Western-style furniture manufactures (Figure 9.8).

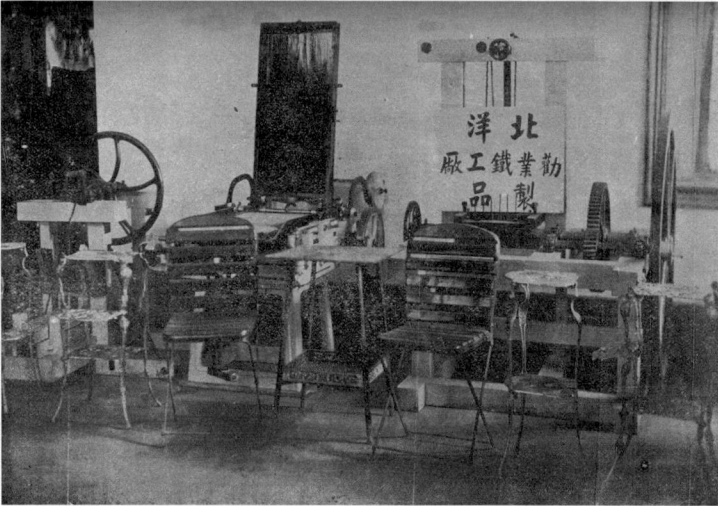

Figure 9.5 From Zhili Province's exhibition displays, manufacture samples provided by the Beiyang Ironworks factory.

Source: Zhili xie zan hui, *Nanyang quanye hui Zhili chu pin lei zuan hebian* (n.p., 1910), vol. 1. Photograph courtesy of the Harvard-Yenching Chinese Library.

Figure 9.6 A textile display from the Zhili Province exhibition hall. The sign indicates products manufactured by a training institute for criminal offenders.

Source: Zhili xie zan hui, *Nanyang quanye hui Zhili chupin lei zuan hebian* (n.p., 1910), vol. 1. Photograph courtesy Harvard-Yenching Chinese Library.

Figure 9.7 "Dragon Best Soap" brand products along with candles and other goods shared as part of the chemical products display in the Zhili Exhibit hall at the Nanyang exposition. In many ways, the displays of goods here seem to echo frames for point of purchase, though the intent was framed as one of study and evaluation.

Source: Zhili xie zan hui, *Nanyang quanye hui Zhili chupin lei zuan hebian* (n.p., 1910), vol. 1. Photograph courtesy of the Harvard-Yenching Chinese Library.

Figure 9.8 Western-style furniture products for the exhibition supplied by a training workshop in Zhili Province.

Source: Zhili xie zan hui, *Nanyang quanye hui Zhili chupin lei zuan hebian* (n.p., 1910), vol. 1. Photograph courtesy of the Harvard-Yenching Chinese Library.

This arrangement of products, tools, and samples—a tour of one corner of the empire's objects—fulfilled one aim of the exposition, namely an "encouragement of industry" to compete with foreign manufactures in both domestic and international markets. As Chen Qi, Xiang Ruikun, and other exposition organizers envisioned, these displays presented a path for China through the presentation of goods produced by a "new people." Ideally visitors would recognize themselves as part of a shared, industrious community while viewing these objects. Xiang Ruikun advocated "an education in material things" for a new citizenry at the exposition. He declared the exposition to be the prime site for this training, a site where visitors could study "which things are those of advanced nations, which things are those of less advanced nation, which are the things made by savage people, and which are those that civilized people are able to make."[33]

San Francisco, 1915: of "besotted and blur eyed opium and morphine fiends"

In the expositions explored above, Chinese elites critiqued representations of China, reclaimed management roles from Western officials, worked to improve and promulgate the tools of display, and staged their own national commodity spectacle aimed at articulating and improving China's position in a global economy. The early years of the Republic of China, founded after the 1911 nationalist revolution and the abdication of the Qing emperor Puyi in February 1912, coincided with both Chinese and American invocations of a rising significance of the Pacific and the celebration of two Republics.

For the Panama Pacific International Exposition of 1915, we find that China's exposition organizers and investors sought to place China in a parallel position with the United States. Chen Qi, the former critic and now presiding president for China's expo contingent set forth a vision of the two nations as standing "like one family" as "the two most rich and populous ... on either coast of the Pacific."[34] The opening of the Panama Canal in 1914 together with San Francisco's Panama Pacific International Exposition one year later amplified concurrent designations by American boosters of the Pacific as a "new Mediterranean," one in which "local interests become world interests," and the interests, particularly, of colonial powers.[35] Amid American invocations of a Pacific Century, China's presentations would also be critiqued by invested audiences both within China and abroad, who each made claims on national representation at the fair in 1915.

China's exposition organizers embraced the new canal and its celebratory Panama Pacific Exposition. Diverse voices spoke to this endeavor and the related colonialist context that the exposition signified. During the course of the event, Chinese-language newspapers in both the United States and China reminded readers of the great impact the Panama Canal would have on world politics as well as the Chinese nation. One author invoked the Canal and the rising competition over the Pacific itself, an arena in which China would need to compete.[36] Feng Ziyou, a writer and revolutionary colleague of Sun Yatsen, also spoke to global competition in the context of the Panama Pacific Exposition. "Now that the canal is built," Feng cautioned,

> the shape of world power has already undergone a change and the international relations of the world's nations become more intricate by the day. Indeed, if one is not hurriedly scheming for prosperity, isn't it difficult to remain independent in this current world?[37]

Feng warned of the fate of Korea and Vietnam, colonial possessions of the Japanese and French respectively, urging his Chinese compatriots to use the opportunity of two events—the construction of the canal and the exposition—to emulate other nations' economic development through improved market competition.

Invocations of a new Pacific arena coincided with the same vision of economic nationalism mobilized for the 1910 Nanyang Exposition. Amid concerns for survival in a competitive global economy, and standing before symbols of progress embodied in the architecture and pageantry of the fair itself, China's exposition contingent staged a celebration to officially open its pavilion at the San Francisco exposition in March, 1915.[38] Led by Chen Qi, China's officials stood before its national hall, a model of the Qing Taihe Palace (now, somewhat ironically, a symbol of the imperial past), wearing modern top hats and tails, sharing speeches and posing for photographs with their wives and attendants. A crowd of guests and onlookers filled the pavilion's grounds while Chinese and American flags flew overhead (Figure 9.9).

A band from the Chinese Six Companies, a leading advocacy organization for Chinese and Chinese Americans in the United States, performed along with an American military band, a drill team from the U.S. Army, and a group of singing Chinese American schoolgirls.[39] In his opening speech, Chen Qi asserted China's own parallel position to the United States. They were two nations that faced each other on the Pacific's coasts, rich in resources, and committed to peace, he argued, announcing that "China and the United States are like one family."[40] Chen's St. Louis sponsor Zhang Jian also added his voice to this effort, now supporting a group of Chinese businessmen who toured the United States in conjunction with the exposition and who produced their own publication, *China's New Industrial Enterprises*. Zhang

Figure 9.9 A mass audience of participant-observers greets opening day speeches at the China Pavilion grounds in San Francisco, 1915. The West Pavilion and Pagoda sit to the right of the frame behind the crowd.

Source: Jiangsu shengzhang gongshu shiye ke, *Jiangsu ban li Banama saihui baogao shu* (Shanghai: Shangwu yinshu guan, 1917).

Jian provided a preface for this work and for their Chinese-language report of the trip. Zhang claimed fundamental similarities between China and the United States at that moment in history. Not only were the two polities now both "brother republics," but they also shared another core promise: "The United States is seen as a great continent—the same. The land is expansive and much remains to be developed—the same. They are full of countless beautiful things—the same." In other words, both the new Chinese Republic and the United States were sites for material exploitation and market development. "Both are indeed heaven," Zhang praised, poetically courting foreign economic investment.[41]

As representatives of a new Republic at the fair, Chen Qi and China's exposition managers offered displays aimed at promoting China's market potential and development. In the exposition's "Universal Halls," China's organizers arranged presentations of commercial goods, art objects, school displays, and surveys of the natural and industrial resources of the nation. In the Palace of Transportation, Xia Changchi (1889–1970), a graduate of Beijing University's Civil Engineering Department, put together a meticulous display designed to demonstrate China's establishment of a diverse and modern transportation system. Visitors encountered photographs, statistical charts and maps, electrical devices as well as a large collection of models that included locomotives, sleeping cars, coal cars, freight cars, trucks, as well as models of railway stations, bridges, steamships, and also a "riksha" offered by the Tung-Chang Bicycle Company of Shanghai.[42] Xia also arranged presentation of the Republic's modern postal service in exhibits of a letter-carrier's uniform and sample postage stamps, as well as a detailed post and railway map of China. To complement these displays, Xia prepared an English language volume that described the exhibit and also offered a detailed report of the history and progress of China's railways, steamships, post and communications infrastructure—all crucial for business and investor-friendly. Xia's volume, illustrated with charts and photographs, also included a lengthy section welcoming overseas tourism, another "export" industry generating income from foreign nations.[43] As Xia noted, his transportation exhibit, like others that the Republic of China offered in the exposition's Universal Halls, was specifically intended to provide the "outside world" with "a glimpse into modern China," and to advertise great "industrial and commercial possibilities" present there.[44] The wealth of China's presentations was seen beyond this Palace, moreover, as the American press celebrated not only the "old kingdom" artwork and porcelains, but also the arrival of a "modern commercial China" and the "industrial significance" of its displays.[45]

As China's exposition managers represented a modern nation and asserted economic aims on this international stage, the constituency making its own claims on the nation continued to broaden. Critiques of Chinese exposition displays now expanded further beyond China's own exposition contingent and diplomatic corps, who had voiced the earlier dissatisfaction with the Customs Service management of Chinese displays at St. Louis. In an era of rising press circulation, criticism of Chen Qi's own management appeared in papers on both sides of the Pacific. Complaints began with a series of articles in San Francisco's local Chinese-language press during the spring of 1915. There, writers disparaged the official China Pavilion as crude and sloppy; within weeks, the same accusations were repeated in newspapers in China itself. The complaints focused on a commercial teahouse that shared grounds with the China Pavilion (Figure 9.10).

One Chinese-language paper based in San Francisco, the *Chung Sai Yat Po* republished a letter of complaint from the Chinese Six Companies to Commissioner-General Chen Qi.[46] Within the same month, newspaper readers in Shanghai also could read similar letters. The Six Companies' letter was lengthy and harsh. Teahouse products, they asserted, were "all inferior in form" and, worse, often consisted of Japanese goods, a direct affront in a time of boycott against Japanese imperialist aggression in China. The Six Companies' letter derided the China Pavilion's design, noting that the verandas flanking the main hall were "crude" and "resemble a horse stable or a pigeon cage." This mattered, the letter asserted, because "overseas Chinese comrades" also "have a stake" in protecting China's reputation.[47]

Chen Qi's reply was formal and polite, opening with a thank you to the Six Companies for their concern, but in his missive he defended the exposition management. Chen explained that he had sent people to investigate and that the teahouse did not, as reported, make use of Japanese goods. While he agreed that the teahouse management would have to improve the appearance of the display, he stressed the importance of the venue to those who would have it shut down. It had been established after a request from Zhejiang Province's product association "so that they might increase their sales of tea and recover economic power in the tea industry." Chen's reference here alluded to markets that had been lost to competition from the Russian and Japanese tea industries. Chen further emphasized the importance of helping national products gain a foothold in the Americas and invited those critical of the teahouse to visit his office to discuss a proper plan to "support the nation, increase trade, and make further progress."[48]

Chen's nationalist exhortations did not achieve the desired result, however, and newspaper attacks on both exposition management and the Chinese Pavilion continued. By the end of March, San Francisco's *Chung Sai Yat Po* reported that "Western newspapers" had begun to comment upon the "laughable" appearance of the pavilion and the crude tables of the teahouse, a situation the *Chung Sai Yat Po* lamented as

巴拿馬博覽會中國政府陳列館平西圖案

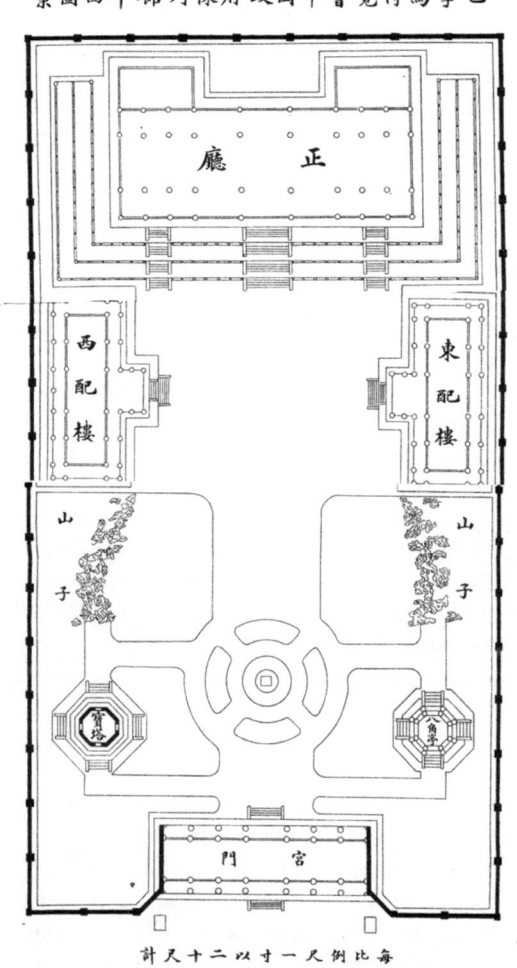

每比例一尺以二十尺計

Figure 9.10 Original plan for the China Pavilion at the Panama Pacific International Exposition, San Francisco, 1915. The main hall sits flanked by west and east pavilions to the left and right, with a pagoda and pavilion towers following before the grand main gate below. The precise location of the controversial teahouse (and other market rental space) is not noted.

Source: Li Xuangong, *Banama Taipingyang wan guo bolan hui yaolan* (Shanghai: Shangwu yin shu guan, 1914). Photograph courtesy of the Harvard-Yenching Chinese Library.

"adding up to more national shame."[49] On the first of March, a front-page editorial blasted "the Commissioner General" and exposition representatives for mismanagement, corruption, and for failing to display the "progress" that China had made both in spirit and in material presentation. Citing an editorial in a "foreigner's paper," the author presented a picture of the China Pavilion as a national travesty; this very same description was republished in the Chinese press in Shanghai. In the complaint, the China Pavilion's exterior was labeled "extremely inferior," while the teahouse resembled a "lower-class restaurant ... entirely insufficient to represent China's progress." The display tables in the Pavilion's exhibit rooms were also declared unworthy of the objects that were placed on them, while a hall that displayed "valuable paintings from Chinese antiquity" was derided for its decoration. "The seats and cushions there are of a red color that differs hardly at all from those that might festoon a black woman's house in a black quarter of town in America's Mississippi," complained the article.[50] Given the hierarchy of races embraced in the expositions of that day, such an association threatened to raise interior decor to the level of national predicament.

A second concern, one that unified rather than divided constituencies invested in images of China, arose with presentations on the Exposition's "Joy Zone," a midway site available for a host of well-prepared concessionaires and impresarios.[51] While at the 1910 national fair in Nanjing, native peoples' goods from Southeast Asia were mobilized by overseas Chinese to display a hierarchy of "civilization" in which China stood on the correct side of history, displays at San Francisco's "Joy Zone" in 1915 threatened this claim. There, at the Chinese Village concession, one attraction known as "Underground Chinatown" drew visitors, but aroused the anger of many Chinese and Chinese Americans. As described in a San Francisco Chinese-language newspaper, *Shao nian Zhongguo* [Young China], visitors to the "Underground Chinatown" saw Chinese opium addicts arrayed in reclined positions, clutching pipes, set forth in dark passages beyond which viewers passed, peering at the addicts through holes in a wall, glancing into the mock opium den and gambling halls. Visitors also saw live performances at one "hazy opium cave" in which "young Western men and women" suddenly appeared as members of the Salvation Army, with police alongside. Reciting scripture, these actors preached to a "pig-tail figure clad in China's old-style garb" meant to be the opium-den's owner, and the police would catch the miscreants fleeing the den itself.[52] The *Chung Sai Yat Po* stridently critiqued "Underground Chinatown" in an editorial warning of the threat that this concession posed, invoking both cultural and racial humiliation as well as the trope of the vanished nation. The editorial spoke of the risks to China's reputation within the international community and urged Chinese people to unite in fighting to have the attraction closed down. "Otherwise," the editorial warned, "[we] will be taken as people of lost nations such as the black people of Africa and the Indians of the Americas, for which managers at the Amusement Zone have devised all sorts of weird displays to make people laugh."[53] These editorials expressed outrage at the performances of racial caricature and their implications for national humiliation; none wanted to be counted among the "lost nations" of that day.

China's expo managers also quickly moved to challenge these caricatures on the Zone. Director of Foreign Affairs for China's exposition contingent Ouyang Qi toured "Underground Chinatown" and met with the Six Companies' attorney to determine possible legal action.[54] As Commissioner-General, Chen Qi composed his own letter of protest to Exposition President C. C. Moore. He declared "Underground Chinatown" a "disgrace to the Exposition and a slander on the Chinese people," noting that it presented an entirely "maudlin and degrading scene" that was "abhorrent to the Chinese government and the people of China in this Country." Chen took his critique a step further, reminding Moore of America's own problems that others might capitalize on in a similarly ugly display, if desired:

Every country has its weak spots. Certainly a concession in a foreign country which would represent as a part of ordinary American civilization the police murders of New York City, the life of the brothel in every port city in America, and other countries, the drinding [sic] of absinthe and its attendant horrors, the dancing of practically nude women as they appear in the Hippodrome of New York City, the fashionable and exclusive gaming clubs of this and other prominent cities and the besotted and blur eyed opium and morphine fiends, and alcoholic degenerates who crowd the dock in Police Court every morning of the week would rouse the American government to instant and justifiable protest.[55]

Similar letters of protest from the Chinese Six Companies, San Francisco's Chinese Chamber of Commerce, the Chinese Consul-General, and the local Chinese community followed. Across the Pacific, in China itself, local newspapers also covered the story closely, publishing copies of protest letters for their own readers to follow.[56]

This concerted effort of a broad Chinese and Chinese American community, press and commercial organizations, Chinese diplomats, and the exposition contingent brought about an end to the "Underground Chinatown" attraction. On March 26, the exposition assistant director of concessions, Jay Bryan, led an escort of exposition guards to shut it down. Though a modified attraction would open again later under the banner of "Underground Slumming" (Figure 9.11), a diverse constituency, including many of the same who had gathered at the opening of the China hall, had rallied to assert both a critique and a claim on the presentations of "China" at the fairgrounds.

Figure 9.11 The "Underground Chinatown" at the Joy Zone concession managed by Sid Grauman was shut down after coordinated protests by Chinese officials as well as a determined Chinese and Chinese American community at the San Francisco Panama-Pacific International Exhibition in 1915. Later that summer, concessionaires offered an "Underground Slumming" attraction in a similar xenophobic manner while removing more overt Chinese imagery.

Source: Photograph courtesy of the Anne T. Kent California Room, Marin County Free Library.

Conclusion

The "Underground Chinatown" protest at the Panama Pacific International Exposition is an indication of the growing significance of a broader audience for "China" as a nation, as well as the increasingly diverse, transnational community invested in representations of China. As this survey of exposition events between 1904 and 1915 reveals, this concern with China's image on the world stage emerges alongside the rise of the exposition as a mass event at the turn-of-the-century, and is bolstered by the increased role of the press and telegraph as agents in the construction of an "imagined community"—or intersecting, identifying *communities*—on both sides of the Pacific.[57] Here we might consider ways in which to study complex models of nationalism itself.[58] People identifying as Chinese, overseas Chinese, or Chinese American, and intersectional combinations of these categories, might all identify with displays set forth in the name of "China" for reasons that reveal a constellation of diverse and shared interests. The nation on display at the fair offered an emblem, a "logoization" (as Benedict Anderson describes the state patrimony at museums) of identity in which many who claimed association with China had a stake.[59] Their reasons for investing in China's displays might include national identification but also could transcend it through links to other definitions of community (e.g., ethnicity, language group or dialect, native place association) and concerns tied to local or regional political contexts (as seen here, San Francisco, the United States, Singapore and Southeast Asia.)

As the 1915 crisis over "Underground Chinatown" echoed an earlier protest by China's provincial attendants at St. Louis in 1904, we see that the stakes for these displays were defined in terms that linked the local to the global and, particularly, to economic nationalism. One of the criticisms of the teahouse at the China Pavilion in San Francisco was that it sold Japanese tea. In San Francisco as in Nanjing, China's exposition managers directed their efforts and the attention of their audience toward international economic competition. They sought to display the nation's strongest goods (including teas, textiles, and porcelain), create opportunities to promote export products to a foreign audience, and, simultaneously, to utilize the same displays of "material things" domestically to educate a "new people" (and a new, cognizant and compliant labor force) to serve the nation in a global economy. As Tony Bennett has described, an embrace of the exhibition was one that might "reverse the orientations of the disciplinary apparatuses in seeking to render the forces and principles of order visible to the populace—transformed, here, into a people, a citizenry."[60] While such modern apparatuses were still developing at the turn of the twentieth century in China (then a nation with a weak central state), Chinese exposition elites embraced the exhibition spectacle as a means to mobilize a national citizenry as loyal producers and consumers. Chen Qi and others linked this endeavor to national survival and to a competition among Chinese individuals in the name of "livelihood" amid global colonialism—one that presages a more recent global modernity.[61] Indeed, as seen in China's recent hosting of the international Expo 2010 in Shanghai, early twentieth-century dynamics of the fair have their successors today in the rise of a renewed Chinese nationalism (if now fostered by a much stronger and authoritarian state under the People's Republic of China), its own global spectacles, and the continued, if ever-intensified, movement of commodities across the planet.

Notes

I thank Ethan Robey and David Raizman for their work in organizing this volume and for their very helpful comments on this essay. Thanks also to Michael G. Chang, Mary Beth Mathews, Krystyn Moon, and Jennifer Neighbors for comments on earlier versions of this paper.

1 For more on these political reformulations, both in relation to notions of citizenship and state formation, see Peter Zarrow, *After Empire: The Conceptual Transformation of the Chinese State, 1885–1924* (Stanford: Stanford UP, 2012).

2 Alcock also sent exhibits from Japan as well. For more on Alcock and his commentary on Japanese and Chinese arts, see Toshio Watanabe, "The Western Image of Japanese Art in the Late Edo Period," *Modern Asian Studies* 18, no. 4 (1984): 667–684, and Anna Jackson, "Imagining Japan: The Victorian Perception and Acquisition of Japanese Culture," *Journal of Design History* 5, no. 4 (1992): 245–256.

3 M. Digby Wyatt and Eliza Paul Kirkbride Gurney, *The Industrial Arts of the Nineteenth Century: A Series of Illustrations of the Choicest Specimens Produced by Every Nation, at The Greatest Exhibition of Works of Industry, 1851*, vol. 1, (London: Day and Son, 1851–1853), Plate LXXXV.

4 For histories of the Customs Service, see Donna Brunero, *Britain's Imperial Cornerstone in China: The Chinese Maritime Customs Service, 1854–1949* (New York: Routledge, 2006) and Chen Shiqi, *Zhongguo jindai haiguan shi* (Beijing: Renmin chuban she, 2002). L. K. Little presents a brief narrative history of the Imperial Customs Service and Robert Hart's own role in his "Introduction" in *The I.G. in Peking: Letters of Robert Hart, Chinese Maritime Customs, 1868–1907*, eds. John King Fairbank, Katherine Frost Bruner, and Elizabeth MacLeod Matheson, vol.1 (Cambridge, MA: Belknap Press of Harvard UP, 1975): 4–7. For further historiographic analysis see Robert Bickers, "Purloined Letters: History and the Chinese Maritime Customs Service," *Modern Asian Studies* 40, no. 3 (July 2006): 691–723.

5 Robert Hart to James Duncan Campbell, 7 March 1873, in John K. Fairbank, Katherine Frost Bruner, and Elizabeth MacLeod Matheson, eds., *The I.G. In Peking...*, vol. 1, 98.

6 James Hevia, *English Lessons: The Pedagogy of Imperialism in Nineteenth-Century China* (Durham, NC: Duke UP, 2003), 212–213.

7 As Edward Said has noted, Europeans utilized diverse forms of knowledge and power to produce an "Orient" associated with a spectrum of ideas and imagery: "Oriental despotism, Oriental splendor, cruelty, sensuality..." See Edward Said, *Orientalism* (New York: Random House, 1978), 4.

8 Robert Rydell, *All the World's a Fair: Visions of Empire at American International Expositions, 1876–1916* (Chicago, IL: University of Chicago Press, 1984), 153.

9 Ibid., 160–161.

10 Timothy Mitchell, *Colonising Egypt* (Berkeley: University of California Press, 1991), xi, xiii.

11 Huang Kaijia (Wong Kai Kah), an 1883 graduate of Yale and official vice-commissioner of China's exposition contingent, oversaw China's exposition space reservations. Lee Toy, meanwhile, managed the American Yee Ging Company, which was co-owned by vaudeville promoters A. Budd and G.B. Hartford. See Mae Ngai, *The Lucky Ones: One Family and the Extraordinary Invention of Chinese America* (Boston: Houghton Mifflin Harcourt, 2010), 99. For descriptions of the Chinese Village concession, see Thomas R. MacMechen, "The True and Complete Story of the Pike and its Attractions," *World's Fair Bulletin* 5, no. 6 (April 1904): 21–22; "Chinese Village," *World's Fair Bulletin* 5, no. 9 (July 1904): 88.

12 Chen Qi and Chen Huide, *Xin dalu Shengluyi bolan hui youji* (1905), 1–2.

13 Cai Kejiao, "Jindai Zhongguo bolan ye de xianqu Chen Qi ji qi zhushu," *Jindai shi yanjiu* 1 (2001): 307–313. The commission Chen accompanied on its tour of Europe and the United States served in preparation for China's later transition from an imperial state to a new constitutional monarchy.

14 Howard Boorman, ed., *Biographical Dictionary of Republican China* (New York: Columbia UP, 1968), vol. 1, 192–196; Chen Yutang, ed., *Zhongguo jinxiandai renwu minghao da cidian* (Hangzhou: Zhejiang guji chuban she, 1993), 530.

15 Karl Gerth, *China Made: Consumer Culture and the Creation of the Nation* (Cambridge, MA: Harvard University Asia Center and Harvard UP, 2003), 59–60.

16 Chen Qi and Chen Huide (1905), 51–52.

17 Ibid., 181–182.

18 Ibid., 188. The location of this display is not made clear in Chen and Chen's account; the majority of China's displays remained in the Liberal Arts Palace due to organizers' limited space reservations ahead of the event.

19 Ibid.

20 The title "Nanyang" is a word with multiple referents, each of which had valuable meaning for Qing officials. The presiding local official for the fair at its proposal and its original sponsor was Duanfang himself, who as Liang-Jiang Governor General held the simultaneous title of *Nanyang da chen* or Superintendent of Trade for the Southern Ports. The "Nanyang" region also refers more broadly to southeast Asia, particularly Singapore, with an active Chinese community from which the Qing officials at this time sought political and entrepreneurial investment. See Michael Godley, "The Late Qing Courtship of the Chinese in Southeast Asia," *The Journal of Asian Studies* 34, no. 2 (February 1975): 361–385.

21 Duanfang, "Chou ban Nanyang quanye hui zhe," (A petition to hold the Nanyang exposition) GX 34.11 (November–December 1908), in Duanfang, *Duan Zhongmin gong zou gao*, 4 vols. (Taipei: Wenhai chuban she, 1967), vol. 4, 1568–1569.

22 Ibid., vol. 4, 157.

23 Chen Qi, "Kaiban shiwu suo yanshuo ci" (A speech on the opening of the general affairs office), *Nanyang quanye hui tonggao*, No. 1 (XT1.2 [February–March 1909]): 42–43.

24 "Fukan quanye hui shuolue" (Summary of the exposition), *Nanyang quanye hui tonggao*, No. 1, XT 1.2 (February–March 1909): 47–48.

25 Chen Qi, "Nanyang quanye hui guan hui zhinan bianyan" (Preface to the guide to the Nanyang Encouraging Industry Exhibition), in *Guan hui zhinan*, by Nanyang quanye hui shiwu suo (Nanyang quanye hui shiwu suo, 1910): 1.

26 *Shenbao* (8 July 1910): 3; *Guide to Nanking and the Nanyang Exposition* (Nanjing: University of Nanking Magazine, May 1910), 7. For more on the first national fair in China, see Susan Fernsebner, "Objects, Spectacle, and a Nation on Display at the Nanyang Exposition of 1910," *Late Imperial China* 27 no. 2 (December 2006): 99–124 and Michael R. Godley, "China's World Fair of 1910," *Modern Asian Studies* 12 (1978): 503–522.

27 "Youlan zhanlan huichang jishi" (An account of a trip to the fairgrounds), in Zhili xie zan hui, *Nanyang quanye hui Zhili chupin lei zuan hebian* (N.p., 1910), vol 4, *Za zhi*, 4.

28 Ibid., 9.

29 Wang Shuyan, *Nanyang quanye hui za yong* (N.p., 1910), vol. 1, 9–9a.

30 For a rich analysis, see Michael R. Godley, *The Mandarin-Capitalists from Nanyang: Overseas Chinese Enterprise in the Modernization of China, 1893–1911* (Cambridge: Cambridge UP, 1981).

31 *Guan hui zhinan* (Nanyang quanye hui shiwu suo, 1910), 51–53.

32 Zhili xie can hui, *Nanyang quanye hui Zhili chupin leicuan hebian*, vol. 1.

33 Xiang Ruikun, "Nanjing yu Nanyang quanyehui" (Nanjing and the Nanyang Exposition), *Nanyang quanye hui tonggao*, No. 2, XT 1.5 (June 1909): 103.

34 *Shishi xinbao* (4 April 1915).

35 Josiah Strong, *Expansion under New World Conditions* (New York: Baker and Taylor, 1900), 163, 246.

36 Xian Wei, "Banama yunhe zhi chengji yu Zhongguo qiantu zhi guanxi" (The relationship between the completion of the Panama Canal and the future of China), *Chung Sai Yat Po* [CSYP] (19 May 1915; 20 May 1915).

37 Feng Ziyou, ed., *Banama Taiping yang wan guo da saihui youji* (San Francisco: Shaonian Zhongguo bao, 1915), 6–7.

38 For more on the imagery and architecture of the fair, as well as the meanings imparted, envisioned, or inscribed, see George Starr, "Truth Unveiled: The Panama Pacific International Exposition and Its Interpreters," in *The Anthropology of World's Fairs*, edited by Burton Benedict (Berkeley: University of California Press, 1983), 134–175; and Robert Rydell (1984), 208–233.

39 The Chinese Six Companies (*Zhonghua zong huiguan*), also known as the Chinese Consolidated Benevolent Association, was an umbrella organization that mirrored and expanded the role of the many smaller regional associations that served Chinese in the Americas. It offered protection and social services to travelers and immigrants, assisting with lodging, help in finding jobs, and extending financial credit. The Six Companies also provided protection amid the xenophobia and violence aimed at ethnic Chinese during that time. Eve Armentrout Ma, *Revolutionaries, Monarchists, and Chinatowns: Chinese Politics in the Americas and the 1911 Revolution* (Honolulu: University of Hawaii Press, 1990), 15–19. At the same time, The Six Companies was dominated by members of the merchant class and may have tended, at times, to favor their own interests. See Renqiu Yu, *To Save China, to Save Ourselves* (Philadelphia, PA: Temple University Press, 1992), chps. 1 and 2, cited in Madeline Y. Hsu, *Dreaming of Gold, Dreaming of Home: Transnationalism and Migration Between the United States and South China, 1882–1943* (Stanford, CA: Stanford UP, 2000), 219n4.
40 CSYB (1 January 1915; 22 January 1915).
41 Zhang Jian, preface to *Zhonghua you Mei shiye tuan baogao*, edited by Nong-Shang bu (Shanghai: Shangwu yinshu guan, 1916).
42 Xia Changchi (a.k.a. C.T. Hsia, 1889–1970), born in Jiangsu Province, graduated from Beijing University's Civil Engineering Department in 1913. Xia's exposition project would be the beginning of a prominent career in framing nationalist imagery; he directed major national construction projects including the Sun Yatsen Mausoleum and Memorial Park in Nanjing during the late nineteen-twenties and, in 1931, the construction of the Republic's National Athletic Stadium. See Chen Yuntang, ed., *Zhongguo jinxiandai renwu minghao da cidian* (Hangzhou: Zhejiang guji chubanshe, 1993), 720.
43 C. T. Hsia [Xia Changchi], *Modern Transportation and Communications in the Republic of China: Report Presented by Mr. C.T. Hsia, Special Commissioner of the Ministry of Communications of Peking, China, to the Panama-Pacific International Exposition, Palace of Transportation* (San Francisco, CA: Marlow Printing Co., 1915).
44 Ibid.
45 Ben Macomber, "Something Old and Something New—China," *San Francisco Chronicle* (7 November 1915): 26.
46 The *Chung Sai Yat Po* (*Zhongxi ribao*) was established in San Francisco in 1900 by Ng Poon Chew (Wu Panzhao) a well-respected Presbyterian minister and immigrant from Guangzhou, China, who also made a name for himself nationally in the United States as a lecturer and writer. The *Chung Sai Yat Po* built a broad readership across the San Francisco area and nationally in over fifty years of printing (1900–1951.) For a detailed study, see Yumei Sun, "From Isolation to Participation: *Chung Sai Yat Po* [China West Daily] and San Francisco's Chinatown, 1900–1920" (Ph.D. diss, University of Maryland, College Park, 1999).
47 "Sang gang da bolan hui zhi chating" (The Great San Francisco Exposition's tea house), *Shishi xin bao* [SSXB] (4 April 1915). See also "Zhonghua Huiguan zhi Chen Qi shu zhao lu" (A reprint of the Chinese Consolidated Benevolent Association's letter to Chen Qi), CSYP (25 February 1915). The Six Companies' letter further suggested that Chen Qi was one who "loves golden cash but does not care about reputation." Others may have differed; Chen Qi had given attention to the issue of foreign goods being used in Chinese exposition displays, warning the Jiangsu Product Association for having used foreign products in the making of supposedly Chinese displays. See *Shibao* (14 August 1914): 7.
48 "Sang gang da bolanhui zhi chating." See also CSYP (3 March 1915).
49 "Di zhi wo guo zhengfu guan zhi xibao tan" (A discussion in Western newspapers of our slanderous Chinese Pavilion), CSYP (30 March 1915).
50 "Banama saihui zhong zhi Zhongguo guan" (The China Pavilion at the Panama Exposition), SSXB (1 May 1915). See also Xian Wei, "Saihui jiandu yu waibao yan lun" (The exposition Commissioner General and the foreign newspaper editorial), CSYP (1 April 1915).
51 For a detailed description, see Rydell (1984), 227–232.
52 "Wo qiaobao jiang he yi xue ci chi ye" (How are we overseas Chinese to clear away this humiliation?) *Shao nian Zhongguo* (19 March 1915). Founded in 1910 in affiliation with the Tong Meng Hui political party, *Shao nian Zhongguo* [Young China] supported

Sun Yatsen and the revolutionary movement in China. For more on both the *Shao nian Zhongguo* and *Chung Sai Yat Po* newspapers in San Francisco as well as their investment in Chinese politics, see Ma (1990), chap. 6.

53 Xian Wei, "Bolan hui zhong zhi chi" (Shame amid the exposition) CSYP (19 March 1915).

54 "Nie zao Hua pu diyu zhe zhi yundong" (The movement against those who have concocted the 'China Hell'), CSYP (19 March 1915).

55 Chen Chi to C. C. Moore, 19 March 1915, Bancroft Archives C-A 190, Box no. 23.

56 See for example *Shenbao* (23 April 1915): 6, 9; (29 April 1915): 6; SSXB (17 April 1915), (23 April 1915), (17 June 1915). Many Chinese exposition guides and report books also provided critical accounts of the "Underground Chinatown" attraction.

57 On the concept of "imagined communities," see Benedict Anderson, *Imagined Communities: Reflections on the Origin and Spread of Nationalism* (New York: Verso, 1983).

58 See Rian Thum, *The Sacred Routes of Uyghur History* (Cambridge, MA: Harvard UP, 2014), 208–209.

59 Anderson (1983), 182.

60 Tony Bennett, *The Birth of the Museum: History, Theory, Politics* (New York: Routledge, 1995), 62–63.

61 See Arif Dirlik, "The End of Colonialism? The Colonial Modern in the Making of Global Modernity," *Boundary* 2, 32, no. 1 (Spring 2005): 1–31.

10 The 1910 Centenary Exhibition in Argentina, Chile, and Uruguay

Manufacturing fine art and cultural diplomacy in South America

M. Elizabeth Boone

When Argentina and Chile commemorated their centenary anniversaries of independence in 1910, they invited a number of other countries, the United States among them, to celebrate with them. Artists in the United States enthusiastically responded to these invitations, despite their distance from Buenos Aires, Santiago, and Montevideo, the three cities that ultimately hosted the single exhibition of 120 paintings and 41 sculptures, which were shipped by sea under the watchful eyes of commissioners John E.D. Trask and Charles Francis Browne.[1] Argentina's centenary, celebrated on May 25, was marked by large agricultural and industrial fairs as well as an international exhibition of fine arts, which opened to the public on July 12, 1910.[2] The Chilean celebration was somewhat smaller, organized to coincide with their national holiday on September 18. Their international art exhibition, which included the same works from the United States that had been shown in Buenos Aires, opened on September 21. Uruguay, a country not yet celebrating its centenary, but eager to present an exhibition from the United States in its own capital city, provided the final venue for the exhibition in February 1911. This traveling exhibition resulted in a number of purchases, and many works from the display—including paintings by Philip Leslie Hale, Edward Redfield and Charles Morris Young—remained in South America, acquired by the sponsoring republics to augment their burgeoning national collections of modern art.[3] Like the international world's fairs that preceded them in Europe and the United States, the Latin American centenary celebrations placed a collection of fine art into the context of industrial and agricultural displays sponsored by manufacturers, farmers, and nations desirous of opening new markets, elevating brand recognition, and promoting political agendas at home and abroad.[4]

This essay uses the collection of paintings sent by the United States to Argentina, Chile, and Uruguay to explore the relationship between the fine arts, agriculture, and the applied arts, as well as to examine the ideological use and political motivations for international exhibitions in the southern cone. The inclusion of original painting and sculpture in exhibition environments dominated by the display of live animals, food products, minerals, and mass-produced goods at first seems odd, but many of the artworks sent by the United States were landscape and genre paintings that incorporated livestock and other signs of agricultural production and industrial wealth within them.[5] In this respect, the fine arts served to illustrate and reinforce the overarching themes of commerce and prosperity at the fair. They also functioned to represent the nation and promote diplomatic policy, contributing to the creation of an international forum in which each could forge its own national identity in relation and reaction to the others.

The relationship between the United States and these three South American nations was by no means analogous.[6] Nonetheless, as an early example of cultural diplomacy, the 1910 display does provide an introduction to some attitudes about the United States and inter-American relations that remain in play even today. A series of popular political cartoons specifically commenting on the United States and the centennial exhibition provides the focus of this investigation. As argued by Benedict Anderson and others, vernacular print culture was a key communicative vehicle in the modern era for the construction of imagined communities, the characteristics of which were both internalized by the local citizenry and exported to promote arguments about similarity and difference abroad.[7] At the centenaries of 1910, mechanically reproduced illustrations in the popular press provided trenchant commentary upon painting and sculpture, that which was handmade and original, in a period of rapidly expanding print communication. They also suggest how agricultural, industrial, and aesthetic trends in the United States and its southern hemispheric neighbors were both similar to and came into competition with each other.

World's fairs as cultural diplomacy

A guiding assumption of cultural diplomacy, in the words of Claire Fox, is "that dialogue and contact among intellectuals lead to greater understanding and mutual respect among their respective societies."[8] The relationship between the United States and Latin America is characterized by collaboration, conflict, and change; and the need for such dialogue increased during the nineteenth century as the many nations of the Americas negotiated their borders and developed their varying national identities. Argentina, Chile, and the United States all began their existence as European colonies, but the United States, achieving independence thirty years before its neighbors to the south, quickly began shedding its position as a colonized nation and assuming the role of colonizer.[9] The Latin American republics followed suit, with Argentina and Brazil wrestling over tiny Uruguay during the course of the nineteenth century, and Chile making territorial incursions to the north at the expense of Bolivia and Peru. The Monroe Doctrine, issued in 1823, characterized any further European colonization in the Americas as an act of aggression. Initially, the policy was favorably viewed as anti-imperialist by the new Latin American republics, and even Simón Bolívar, one of the fathers of Latin American independence, imagined a pan-American union that might bring together all the nations of the hemisphere.

Sixty years later, however, when the United States began promoting the idea of an Inter-American alliance, many Latin American nations—Argentina, Chile, and Uruguay among them—saw this new manifestation of *Pan-Americanism* as an imperialistic gesture led by a government determined to develop markets and exploit resources for its own profit.[10] U.S. Secretary of State James G. Blaine began promoting the idea of an inter-American conference in 1881, and the Bureau of American Republics held the First International Conference of American States in 1890 in Washington, DC. *Pan-Americanism* simultaneously developed with the emergence of *Latin Americanism*, which sought to unite the Spanish- and Portuguese-speaking nations of the Americas into a coalition in defense of its own interests and independent of the United States. In 1904, six years before the centennial year, Theodore Roosevelt delivered his Corollary to the Monroe Doctrine, which sanctioned the use of armed force in Latin America should events in the region threaten U.S. economic interests. Roosevelt's successor William Taft, elected in 1909, softened this stance somewhat with Dollar Diplomacy, using foreign loans and other forms of financial support, rather than the military, to assert and protect United States interests abroad. For all intents and purposes, however, the goal remained the same: to maintain influence in the south. *Caras y Caretas*, a weekly magazine of political satire, news, and literature, addressed this situation from an Argentine perspective (Figure 10.1).

Caras y Caretas—the title translates roughly as *Faces and Masks*—was a weekly magazine founded in Montevideo that moved to the larger metropolis of Buenos Aires in 1898. José María Cao, who had immigrated to Argentina from Spain in 1882, provided a large number of illustrations for the journal.[11] In a cover from 1909, the first year of Taft's presidency, Cao shows the corpulent leader leaning casually against the top of the globe, his large bulk dominating South America and his expressive face widening into a satisfied grin. The president appears a second time to the right,

Figure 10.1 José María Cao, "El Humor Yankee."
Source: *Caras y Caretas* 12, 581 (November 20, 1909), cover.

with top hat in hand and a roll of documents tucked under his arm. He now stands atop Brazil, which hosted the third Inter-American conference in Rio de Janeiro in 1906, and faces south toward Argentina, which would host the fourth in 1910, the centenary year.

The caricature, titled *El Humor Yankee (Yankee Humor),* was produced in the context of the international military build-up that preceded the outbreak of World War I. Following England and Germany, the United States was expanding its navy, known as the "Great White Fleet" for the white paint used to protect the ships' modern steel hulls. In fact the vessels had just returned from an around- the-world tour that included stops in Brazil, Chile, and Peru, among other countries in Latin America. Presented as both good-will visit and veiled threat, the journey provided U.S. naval commanders and sailors with hands-on experience in maneuvering, fueling, and maintaining these ships in ports far from home. The curiously shaped shadow cast by Taft's bulbous body—the darkest section vaguely resembles the inverted outline of the continental United States, with his two legs suggesting the states of Florida and Texas—is explained by the caption below: "His gigantic shadow covers half a continent; and it will project even more forcefully once he gets the new searchlight being built for him in the shipyards of England." With considerable coastline and access by sea to North America, Europe and Asia, the Latin American republics were likewise engaged in a pre-war naval buildup. The contest was both external, with the South American countries uniting to position themselves against their rivals to the north and in Europe, and internal, as shifting borders, economic competition, as well as ethnic and cultural differences, continued to lead the various neighbors into a series of political disputes and military encounters during the course of the late nineteenth and early twentieth centuries.

The 1910 travelling art exhibition was, in fact, an early example of governmental use of culture to support economic and political goals. Scholarly discussions of soft power and cultural diplomacy have primarily focused on the mid-twentieth century, a period marked by Franklin D. Roosevelt's Good Neighbor Policy, World War II and the Cold War. Nelson Rockefeller and the Office of the Coordinator of Inter-American Affairs used popular singers and Hollywood actors as goodwill ambassadors to counter German influence in Latin America during World War II, and the CIA ran a covert campaign to turn post-war Europeans away from Communism by supporting a vast network of unsuspecting artist and writers through the Congress for Cultural Freedom in the 1950s.[12] The use of art exhibitions, particularly those featuring abstract expressionism, by the U.S. government as a political tool in the Cold War is now also well known.[13] The Taft administration's support for the 1910 centenary exhibition, and participation by the United States and other nations in world's fairs more generally, were effectively the precursors to this sort of surreptitious promotion of United Stated interests abroad.

The values of the nation

Argentina and Chile were working independently when they decided to include collections of fine art at their centenary celebrations. Both sent invitations through diplomatic channels to potential participants, and displays were organized by artists and arts administrators from a number of nations—England, France, Germany, and Spain among them. The United States was the only country to provide government funding for its display, sending one body of work to both countries.[14] Responding to the lobby of John Barrett, Director of the Bureau of American Republics (later reconstituted as the Organization of American States), Congress allocated a modest sum to support the cost of participation.[15] In keeping with the policy of Dollar Diplomacy, the United States would use money (rather than weapons) to influence its southern neighbors and promote goodwill in the region. The works selected, which consisted of a large number of late Impressionist and Tonalist land- and cityscapes, along with a few academic nudes, figure and still life paintings, were mostly for sale. Many displayed the refined subjects and stylistic sensibilities of the Aesthetic Movement. Following the art display in Argentina and Chile, Edwin V. Morgan, U.S. Minister to Uruguay, arranged for those works that remained unsold to be exhibited a third time in Montevideo. Each nation produced its own exhibition catalogue; the publications produced by Argentina and Chile are illustrated and include essays by prominent critics, while the Uruguayan catalogue is a modest pamphlet that consisted simply of a list of works on display.[16]

John Trask explained the selection of paintings and sculpture sent to Latin America in a final report prepared for government officials.[17] Announcements appeared in targeted periodicals such as *American Art News*, and artists were invited to submit their entries to Philadelphia, where a jury consisting of sculptors Karl Bitter and Charles Grafly and painters Francis Coates Jones, Lewis Henry Meakin, Edward Redfield, Edmund Tarbell and Irving Wiles made the final selection.[18] Each artist was allowed to submit only one work; each work had to conform to a conventional size for ease of transport; and to ensure the national character of the display, only artists residing in the United States could participate.

Trask was secretary of the Pennsylvania Academy of the Fine Arts, and Charles Francis Browne, his assistant, taught at the Art Institute of Chicago. These were institutional men with conservative tastes. The United States, Trask explained in an essay translated into Spanish for the Argentine catalogue, had liberated itself from England and by the late nineteenth century had developed a mature culture that expressed "the values of the nation."[19] The paintings selected were primarily by artists working in the Northeast, Mid-Atlantic, and Midwest regions of the country. For example, John Franklin Stacey's *Church Spires of a New England Town*, which was purchased for Chile's National Museum of Fine Arts, depicts a part of the nation associated with the arrival of English colonization and the development of transatlantic trade. The steeple of a white church rises above the coastal town, conveniently located to guide and support ship captains traveling back and forth across the Atlantic. Few artists working in the Southeast, Florida, the Southwest, and California—regions associated with contentious issues like slavery and the wresting of land from Spain and Mexico—bothered to send their work to Philadelphia, and these regions of the country were poorly represented as a result.[20]

The 1910 exhibition provided the South American public with its first opportunity to see a large number of works from the United States, and art critics who reviewed the show expressed their surprise that a country they associated with industry, mass production and economic competition might also produce fine art. Pedro Lira, director of the School of Fine Arts in Santiago de Chile, called special attention to the art works sent by the United States in his review of the exhibition:

> After quickly reviewing the numerous rooms of this complicated survey of world art, we salute the great republic of the United States for its surprising achievements, which allow its art to be hung with dignity alongside the great European schools of painting and sculpture.[21]

The critic for a Montevideo daily concurred, observing that "this gallery of art is not only interesting for the truly wonderful works exhibited in it, but also because it allows us to appreciate a new aspect of the proud and magnificent activity found in North America."[22] Only ten years earlier, in 1900, Uruguayan literary scholar José Enrique Rodó had published his famous essay *Ariel*, artificially contrasting art with industry to differentiate Latin America from the United States. Rodó encouraged the young people of Latin America to reject the utilitarianism of the United States and embrace the cultural and spiritual values of Latin America instead.[23] However, the 1910 exhibition of painting and sculpture led some in South America to suggest that the United States could reconcile the poetic sensibility of the arts with the prosaic concerns of economic expansion: cultural diplomacy was clearly at work.

The United States shared with the nations of the southern cone a deep appreciation for European culture. Images such as John Christen Johansen's *Child Bathing*, which was also purchased for the Chilean national collection, convey the New York-based painter's admiration for the academic tradition and facility in painting the nude. The canvas, with a Whistlerian palette of blue and white imparting a touch of aestheticism to the subject, was appreciated by those who, like Lira, found it to be an example worthy of emulation by the younger artists of Chile. Painting academic nudes, however, forced artists from the Americas into an awkward position. Because their work bore direct relation to French mentors, it was also considered derivative and lacking in originality by contemporary critics.[24] While painters in Mexico, Ecuador and Peru were able to differentiate themselves from the European tradition by depicting the large indigenous populations in their own countries, artists in the United States, Argentina, and Chile, nations in which native populations were smaller in number and had been brutally exterminated and removed from their land, only infrequently saw, encountered, and painted these less visible populations by the early twentieth century. For these countries, native culture did not become a national style and remained essentially an anthropological curiosity. Only two paintings sent by the United States to the Latin American Centenaries depicted Native American subjects—one by George De Forest Brush and the other by Henry Farny.[25] Historian Frederick Jackson Turner famously declared at the 1893 World's Columbian Exposition in Chicago that the frontier, which had played a key role in the development of U.S. national identity, was closed. Twenty years later, American artists in the United States, as well as in Argentina and Chile, were looking for other ways to distinguish themselves from Europe.

Trask claimed that landscape, rather than figure painting, was the most important genre in the United States, calling its distinguishing national characteristic a "clarity of vision and firmness of purpose."[26] His defense of landscape appeared during a period when the United States was becoming increasingly urbanized. Many were leaving individually owned farms to work in factories, and the thousands of immigrants who arrived during the first decade of the twentieth century primarily settled in cities like New York, Philadelphia, and Chicago rather than in the countryside.[27] Agriculture was becoming agribusiness, and factories organized according to new theories of scientific management were multiplying profits. Argentina was beginning to experience a similar phenomenon, and the paintings purchased for the Argentine national collection—like Edward Redfield's *Hillside Village* in Pennsylvania and Colin Campbell Cooper's view of New York in winter—depicted scenes of both rural as well as urban regions of the

United States. Rural landscapes, which predominated in the exhibition, nostalgically evoked the past, and cityscapes nodded optimistically toward a man-made future.

Argentine critic Godofredo Daireaux suggested that painters in his country use the United States as a model for the development of their own national school, arguing that painters from the U.S. had produced a rich artistic tradition despite their commercial instincts by turning their attention to the land.[28] Like Argentina, the United States was a relatively young republic. By employing "clear vision and practical energy," in Daireaux's words, artists in the U.S. had succeeded in creating a national school, which was becoming with each passing day more and more original. "What more deserves to be painted by the artists of a nation than their own landscapes and the people who populate it?" he concluded.

Caras y Caretas, with Cao serving as illustrator, took the juxtaposition of agricultural richness and urban sophistication in a humorous direction, while also making visible an underlying fear of cultural inferiority in the Americas (Figure 10.2). Manuel

Figure 10.2 José María Cao, "El Indendente Ambidextro."
Source: *Caras y Caretas* 13, 616 (July 23, 1910), cover.

José Güiraldes Guerricó, who appears on the cover of the July 23, 1910 issue, was a wealthy *hacendado* (rural landowner) turned big city mayor. Güiraldes had spent his childhood and been educated in Europe, then returned to Argentina to raise horses on land inherited from his parents in the provinces. Selected by President José Figueroa Alcorta as *intendente* (mayor) of Buenos Aires in 1908, Güiraldes dedicated himself to beautifying the capital city through the widening of streets and boulevards, the development of public parks and recreational areas, and erecting sculptural monuments in honor of heroes and events from Argentine history. He also served as chief administrator for the national centenary celebrations, overseeing horse and bicycle races, agricultural exhibits, industrial displays and the exhibition of fine art. Each of the various immigrant communities resident in Argentina—the French, Italian, German, and Spanish *colonias*—promoted their history and presence in the new home as well, with the Spanish receiving particular attention when Princess Isabel, aunt of King Alfonso XIII, made an official state visit timed to mark a moment of rapprochement between the European monarchy and its former colony.

Güiraldes appears two times in the cartoon titled *El Intendente Ambidextro* (*The Ambidextrous Mayor*), first as president of the agricultural exposition and a second time as president of the exhibition of fine arts. On the left, the debonair mayor gestures at a slightly knock-kneed mount and smiles back toward the viewer. "That horse is blurry and his horseshoes are lacking in perspective," he states with the assurance of a man who knows his livestock. The *intendente* then turns his attention to the paintings on the right: "Che," he begins in excitement, using a colloquial Argentine interjection and now pointing with the opposite hand. "I don't know whether to choose this overo landscape or roan portrait." The jokes revolve around the mayor's use of specialized agricultural and artistic vocabulary to describe objects of the opposite kind; the term "overo" is used in both English and Spanish for a pinto horse with white markings over a dark background color, while a "roan" (*rosillo* in Spanish) is a horse that has white hair mixed into its primary color. The cartoonist furthers the ambidextrous nature of the mayor's action by facing his bewhiskered protagonist in opposing directions, directing the viewer's eye away from the center of the composition and out of the picture frame.

Cartoonists in Chile, a nation likewise characterized by bountiful farmland, mineral resources, and, to an even greater degree than Argentina, distance from European culture, produced their own caricatures of people using an agricultural eye to appraise fine art (Figure 10.3). Such images point to the ironies and complications in building a sophisticated national identity related to the cultural achievements of Europe, but also based on nature.

Julio Bozo, a comic artist known by his pseudonym Moustache, produced depictions of local people and customs for the Santiago-based magazine *Zig-Zag*.[29] In the top panel of his image is a prosperous landowner visiting the galleries of the International Exhibition of Fine Art. He is a well fed man of the earth, who has come in to the city from the countryside and now hopes to become an art collector. "I like this little picture of oxen. How much is it?" he asks an assistant in the gallery. "Three thousand pesos," replies the lackadaisical employee. "And in two months, how much will they ask?" continues the rancher, surprised by the painting's high price. "The same" is the disappointing reply. "Then I shall come back in two months," he concludes, "because these oxen are only half fattened up."

Figure 10.3 Moustache (Julio Bozo), "Bellas Artes."
Source: *Zig-Zag* 6, 299 (November 12, 1910), unpaginated.

The cartoon focuses on the agriculturalist's rude formation; he is a simple man who does not completely comprehend the difference between a painting and an animal, illusion and reality. A Chilean of European descent, who has probably never visited Europe, he is better able to judge cattle than a fine work of art. Turning his back on the painting of a nude reading a book hung on the adjoining wall, he may not even be literate. Below this cartoon is a second image, with an aspiring collector intently peering at the portrait he has commissioned from a young Bohemian painter. The artist, practicing one of the latest styles from Paris, has represented his model in abstract form, and the unsuspecting patron does not recognize his likeness. "What are you painting?" he asks politely. "Guess," replies the scruffy painter with a sly grin. The patron—who is balding (*pelón*)—mistakes the shape of his head for a cantaloupe (*melón*), and the mischievous artist can barely contain his mirth. A near homonym for *pelón*, *melón* is also a Spanish word used colloquially to refer to the head, as well as to an idiot. In Chile, as in the United States, most viewers preferred recognizable people and places to abstract styles of art. The academic nude on the wall of the painter's studio likewise turns away in disgust.

A transnational dialogue took place on the walls of the 1910 exhibitions as well as in the press. Lira's positive response to painting from the United States is not surprising, for he too trained in the ateliers of Paris and, like the United States, his country boasted a varied topography and rich agricultural terrain.[30] Chile, wrote one critic of Lira's work, "must be a nation of landscapes."[31] The specificity of land, moreover, was a way to express national pride and assert difference from Europe. Lira won a first place medal at the 1910 International Exposition of Fine Art. In addition to figure painting and portraiture, demonstrating his facility in painting the body, he also exhibited bucolic images of oak trees in La Quinta Normal, a large urban park developed as an agricultural reserve. Other landscapes by Lira from this time depict old country farms in Chile. The wispy brushwork and subtle coloring of his paintings appealed to urbanized Chilean collectors who regarded rural life with nostalgia and had little actual experience working the land.

Charles Morris Young exhibited *The Brandywine in Winter*, which was purchased by the Chilean government at the end of the show. Young was from Gettysburg, Pennsylvania, a town made famous during the Civil War, and he painted several views of the Brandywine River, the site of an important battle that took place during the Revolutionary War.[32] Chilean painters were likewise painting landscapes with historical resonance to exhibit at the centenary exhibition. Alberto Valenzuela Llanos submitted *Banks of the Mapocho River*, which shares with Young's work a delicate approach to color, light, and atmospheric effect as well as a connection to history: Spanish explorer Pedro de Valdivia founded the city of Santiago on the banks of the Mapocho River in 1541. In both the United States and Chile, the land could incarnate events of the past and communicate national values to present day inhabitants. Moreover, the city of Santiago had recently canalized the Mapocho River, and Valenzuela's painting evoked contemporary ecological concerns about the potential loss of the nation's natural beauty.[33] Like the landscapes that dominated the U.S. exhibition, paintings of the Chilean land provided artists and critics alike with images that resonated with memories of the past and nationalistic feeling.

Art applied to industry and industry applied to art

Complementing notions that equated Argentina, Chile, and the United States with agricultural riches were aspirations in the industrial arts, and cartoonists were quick to find humor in this theme as well (Figure 10.4). Moustache again was responsible for two images reproduced in the pages of *Zig-Zag* that address mass reproduction at the exhibition of fine art.

The first depicts a well-dressed couple, accompanied by an unassuming dog, staring intently at two stacks of paintings that have arrived from abroad. Thousands of canvases have arrived for display, and the walls of the Museo de Bellas Artes will never contain so many works. As a result, organizers have stacked them horizontally on the floor, one atop the other, creating an overriding impression of repetition and abundance. One painting is indistinguishable from the next, and they arrive *en masse*, like identical mass-produced objects from a factory. The addition of an historical exhibition has reduced the space available for the exhibition of fine art, explains

Figure 10.4 Moustache (Julio Bozo), "Exposición de Bellas Artes."
Source: *Zig-Zag* 6, 286 (August 13, 1910), unpaginated.

the caption, and the unusual display scheme will make no difference to the public, which knows little about painting anyway. Installation photographs of the galleries in Argentina and Chile reveal a more conventional display, yet an overall impression of sameness remains (Figure 10.5). Double-sided benches and sculpture pedestals were placed in the center of the room, and long carpet runners kept viewers moving along what looks like an assembly line of paintings.

Below the cartoon of paintings stacked on the floor is a second image depicting a disheveled painter, struggling to make ends meet by painting several small marine paintings at once. The room is sparsely furnished and poorly heated; the single bed is neatly made, but the window pane is broken and a candle stuck into the neck of a wine bottle provides the makeshift studio with its only source of nighttime illumination. "The Exposition of Fine Arts," explains the text, "includes a section of 'art

UNITED STATES SECTION.   SANTIAGO EXPOSITION.

Figure 10.5 "United States Section: Santiago Exposition."
Source: John E.D. Trask, *The United States Section; International Fine Arts Expositions at Buenos Aires and at Santiago* (Philadelphia: Office of the Commissioner General, 1910), unpaginated.

applied to industry,' but it should also have had a section of 'industry applied to art.' Here in Chile," it continues, "they sell beautiful seascapes for five pesos a piece. As artists don't earn enough to eat by painting one a day, they have invented some ingenious instruments to paint more than one at a time. This is industry applied to art." The painter's tool, a forked extension that attaches to three separate brushes, allows the aspiring industrialist to dip into multiple buckets of house paint and concurrently produce three delicate marine paintings, thereby tripling his earnings. The techniques of scientific management, derived from the theories of Frederick Winslow Taylor and famously implemented by Henry Ford, were widening the division between fine art and design in the early twentieth century. As with art and agriculture, placing art and industry into intimate conversation now led to humorous, even ridiculous, situations that questioned traditional hierarchies of value.

Playing the game

Following sales in Argentina and Chile—12 paintings and 8 sculptures remained in Buenos Aires, while 11 paintings and 2 sculptures stayed in Santiago—the rest of the collection was shipped from Chile around the bottom of the continent and through the Straits of Magellan to Uruguay, where the unsold works were hung in Montevideo's Municipal Pavilion, soon to become the site of that country's new art museum. Montevideo is smaller than either Buenos Aires or Santiago, but a national collection only becomes national in the context of an international display. Ambitious government officials in Uruguay therefore wanted their nation to have an international collection of art as well. They selected three works, a landscape and marine by Walter Elmer Schofield and George Albert Thompson as well as Philip Leslie Hale's spectacular life-sized nude, *The Spirit of Antique Art*, to represent the United States in Uruguay. Like many of the works now in Latin American collections, this painting has an impressive exhibition history. Hale had sent it to the annual exhibition held at the Pennsylvania Academy of the Fine Arts in 1907 and to the Corcoran Gallery the following year. The first acquisitions made in the southern cone were conservative, institutionally sanctioned works of art; there were no modernist paintings displayed at the Latin American centenary exhibitions. Robert Henri and three of his rebellious followers, George Bellows, William Glackens, and Ernest Lawson, did send one painting each, but no works by these Ashcan School artists were purchased by collectors in South America.[34]

The political and economic conflicts that kept U.S., Argentine, Chilean, and Uruguayan diplomats busy throughout the nineteenth and early twentieth centuries were mostly avoided in reviews of the fine art displayed at the 1910 exhibition. Similarities were emphasized instead. In this respect, the guiding assumption of cultural diplomacy—"that dialogue and contact among intellectuals lead to greater understanding and mutual respect among their respective societies"—was affirmed, but in an odd manner. Is an insistence on similarity always friendly? And is avoidance of difference the best way to foster dialogue resulting in deeper knowledge? In sending a group of landscapes, marines and figure paintings to Argentina, Chile and Uruguay, the United States presented itself as like the southern-most countries of Latin America. Implicitly it also presented a challenge, for paintings that are similar, like industrial products manufactured and exhibited at a world's fair, offer consumers a choice. For every work purchased from a foreign artist, a local one loses the opportunity for a sale. The exhibition of fine art at an industrial fair provides a rich opportunity for humorous reflection and brings into question assumptions of difference that insist on keeping these two realms apart.

The current exhibition context of a fourth painting from the United States that remained in Montevideo suggests the continued relevance of these questions for the twenty-first century. Edwin Morgan, the career diplomat who successfully negotiated the final stop in the South American tour, purchased Frederick Waugh's marine *On the High Seas* in 1911 and donated it to the nation's new museum the following year. Since 1980 the painting has been on long-term loan to Uruguay's Ministry of Foreign Affairs, where it hangs in the corner of the dining room, one painting amid an international company of many, performing the work of cultural diplomacy today. Thompson's *Nocturne in Quinnipiack*, which depicts several boats pulled up at dock, is also on long-term loan to the Uruguayan government; it decorates the walls of the Navy's Division of Oceanography, Hydrography and Meteorology, which provides technical assistance to the Armed Forces.[35] The paintings, hung as part of an innocuous interior design scheme, serve as gentle reminders that our nations share common values, competitive economies, and potentially explosive disagreements.

The cover of another Chilean magazine, *Sucesos*, suggests the tone of diplomatic dialogues that did occur—or could have occurred—in 1910 between the United States and its South American neighbors (Figure 10.6). *Sucesos* was published in Valparaíso, the port city from whence President Pedro Montt embarked

Figure 10.6 Carlos Wiedner, "El Afecto Yankee."
Source: *Sucesos* 8, 413 (August 4, 1910), cover.

during the centenary year on a trip to the United States and Germany. Montt was
suffering from ill health, and the journey was prompted by an unfulfilled hope
that time in the spas of Europe would restore his strength. The cover illustration, signed by German-Chilean caricaturist Carlos Wiedner, imagines a reunion
between Taft and Montt that occurred in the course of this trip.[36] A diminutive
Montt, wearing a tuxedo, top hat and white gloves, looks up at the portly U.S.
president, dressed in a wrinkled suit and bedroom slippers. A huge diamond ring
shines from Taft's corpulent finger. The meeting takes place in August, summer
holidays in the United States, and the U.S. president is vacationing in the coastal
town of Beverly, Massachusetts. Montt has therefore traveled up from New York

to meet Taft in a greenhouse, where he is busy watering a tropical flower, perhaps imported from Latin America. Taft tends his exotic plants just as he cares for the countries of South America. "But where the hell is Chile?" he asks after receiving cordial greetings from the Chilean president, who has arrived to suggest a friendly understanding (*entente cordiale*).

Montt begins to explain the location of his country, only to be rudely interrupted: "Aren't you the ones involved in the Alsop Claim?" asks Taft, in reference to a diplomatic conflict that had arisen in the late-nineteenth century from economic demands made by U.S. citizens who had invested in a Bolivian mining company. Bolivia had recognized the debt, but passed it on to Chile after losing the land in which it had been incurred at the end of the War of the Pacific in 1883. Thirty years later, the United States and Chile were still trying to reach agreement, and Taft responds to Montt's overture with disdain. He dismisses the frail Chilean, abruptly suggesting he go talk with Philander C. Knox, his Secretary of State, instead.[37] Knox, who promoted such Latin American development projects as the Panama Canal, was a staunch advocate of Dollar Diplomacy. Supported by a grant from the U.S. government, the exhibition of painting and sculpture in Argentina, Chile and Uruguay functioned as a tool of cultural diplomacy in the centenary year to smooth over economic disagreements like the Alsop Claim and defer attention from investments in Latin America such as the Panama Canal (completed in 1914), that would so obviously serve U.S. interests. Even if politicians could not productively talk and agree, art might fulfill the commitment, or at least divert attention from the problems.

Claire Fox accurately notes that

> the Americas have their own centers and peripheries, they are often better acquainted with global *metropoli* than they are with one another, and internecine competition and mutual distrust is as common as regional solidarity among the citizens of greater America.[38]

The reality of this observation is illustrated in a second Wiedner cartoon reproduced in the Chilean press (Figure 10.7). Uncle Sam, who wears a sparkling diamond on his ring finger to indicate his wealth, grabs his hat in disbelief while a Chilean *huaso* and Argentine *gaucho*—Latin American cowboys who work on the land—negotiate a balance of power by placing wooden blocks representing their countries on a map of South America. A third figure, an Afro-Brazilian dandy, is slower to act, but as the caption makes clear, he too will learn the rules and shut the United States out of the game.[39] The three countries remain ethnically and culturally differentiated from one another; at the same time they are positioned together, in opposition to the United States. The fine art works in the 1910 exhibition were selected and viewed in this context, and the responses by critics demonstrate some of the complex ways *Pan-Americanism*, *Latin Americanism*, and shifting political alliances contradict, reinforce and develop in relation to each other. Cartoons and caricatures penned by commercial artists in Argentina and Chile, moreover, bring out the ironies (and humor) of allowing fine art, a medium characterized by aesthetics, connoisseurship and individual craftsmanship, to speak, when wealth in the Americas was primarily generated and valued through the lenses of agriculture and industry.

Figure 10.7 Carlos Wiedner, "ABC Internacional."
Source: *Sucesos* 9, 428 (November 17, 1910), cover.

Notes

Many thanks to David Raizman and Ethan Robey for their productive suggestions in relation to this essay. Alex Taylor graciously invited me to present some of these ideas at the 2015 College Art Association Annual meeting. Thanks also to Roberto Amigo and María José Herrera in Argentina; Soledad Novoa and Pedro Zamorano in Chile; and Willy Rey and Enrique Aguerre in Uruguay. Luciana Erregue and Marco Katz Montiel helped me work through some of the subtleties of Argentine and Chilean humor. All translations are my own.

   1 The countries responding to Argentina's call included, in addition to the United States, Germany, Belgium, Chile, England, France, Italy, Norway, the Netherlands, Spain, and Sweden. Argentina also produced a national display of fine art. Those that joined the United States and Chile in organizing displays for the Chilean centenary included Argentina, Austria, Belgium, Brazil, England, France, Germany, Italy, the Netherlands, Portugal, Spain, and Uruguay. Argentina, with the largest population and biggest economy

of the three countries to host the U.S. display, organized what came closest to a world's fair, with pavilions spread throughout the city that brought together an impressive selection of agricultural and industrial displays.

2 Although the fine arts exhibition in Argentina remained open until November 13, the works from the United States were removed on September 1 in order to travel to Chile in time for the opening in Santiago.

3 The exact constitution of this exhibition and the paintings that remained in South America form the subject of another article that I am preparing for publication. For further discussion of the artistic conversation taking place through paintings between the United States and Chile, see my essay "'Una cualidad lírica de un encanto duradero': El diálogo de pintores estadounidenses y chilenos en el Centenario de 1910," in *Relaciones Culturales entre Estados Unidos y América Latina*, eds. Daniel Expósito Sánchez and Juan Ignacio Guijarro González (Valencia [Spain]: Biblioteca Javier Coy de la Universitat de València, forthcoming). I curated an exhibition with this same title for the Museo Nacional de Bellas Artes, Santiago de Chile, which was on view from March 18 to May 18, 2014.

4 The literature on the 1910 centenaries is rich. See especially the series of essays published in *Apuntes* (Pontificia Universidad Javeriana, Bogotá) 19, 2 (July–December 2008); a second series of essays in *Historia Mexicana* 60, 1 (July–September 2010); and Javier Moreno Luzón and Rodrigo Gutiérrez Viñuales, eds. *Memorias de la independencia: España, Argentina y México en el primer centenario (1908–1910–1912)* (Madrid: Acción Cultural Española, [2012]).

5 Paul Greenhalgh points out the inflated reputation of the fine arts exhibitions in what were essentially agricultural and industrial fairs. See *Ephemeral Vistas: The Expositions Universelles, Great Exhibitions and World's Fairs, 1851–1939* (Manchester: Manchester University Press, 1988), 198.

6 An introduction to the particularities of each relationship may be found in Thomas M. Leonard, ed. *United States-Latin American Relations: Establishing a Relationship, 1850–1903* (Tuscaloosa: University of Alabama Press, 1999). See especially the essays by Joseph S. Tulchin, "Argentina: Clash of Global Visions I," William F. Sater, "Chile: Clash of Global Visions II," and José B. Fernández and Jennifer M. Zimnoch, "Paraguay and Uruguay: On the Periphery II."

7 Benedict Anderson, *Imagined Communities: Reflections on the Origin and Spread of Nationalism* (Rev. ed., London: Verso Books, 1991), 9–36. For the usefulness of Anderson's thesis for the study of Latin America, see John Charles Chasteen, "Introduction: Beyond Imagined Communities," in *Beyond Imagined Communities: Reading and Writing the Nation in Nineteenth-Century Latin America*, ed. Sara Castro-Klarén and Chasteen (Baltimore, MD: Johns Hopkins University Press, 2003), ix–xxv.

8 Claire F. Fox, *Making Art Panamerican: Cultural Policy and the Cold War* (Minneapolis: University of Minnesota Press, 2013), xvi. For a further introduction to cultural diplomacy, see Jessica C. E. Gienow-Hecht and Mark C. Donfried, eds. *Searching for a Cultural Diplomacy* (New York: Berghahn Books, 2010), 3–29.

9 Michele Greet makes this point in *Beyond National Identity: Pictorial Indigenism as a Modernist Strategy in Andean Art, 1920–1960* (University Park: Pennsylvania State University Press, 2009), 7.

10 Benjamin A. Coates provides a succinct explanation of the economic and cultural interpretations of *Pan-Americanism* in "The Pan-American Lobbyist: William Eleroy Curtis and U.S. Empire, 1884–1899," *Diplomatic History* 38, no. 1 (2014): 23–24. See also Alan McPherson, "Introduction: *Antiyanquimo*: Nascent Scholarship, Ancient Sentiments," in *Anti-Americanism in Latin America and the Caribbean* (New York: Berghahn Books, 2006), 1–34.

11 Cao formed the subject of a comprehensive exhibition and catalogue by Julio Neveleff, *José María Cao: Padre de la caricatura argentina* (Exh. cat., Buenos Aires: Museo de Artes Plásticas Eduardo Sívori, 2011).

12 Darlene J. Sadlier focuses on Rockefeller and the 1940s in *Americans All: Good Neighbor Cultural Diplomacy in World War II* (Austin: University of Texas Press, 2012). On the CIA and the 1950s, see Frances Stonor Saunders, *The Cultural Cold War: The CIA and World of Arts and Letters* (New York: The New Press, 1999).

13 Several key essays on this topic are reprinted in *Pollock and After: The Critical Debate*, ed. Francis Frascina, 2nd ed. (London: Routledge, 2000). See also Alex J. Taylor, "Unstable Motives: Propaganda, Politics, and the Late Work of Alexander Calder," *American Art* 26, no. 1 (spring 2012): 24–47.

14 The means by which the displays were selected in other countries varied widely. The Spanish displays in Argentina, for example, were organized primarily by Spanish businessmen living in Buenos Aires. José Artal, who imported paintings from Spain for his gallery in Buenos Aires, was especially active in this instance. See my essay, "'A Renewal of the Fraternal Relations that Shared Blood and History Demand': Latin American Painting, Spanish Exhibitions, and Public Display at the 1910 Independence Celebrations in Argentina, Chile, and Mexico," *Revue d'art canadien/Canadian Art Review (RACAR)* 38, no. 2 (Fall 2013): 90–108.

15 "American Art in South America," *Art and Progress* 1, no. 7 (May 1910): 207. The Bureau of American Republics was renamed the Pan American Union in 1910, the same year the Pan American Union Building was constructed in Washington, DC, in 1910. The organization was reconstituted in 1948 as the Organization of American States (OAS).

16 The three catalogues are *Exposición Internacional de Arte del Centenario; Buenos Aires 1910; Catálogo Ilustrado* (Buenos Aires: Est. Gráfico M. Rodríguez Giles, 1910); *Exposición Internacional de Bellas Artes, Santiago de Chile: Catálogo Oficial Ilustrado* (Santiago: Imprenta Barcelona, 1910); and *Exposición de pinturas y bronces por artistas norte-americanos; Catálogo* (Montevideo: Talleres gráficos A. Barreiro y Ramos, 1911). The Argentine and Chilean catalogues are readily available, but I have only located one example of the Uruguayan catalogue, at the Biblioteca Nacional in Montevideo.

17 John E.D. Trask, *The United States Section; International Fine Arts Expositions at Buenos Aires and at Santiago* (Philadelphia, PA: Office of the Commissioner General, 1910), n.p.

18 Exhibition announcements specified delivery of entries by March 31. *American Art News* 8, 23 (March 19, 1910) and *American Art News* 8, 24 (March 26, 1910). The exhibition was also announced in *Art and Progress* 1, 4 (February 1910): 106.

19 [John E.D. Trask], "Estados Unidos," in *Exposición Internacional de Arte del Centenario; Buenos Aires 1910*, 101–104.

20 Only a few paintings appear to have depicted the West: Albert Groll, who moved to Arizona in 1905, sent a painting called *Lloviznando en Arizona* (*Rainy Day in Arizona*) and California painter William Keith sent *El regreso al hogar* (*The Return Home*). Francis S. Mollhenny sent a painting of a miner by Frederic Remington, which may have depicted the West as well. Because the catalogues in the three venues were produced independently, the Spanish titles vary from list to list. These titles are from the Argentine catalogue.

21 Pedro Lira, "La Exposición Internacional de Bellas Artes," *Selecta* 2, no. 7 (octubre 1910): 271–273.

22 "La exposición artística norteamericana," *La Tribuna Popular* (Montevideo), 6 febrero 1911, 4.

23 Rodó's essay appeared in Spanish in 1900 and was translated into English shortly after. See José Enrique Rodó, *Ariel*, trans. F.J. Stimson (Boston, MA: Houghton Mifflin Company, 1922).

24 Natalia Majluf discusses this problematic in the context of Peruvian art displayed at the 1855 Exposition Universelle. See "'Ce n-est pas le Pérou', or, the Failure of Authenticity: Marginal Cosmopolitans at the Paris Universal Exhibition of 1855," *Critical Inquiry* 23, no. 4 (1997): 868–893.

25 Neither painting was sent by the artist: George De Forest Brush's *Un piel roja con un lirio o un indiano* (probably *The Indian and the Lily*, 1887; lent by Dr George Woodward and now in the Crystal Bridges Museum) and Henry Farny's *Apaches traidores a su causa* (*Renegade Apaches*, 1892, lent by the Cincinnati Museum).

26 [Trask], "Estados Unidos," 103.

27 Among the many sources that discuss the complex changes occurring on farms, in factories, and in city, see Susan G. Larkin, *American Impressionism: The Beauty of Work* (Greenwich, CT: Bruce Museum of Arts and Science, 2005) and Rebecca Zurier, *Picturing the City: Urban Vision and the Ashcan School* (Berkeley: University of California Press,

2006). The classic introduction to this theme is Alan Trachtenberg, *The Incorporation of America* (New York: Hill and Wang, 1982).

28 Godofredo Daireaux, "La Exposición I. de Arte: Impresiones," *Athinae: Revista Argentina de Bellas Artes*, año III, no. 2 (julio 1010): 11.

29 See "Julio Bozo (Moustache) (1879–1942)," in *Memoria chilena*, Biblioteca Nacional de Chile, at www.memoriachilena.cl/602/w3-article-4927.html, accessed July 18, 2017.

30 On Pedro Lira and painting in Chile more generally, see Ana Francisca Allamand, *Pedro Lira: El maestro fundador* (Santiago: Origo Ediciones, 2008) and Ricardo Bindis, *Pintura chilena: Doscientos años* (Santiago: Origo Ediciones, 2006).

31 "Chile, por su naturaleza, tiene que ser tierra de paisajistas. Si alguna vez en pintura tenemos escuela propia, será escuela de paisaje" (Chile, because of its nature, must be a nation of landscapes. If we ever have a painting school of our own, it will be a landscape school.) Benjamín Vicuña Subercaseaux, "Don Pedro Lira," *Revista Selecta* 2 (May 1909): n.p.

32 See Charles Teaze Clark, "In Pursuit of Higher Truth: The Landscape Paintings of Charles Morris Young," *Antiques* 168 (November 2005): 163–169.

33 Gloria Cortés Aliaga, "El paisaje habitado: de geografía humana a objeto de deseo," in *Puro Chile: Paisaje y territorio* (Santiago: Centro Cultural de La Moneda, 2014), 201. See also Ana Francisca Allamand, *Alberto Valenzuela Llanos: Visión entrañable del mundo rural* (Santiago: Origo Ediciones, 2008). Ecocritical studies in the history of art are just beginning to develop. For an anthology of essays addressing the topic in the context of art in the United States, see Alan C. Braddock and Christoph Irmscher, eds. *A Keener Perception: Ecocritical Studies in American Art History* (Tuscaloosa: University of Alabama Press, 2009).

34 Robert Henri did receive a silver medal for *Willie Gee* (1904; Newark Museum). George Bellows sent a painting called *La lluvia en el río* (*Rain on the River*), William Glackens *El Puerto de Portsmouth* (*Portsmouth Harbor*), and Ernest Lawson *Sepulcro del General Grant* (*Grant's Tomb*). Thanks to Valerie Leeds for confirming that the Newark painting was indeed the painting sent to South America.

35 I appreciate the assistance of Sr Alejandro Mongrel and Sr Oscar de María at the Ministerio de Asuntos Experiores and of Sra María del Carmen Harriett Giúdice and Capt Claudio López at the Servicio de Oceanografía, Hidrografía y Meteorología de la Armada for facilitating my visit to these government buildings in Montevideo.

36 See "Carlos Wiedner," in *Memoria chilena*, Biblioteca Nacional de Chile, at www.memoriachilena.cl/602/w3-article-94663.html, accessed July 18, 2017.

37 Montt did indeed meet with Taft and Knox in Beverly, MA. A photograph of the two presidents in the garden of Taft's summer home was reproduced in *Zig-Zag*, October 15, 1910. Among the dignitaries present in the photograph are the wives of both men, as well as Knox.

38 Fox, *Making Art Panamerican*, xiv.

39 "Tío Sam. —Chile y la Argentina ya han colocado sus letras; como encuentren la otra aprenden el juego y me la juegan estos muchachos." (Uncle Sam. —Chile and Argentina have already placed their letters; look how these boys play each other, learn to play the game, and play against me!).

Bibliography

Introduction

Adamson, Glenn, Giorgio Riello, and Sarah Teasley. *Global Design History*. London: Routledge, 2011.

Anderson, Benedict. *Imagined Communities: Reflections on the Origin and Spread of Nationalism*. Revised edition. London and New York: Verso, 1983, 2006.

Banham, Reyner. *Design by Choice/Reyner Banham*, edited by Penny Sparke. New York: Rizzoli, 1981.

Bennett, Tony. "The Exhibitionary Complex." *New Formations*, no. 4 (spring 1988): 73–102.

Cole, Henry. *Lectures on the Results of the Great Exhibition of 1851*. Second Series. London: David Bogue, 1853.

Filipová, Marta. *Cultures of International Exhibitions, 1840–1940: Great Exhibitions in the Margins*. Farnham and Burlington: Ashgate, 2015.

Gellner, Ernest. *Nations and Nationalism*. Ithaca, NY: Cornell University Press, 1983.

Geppert, Alexander C.T., Jean Coffey, and Tammy Lau. *International Exhibitions, Expositions Universelles and World's Fairs, 1851–2005: A Bibliography*. 3rd ed. (November 2006). http://fresnostate.edu/library/subjectresources/specialcollections/worldfairs/ExpoBibliography3ed.pdf.

"The Great Exhibition and its Results." *Exhibition Supplement to The Illustrated London News* 19, no. 523 (11 October 1851): 457–458.

Greenhalgh, Paul, ed. *Art Nouveau: 1890–1914*. London: V & A Publications, 2000.

Greenhalgh, Paul. *Ephemeral Vistas: The Expositions Universelles, Great Exhibitions, and World's Fairs, 1851–1939*. Manchester: Manchester University Press, 1988.

Goury, Jules. *Plans, Elevations, Sections, and Details of the Alhambra*. London: Owen Jones, 1842–1845.

Heskett, John. *Industrial Design*. New York: Oxford University Press, 1980.

Huppatz, Daniel. "Globalizing Design History and Global Design History." *Journal of Design History* 28, no. 2 (May 2015): 182–202.

Jones, Owen. *The Grammar of Ornament, Illustrated by Examples from Various Styles of Ornament*. London: Day and Son, 1856.

Mainardi, Patricia. *Art and Politics of the Second Empire: The Universal Expositions of 1855 and 1867*. New Haven, CT: Yale University Press, 1987.

Mandell, Richard D. *Paris 1900: The Great World's Fair*. Toronto, ON: University of Toronto Press, 1967.

Mitchell, Timothy. "The World as Exhibition." *Comparative Studies in Society and History* 31, no. 2 (April 1989): 217–236.

Pevsner, Nikolaus. *Pioneers of Modern Design: From William Morris to Walter Gropius*. London: Faber and Faber, 1936. 4th edition, revised and expanded. New Haven, CT: Yale University Press, 2005.

Purbrick, Louise, ed. *The Great Exhibition of 1851: New Interdisciplinary Essays*. Manchester and New York: Manchester University Press, 2001.

Raizman, David. "Giuseppe Ferrari's Carved Cabinet for the 1876 Centennial Exhibition: Presentation Furniture in the Cultural Context of World's Fairs." *West 86th: A Journal of Decorative Arts, Design History, and Material Culture* 20, no. 1 (2013): 62–91.

Richards, Thomas. *The Commodity Culture of Victorian England: Advertising and Spectacle 1851–1914*. Stanford, CA: Stanford University Press, 1990.

Rydell, Robert W. *All the World's a Fair: Visions of Empire at American International Expositions, 1876–1916*. Chicago, IL: University of Chicago Press, 1984.

Rydell, Robert. *The Books of the Fairs-Materials about the World's Fairs, 1834–1916, in the Smithsonian Institution Libraries*. Chicago: The American Library Association, 1992.

Silverman, Debora L. "Art Nouveau, Art of Darkness: African Lineages of Belgian Modernism." *West 86th: A Journal of Decorative Arts, Design History, and Material Culture* 18, no. 2 (Fall-Winter 2011): 139–181; 19, no. 2 (Fall-Winter 2012): 175–195; 20, no. 1 (Spring-Summer 2013): 3–61.

Snodin, Michael and John Styles. *Design & the Decorative Arts: Victorian Britain 1837–1901*. London: V & A Publications, 2004.

Sparke, Penny. *An Introduction to Design and Culture in the Twentieth Century*. New York: Harper and Row, 1986.

Whiteley, Nigel. *Reyner Banham, Historian of the Immediate Future*. Cambridge, MA: MIT Press, 2002.

Chapter 1

AlSayyad et al., eds. *Making Cairo Medieval*. New York: Lexington Books, 2005.

The Art-Journal Illustrated Catalogue of the Exhibition of the Works of Industry of All Nations. London: George Virtue, 1851.

Auerbach, Jeffrey. *The Great Exhibition of 1851: A Nation on Display*. New Haven, CT: Yale University Press, 1999.

Auerbach, Jeffrey and Peter Hoffenberg, eds. *Britain, the Empire and the World at the Great Exhibition of 1851*. Aldershot: Ashgate, 2008.

Benjamin, Walter. *Reflections: Essays, Aphorisms, Autobiographical Writings*, edited by Peter Jernetz, and translated by Edmund Jephcott. New York: Harcourt Brace Jovanovich, 1978.

Bennett, Tony. *The Birth of the Museum: History, Theory, Politics*. London: Routledge, 1995.

Blake, William P., ed. *Reports of the U.S. Commissioners to the 1867 Paris Universal Exposition*. Washington, DC: U. S. Government Printing Office, 1870.

Brain, David. "Discipline and Style: The École des Beaux Arts and the Social Production of American Architecture." *Theory and Society* 198, no. 6 (November 1989): 807–868.

Brown, Carl. *The Tunisia of Ahmed Bey 1837–1855*. Princeton, NJ: Princeton University Press, 1974.

Buel, James W. *The Magic City: A Massive Portfolio of Original Photographic Views of the Great World's Fair*. St. Louis, MO and Philadelphia, PA: The Historical Publishing Co., 1894.

Busch, Jason and Catherine Futter, eds. *Inventing the Modern World Decorative Arts at the World's Fairs, 1851–1939*. New York: Rizzoli, 2012.

Buzard, James, Joseph W. Childers, and Eileen Gillooly, eds. *Victorian Prism: Refractions of the Crystal Palace*. Charlottesville, VA: University of Virginia Press, 2007.

Catalogue of the Turkish Section of the Great Exhibition. London: McKewan & Co., 1851.

Çelik, Zeynep. "Bouvard's Boulevards: Beaux-Arts Planning in Istanbul." *Journal of the Society of Architectural Historians* 43, no. 4 (December 1984): 341–355.

Çelik, Zeynep. *Displaying the Orient: Architecture of Islam at the Nineteenth-Century World's Fairs*. Berkeley: University of California Press, 1992.

Çelik, Zeynep and Leila Kinney. "Ethnography and Exhibitionism at the Expositions Univer-selles." *Assemblage* no. 13 (1990): 34–59.

Chatty, Dawn. *From Camel to Truck: The Bedouin in the Modern World.* New York: Vantage Press, 1986.

Chatty, Dawn, and William Young. "Bedouin." In *Encyclopedia of World Cultures,* edited by David Levinson et al., vol. 9, "Africa and the Middle East," 42–46. New York: Macmillan Reference, 1996.

Coleman, Peter. *Shopping Environments-Evolution, Planning, and Design.* London: Routledge Architectural Press, 2006.

Cooke, Miriam. *Tribal Modern: Branding New Nations in the Arab Gulf.* Berkeley: University of California Press, 2014.

Cowper, Charles and Charles Downes. *The Building Erected in Hyde Park for the Great Exhibition of 1851.* London: Woodfall and Sons, 1852.

Dickinson Brothers, *Dickinsons' Comprehensive Pictures of the Great Exhibition of 1851.* London: Dickinson Bros., 1852.

Dilke, Charles Wentworth. *Reminiscences of the Crystal Palace.* London: George Routledge & Co., 1852.

Dyce, William. "Universal Infidelity in Principles of Design." *Journal of Design and Manufactures* 5, no. 28 (1851): 158–161.

Edwards, Holly, et al. *Noble Dreams, Wicked Pleasures Orientalism in America 1870–1930.* Princeton, NJ: Princeton University Press, 2000.

Ersoy, Ahmet. "Osman Hamdi Bey and the Historophile Mood: Orientalist Vision and the Romantic Sense of the Past in Late Ottoman Culture." In *The Poetics and Politics of Place: Ottoman Istanbul and British Orientalism,* edited by Zeynep Inankur, Reina Lewis, and Mary Roberts, 144–155. Istanbul: Pera Muzesi, 2011.

Glassner, Martin. "The Bedouin of the Southern Sinai." *Geographic Review* 64, no. 1 (1974): 31–60.

Great Exhibition of the Works of Industry of All Nations, 1851, Official Descriptive and Illustrated Catalogue. London: Spicer Bros., 1851.

Haddawy, Husain, trans. *The Arabian Nights.* New York: Alfred A Knopf, 1990.

"How We Hunted the Prince." *Punch,* no. 20 (June 1851): 222.

Irwin, Robert. *For the Lust of Knowing: The Orientalists and their Enemies.* London: Penguin Books, 2006.

Jones, Owen. "Gleanings from the Great Exhibition of 1851." *Journal of Design and Manufactures* 5, no. 28 (1851): 89–93.

Jones, Owen. "The Interior Decoration of the Crystal Palace." *Bulletin of the American Art Union* no. 2 (May 1851): 28–29.

Mitchell, Timothy. "Orientalism and the Exhibitionary Order." In *The Visual Culture Reader,* edited by Nicholas Mirzoeff. London: Routledge, 1998.

Muravchik, Joshua. "Enough Said: The False Scholarship of Edward Said." *World Affairs* 175, no. 6 (March 2013): 9–21.

Nance, Susan. *How the Arabian Nights Inspired the American Dream 1790–1935.* Chapel Hill: University of North Carolina Press, 2009.

Nochlin, Linda. "The Imaginary Orient," In *The Politics of Vision: Essays on Nineteenth Century Art and Society,* 33–59. New York: Harper and Row, 1983.

Ormos, István. "The Cairo Street at the World's Columbian Exposition, Chicago, 1893." In *L'Orientalisme architectural entre imaginaires et saviors,* edited by Nabila Oulebsir and Mercedes Volait, 195–214. Paris: Picard, 2009.

Quateret, Donald. "Clothing Laws, State, and Society in the Ottoman Empire." *International Journal of Middle Eastern Studies* 29 (1997): 403–425.

Ralph, Julian. *Harper's Chicago and the World's Fairs.* New York: Harper and Bros., 1893.

Revault, Jacques. *Designs and Patterns in North African Textiles.* New York: Dover Press, 1973.

Said, Edward. *Orientalism.* New York: Vintage Books, 1978.

Sanders, Paula. "Islam for the Modern World: Medieval Cairo between Egyptian Reforms and British Critics." In *Creating Medieval Cairo.* Cairo: American University in Cairo Press, 2008.

Shorter, Clement. *The Brontes: Life and Letters*. London: Haddon Co., 1908.

Spring, Christopher and Hudson, Julie. *North African Textiles*. London: The British Museum Press, 1995.

Tallis, John. *Tallis's History and Description of the Crystal Palace*. London: Tallis & Sons, 1852.

Warren, Samuel. *The Lily and the Bee: An Apologue of the Crystal Palace*. London: Blackwood, 1851.

Young, Paul. *Globalization and the Great Exhibition*. New York: Palgrave Macmillan, 2009.

Chapter 2

Abu-Lughod, Janet. *Cairo: 1001 Years of the City Victorious*. Princeton, NJ: Princeton University Press, 1971.

AlSayyad, Nezar. *Cairo: Histories of a City*. Cambridge, MA: Belknap Press, 2011.

AlSayyad, Nezar, Irene A. Bierman, and Nasser Rabbat, eds. *Making Cairo Medieval*. New York: Lexington Books, 2005.

Amtlicher Bericht über die Industrie-und Kunst-Ausstellung zu London im Jahre 1862, erstattet nach Beschluß der Kommissarien der Deutschen Zollvereins-Regierungen, XVI. Heft. Berlin: Verlag der Königlichen Geheimen Ober-Hofbuchdruckerei, 1864.

Berman, Nina. *German Literature on the Middle East: Discourses and Practices, 1000–1989*. Ann Arbor, MI: University of Michigan Press, 2011.

Berman, Nina. "Historische Phasen orientalisierender Diskurse in Deutschland." In *Orient-und Islam Bilder: interdisziplinäre Beiträge zu Orientalismus und antimuslimischem Rassismus*, edited by Iman Attia, 71–84. Muenster: Unrast, 2007.

Berque, Jacques. *Egypt: Imperialism & Revolution*. Translated by Jean Stewart. New York: Praeger, 1972.

Bohle-Heintzenberg, Sabine. "Die Dampfkraft in der Parklandschaft." In *Ludwig Persius, Architekt des Königs: Baukunst unter Friedrich Wilhelm IV*, edited by Stefan Gehlen, 73–80. Potsdam: SPGS, 2003.

Bohle-Heintzenberg, Sabine and Manfred Hamm. *Architektur & Schönheit: Die Schinkelschule in Berlin und Brandenburg*. Berlin: Transit Buchverlag, 1997.

Börsch-Supan, Eva. *Berliner Baukunst nach Schinkel 1840–1870*. Munich: Prestel, 1977.

Börsch-Supan, Eva, ed. and Ludwig Persius. *Das Tagebuch des Architekten Friedrich Wilhelms IV., 1840–1845*. Munich: Deutscher Kunstverlag, 1980.

Brose, Eric Dorn. *The Politics of Technological Change in Prussia: Out of the Shadow of Antiquity, 1809–1848*. Princeton, NJ: Princeton University Press, 1993.

Channing Downs, Arthur, Jr. "Zinc for Paint and Architectural Use in the 19th Century." *Bulletin of the Association for Preservation Technology* 8, no. 4 (1976): 81–82.

Coste, Pascal-Xavier. *Architecture arabe, ou Monuments du Kaire, mesurés et dessinés, de 1818 à 1826*. Paris: Typ. de Firmin Didot frères et compagnie, 1839.

Crinson, Mark. *Empire Building: Orientalism & Victorian Architecture*. London: Routledge, 1996.

Davis, John R. *Britain and the German Zollverein, 1848–66*. New York: St. Martin's Press, 1997.

Dishon, Dale. "South Kensington's Forgotten Palace: The 1862 International Exhibition Building." In *Die Weltausstellung von 1851 und ihre Folgen*, edited by Franz Bosbach and John R. Davis, 315–324. Munich: Saur, 2002.

Dobbert, Eduard. *Chronik Der Königlichen Technischen Hochschule Zu Berlin, 1799–1899*. Berlin: Verlag von Wilhelm Ernst & Sohn, 1899.

Douin, Georges. *Histoire du règne du khédive Ismaïl*. Roma: Stampata… nell Ìstituto poligrafico dello stato per la Reale società di geografia d'Egitto, 1933.

"Dr. L. zu Berlin." "Bauwissenschaftliche und Kunst-Nachrichten: Siebente Versammlung deutscher Architekten und Ingenieure zu Braunschweig vom 26. Bis 29. Mai 1852. Bericht von Herrn Dr. L. zu Berlin." *Zeitschrift für Bauwesen* 2 (1852): 334.

Ebers, Georg. *Ägypten in Bild und Wort: Dargestellt von unseren ersten Künstlern*. Stuttgart and Leipzig: E. Hallberger, 1879–1880.

Farag, Maged M. *The Palace Since 1869*. Researched, compiled and designed by Maged M. Farag; additional research by Fayza Hassan and Samir Raafat. Cairo: Max Group, 1999.

Fehle, Isabella. *Der Maurische Kiosk in Linderhof von Karl von Diebitsch: ein Beispiel für die Orientmode im 19. Jahrhundert*. Munich: Kommissionsverlag UNI-Druck, 1987.

Flores, Carol A. Hrvol. *Owen Jones: Design, Ornament, Architecture, and Theory in an Age in Transition*. New York: Rizzoli, 2006.

Francastel, Pierre. *Art & Technology in the Nineteenth and Twentieth Centuries*. Translated by Randall Cherry. New York: Zone Books, 2003. Originally published as *Art et technique au XIXe et XX siècle*. (Paris: Éditions de Minuit, 1956.)

Franz, Julius. "Cairo's Neubauten. Mitgetheilt vom Ober-Baudirektor des Khedive J. Franz Bey in Cairo." *Zeitschrift für praktische Baukunst* 31 (1871): 193–197 and 325–330.

Franz, Julius. *Kairo*. Leipzig: E. A. Seemann, 1903.

Gautier, Hypolyte. *Les Curiosités de L'Exposition universelle de 1867*. Paris: CH. Delagrave et Cie, 1867.

Grabar, Oleg. *The Alhambra*. Cambridge, MA: Harvard University Press, 1978.

Green, Abigail. "The Representation of the German States at the Great Exhibition." In *Die Weltausstellung von 1851 und ihre Folgen*, edited by Franz Bosbach and John R. Davis, 267–278. Munich: Saur, 2002.

Greenhalgh, Paul. *Fair World: A History of World's Fairs and Expositions, from London to Shanghai, 1851–2010*. Winterbourne: Papadakis, 2011.

Guémard, Gabriel. "Le Tombeau et les 'armes parlantes' de Soliman Pacha." *Bulletin de l'Institut d'Egypte* 9 (1927): 72–73.

Guindi, Georges and Jacques Tagher, eds. *Ismail d'après les documents officiels: avec avant-propos et introduction historique*. Cairo: Imprimerie de l'Institut français d'Archéologie orientale, 1946.

Hafemann, Ingelore, ed. *Preußen in Ägypten & Ägypten in Preußen*. Berlin: Kulturverlag Kadmos, 2010.

Hagedorn, Annette. "'Thier weiß u braun' Die Zeichnung der nasridischen Gazellenvase von Carl von Diebitsch, 1846–1847." In *Von Gibraltar bis zum Ganges: Studien zur islamischen Kunstgeschichte in memoriam Christian Ewert*, edited by Christian Ewert, Martina Müller-Wiener, and Marion Frenger, 59–74. Berlin: EB-Verlag, 2010.

Hahn, Hans-Werner and Marko Kreutzmann. *Der deutsche Zollverein: Ökonomie und Nation im 19. Jahrhundert*. Cologne: Böhlau, 2012.

Haubrich, Rainer. "Kunst am Bau: Die Welt durch Schönheit veredeln: Karl Friedrich Schinkel beeinflusste Generationen von Architekten. Jörg Trempler verrät, warum." *Die Welt*. September 18, 2012.

Hawwas, Sohayr Zaki. *Khedivian Cairo: Identification and Documentation of Urban-Architecture in Downtown Cairo*. Al-Muhandisn [Cairo]: Markaz al-Tammt al-Mimryah, 2002.

Henderson, W.O. *The Rise of German Industrial Power, 1834–1914*. Berkeley: University of California Press, 1975.

Henderson, W.O. *The Zollverein*. Cambridge [Eng.]: University Press, 1939.

Hodkinson, James R, and Jeffrey Morrison, eds. *Encounters with Islam in German Literature and Culture*. Rochester, NY: Camden House, 2009.

Holynski, Alexandre. *Nubar Pacha, devant l'histoire*. Paris: E. Dentu, 1885.

Hübsch, Heinrich. *In Welchem Style Sollen Wir Bauen?* Karlsruhe: C.F. Müller, 1828.

Hunter, Robert. *Egypt under the Khedives, 1805–1879: From Household Government to Modern Bureaucracy* [1984]. Cairo: American University in Cairo Press, 1999.

Ilbert, Robert and Mercedes Volait. "Neo-Arabic Renaissance in Egypt, 1870–1930." *Mimar Architecture in Development* 13 (1984): 26–34.

International Exhibition, 1862. Medals and Honourable Mentions awarded by the International Juries with a list of Jurors, and the Report of the Council of Chairmen. 2nd edition. London: Geo. E. Eyre and Wm. Spottiswoode Printers, 1862.

International Exhibition 1862: Official Catalogue of the Industrial Department. London: Truscott Son & Simmons, 1862.

Koppelkamm, Stefan. *Der Imaginäire Orient: Exotische Bauten des achtzehnten und neuntzehen Jahrhunderts in Europa.* Berlin: Ernst & Sohn, 1987.

Kontje, Todd. *German Orientalisms.* Ann Arbor, MI: University of Michigan Press, 2004.

Köster, Cornelia and Wolfgang Schwanitz. "Kunsthistoriker, Architekten, Ingenieure und Manager: Ein Gespräch mit Elke Pflugradt-Abdel Aziz." In *125 Jahre Sueskanal: Lauchhammers Eisenguss am Nil*, edited by Wolfgang G. Schwanitz, 34–54. Hildesheim: G. Olms, 1998.

"Kunst-Chronik." *Die Dioskuren: Deutsche Kunstzeitung* 7, no. 17 (27 April 1862): 132.

Landes, David. *Bankers and Pashas: International Finance and Economic Imperialism in Egypt.* Cambridge, MA: Harvard University Press, 1979.

Lübke, Wilhelm, "Wilhelm Stier. Nekrolog." *Zeitschrift für Bauwesen* 7 (1857): 92.

Marchand, Suzanne L. *German Orientalism in the Age of Empire: Religion, Race, and Scholarship.* Washington, DC: German Historical Institute, 2009.

Mangold, Sabine. *Eine "weltbürgerliche Wissenschaft" – Die deutsche Orientalistik im 19. Jahrhundert.* Stuttgart: Steiner, 2004.

Morgan, Ihab. *Kairo: Die Entwicklung des modernen Stadtzentrums im 19. und frühen 20. Jahrhudert.* Bern: Peter Lang, 1999.

Nägelke, Hans-Dieter. "Carl von Diebitsch (1819–1869) Entwurf für eine Villa in Kairo." In *Architekturbilder 125 Jahre Architekturmuseum der Technischen Universität Berlin*, edited by Hans-Dieter Nägelke and translated by Catherine Framm, 64–65. Kiel: Verlag Ludwig, 2011.

Official Catalogue of the Fine Art Department (International Exhibition). London: Truscott, Son & Simmons, 1862.

Ohnesorge, Carl. *Orientalische Skizzen: Unsers Vaters Erinnerungen an sein Arbeiten un Wandern im Orient 1863–65 zu seinem siebzigsten Geburtstage dem 17. Juli 1908.* Collected and presented by his children. Magdeburg: A Wohlfeld, 1908.

Osterhammel, Jürgen. *The Transformation of the World: A Global History of the Nineteenth Century.* Princeton, NJ: Princeton University Press, 2014.

Oulebsir, Nabila and Mercedes Volait. *L'orientalisme architectural entre imaginaires et saviors, textes réunis par Nabila Oulebsir et Mercedes Volait.* Paris: Picard, CNRS, 2009.

Phillips-Matz, Mary Jane. *Verdi: A Biography.* London and New York: Oxford University Press, 1993.

Pflugradt-Abdel Aziz, Elke. "Der Preußische Palast in Ägypten." In *125 Jahre Sueskanal: Lauchhammers Eisenguss am Nil*, edited by Wolfgang G. Schwanitz, 55–77. Hildesheim: G. Olms, 1998.

Pflugradt-Abdel Aziz, Elke. *Islamisierte Architektur in Kairo: Carl von Diebitsch und der Hofarchitekt Julius Franz-Preußisches Unternehmertum im Ägypten des 19. Jahrhunderts.* Bonn: Universität Bonn, 2003.

Pflugradt-Abdel Aziz, Elke. "The Mausoleum for Soliman Pasha 'el-Faransawi' in Cairo." *Mitteilungen des deutschen archaologischen Instituts: Abteilung Kairo* 44 (1988): 210–212.

Pflugradt-Abdel Aziz, Elke. "Orientalism as an Economic Strategy: The Architect Carl von Diebitsch in Cairo (1862–1869)." In *Le Caire – Alexandrie: Architectures européennes 1850–1950*, edited by M. Volait, 3–22. Cairo: Institut Français d'Archéologie Orientale, 2001.

Pflugradt-Abdel Aziz, Elke. "A Proposal by the Architect Carl von Diebitsch (1819–1869): Mudejar Architecture for a Global Civilization." In *L'Orientalisme architectural entre imaginaires et saviors*, edited by Nabila Oulebsir and Mercedes Volait, 69–88. Paris: Picard, 2009.

Pflugradt-Abdel Aziz, Elke. *A Prussian Palace in Egypt*. Cairo: Museum of Egyptian Modern Arts, 1993.

Polaschegg, Andrea. *Der andere Orientalismus: Regeln deutsch-morgenländischer Imagination im 19. Jahrhundert*. Berlin: W. de Gruyter, 2005.

Prangey, Girault de. *Souvenirs de Grenade et de l'Alhambra: monuments arabes et moresques de Cordoue, Séville et Grenade, dessinés et mesurés en 1832 et 1833*. Paris: Veith et Hauser, 1837.

Rabbat, Nasser O. "The Formation of the Neo-Mamluk Style in Modern Egypt." In *The Education of the Architect: Historiography, Urbanism, and the Growth of Architectural Knowledge*. Essay Presented to Stanford Anderson, edited by Martha Pollak, 363–386. Cambridge, MA: MIT Press, 1997.

Raymond, André. *Cairo*. Translated by Willard Wood. Cambridge, MA: Harvard University Press, 2000.

Rohde, Elisabeth. "Lehrer und Schüler der Schinkelschen Bauakademie: Ein Beitrage zur Stadtgeschichte Berlins." In *Karl Friedrich Schinkels Berliner Bauakademie*, edited by Elke Blauert, 87–108. Berlin: Nicolaische Verlagsbuchhandlung Beuermann GmbH, und Kunstbibliothek Staatliche Museen zu Berlin – Preußischer Kulturbesitz, 1996.

Scharabi, Mohamed. *Kairo: Stadt und Architektur im Zeitalter des europaischen Kolonialismus*. Tuebingen: Wasmuth, 1989.

Schmitz, Hermann. *Berliner Baumeister vom Ausgang des Achtzehnten Jahrhunderts*. Berlin: Gebr. Verlag, 1981.

Schmuttermeier, Elisabeth. *Cast Iron from Central Europe, 1800–1850*. New York: Bard Graduate Center for Studies in the Decorative Arts, 1994.

Schütz, Christiane. *Preussen in Jerusalem (1800–1861): Karl Friedrich Schinkels Entwurf der Grabeskirche und die Jerusalempläne Friedrich Wilhelms IV*. Berlin: Mann, 1988.

Schwanitz, Wolfgang G., ed. *125 Jahre Sueskanal: Lauchhammers Eisenguss am Nil*. Hildesheim: G. Olms, 1998.

Senn, Rolf Thomas. *Orientalisierende Baukunst in Berlin im 19. Jahrhundert*. PhD Diss., Berlin: Freie Universität Berlin, 1990.

Senn, Rolf Thomas. *Schätze der Alhambra-Orientrezeption in Berlin und Brandenburg* [eine Ausstellung in den Sonderausstellungshallen am Kulturforum Berlin,... 29. Oktober 1995 bis 3. März 1996]. Berlin: Haus der Kulturen der Welt, 1995. Exhibition Catalog.

Stier, Hubert. "Karl von Diebitsch." *Deutsche Bauzeitung* 3 (August 26, 1869): 418–436.

Stier, Wilhelm. *Architektonische erfindungen*, edited by Hubert Stier. Berlin: Stier, 1867.

Stier, Wilhelm. "Beiträge zur Feststellung des Principes der Baukunst für das vaterländische Bauwesen der Gegenwart: Architrave und Bogen." *Allgemeine Bauzeitung* 8 (1843): 309–339.

Sweetman, John. *The Oriental Obsession: Islamic inspiration in British and American Art and Architecture, 1500–1920*. Cambridge, [Eng]: Cambridge University Press, 1988.

Toledano, Ehud R. *State and Society in Mid-Nineteenth-Century Egypt*. Cambridge [Eng]: Cambridge University Press, 1990.

Vatikiotis, P.J. *The History of Egypt*. 3rd ed. Baltimore, MD: The Johns Hopkins University Press, 1985.

Vingtrinier, Aimé. *Soliman-pacha, colonel Sève, généralissime des armées égyptiennes: ou, Histoire des guerres de l'Égypte de 1820 a 1860*. Paris: Didot, 1886.

Mercedes Volait, *Architectes et architectures de l'Égypte moderne (1830–1950): genèse et essor d'une expertise locale*. Paris: Maisonneuve et Larose, 2005.

Wise, M. Norton. "Architectures for Steam." In *The Architecture of Science*, edited by Peter Gallison and Emily Thompson, 107–140. Cambridge, MA: MIT Press, 1999.

Wise, M. Norton and Elaine M. Wise. "Staging an Empire." In *Things That Talk: Object Lessons from Art and Science*, edited by Lorraine Daston, 101–145. New York: Zone Books, 2004.

Woköck, Ursula. *German Orientalism: the Study of the Middle East and Islam from 1800 to 1945*. London: Routledge, 2009.

Wyatt, Matthew Digby. "Orientalism in European Industry." *Macmillan's Magazine* 21 (1870): 551–556.

Chapter 3

Amari, Monica. *I musei delle aziende: la cultura della tecnica tra arte e storia*. Milan: Franco Angeli, 2011.

Baes, Jean. "Conseil de perfectionnement de l'enseignement des arts du dessin. Réponse aux questions posées par le ministre de l'Agriculture, de l'Industrie et des Travaux publics. Reproduction ordonnée par le Conseil." *L'Emulation*, (1893), 5, col. 65–7; 6, col. 81–5; 7, col. 97–103; 8, col. 113–20.

Bamps, Anatole. "L'exposition universelle d'Anvers." *Revue Internationale* 8 (1885): 223–224.

Barzaghi, Ilaria M.P. *Milano 1881: tanto lusso e tanta folla: rappresentazione della modernità e modernizzazione popolare*. Cinisello Balsamo: Silvana Editoriale, 2009.

Bencivenni, Mario, Riccardo Dalla Negra, and Paola Grifoni. *Monumenti e istituzioni. Parte I. La nascita del servizio di tutela dei monumenti in Italia 1860–1890*. Florence: Alinea, 1987.

Bencivenni, Mario, Riccardo Dalla Negra, and Paola Grifoni. *Monumenti e istituzioni – Parte II. Il decollo e la riforma del servizio di tutela dei monumenti in Italia 1880–1915*. Florence: Alinea, 1992.

Bendiscioli, Mario. "La sinistra storica e la scuola." *Studium* 4 (1977): 447–466.

Boito, Camillo. "Cenni di recenti pubblicazioni sul nuovo stile e sullo stile della natura." *AIDI*, (1901): 83–84; 96–98.

Boito, Camillo. "L'industria delle suppellettili sacre." In *Oreficerie, stoffe, bronzi, intagli all'esposizione di arte sacra in Orvieto*, edited by Raffaele Erculei, VI–IX. Milan: Hoepli 1898.

Boito, Camillo and Moritz Meurer. "Il professore M. Meurer e i suoi studi sulle piante e sull'ornamento/Lo studio ornamentale della pianta." *Arte italiana decorativa e industriale* 7 (1901): 53–57.

Bossaglia, Rossana. *Torino 1902: le arti decorative internazionali del nuovo secolo*. Milan: Fabbri 1994.

Bulegato, Fiorella. *I musei d'impresa. Dalle arti industriali al design*. Rome: Carocci, 2008.

Buti, Sandra. *La Manifattura Ginori. Trasformazioni produttive e condizione operaia (1860–1915)*. Florence: Olschki, 1990.

Carocci, Giampiero. *Agostino Depretis e la politica interna italiana dal 1876 al 188*. Turin: Einaudi, 1956.

Carraro, Martina. "I Belgi e la Biennale: premesse e protagonisti del primo padiglione nazionale ai giardini (1895–1914)." PhD diss., Università Ca' Foscari Venezia, 2010.

Carta, Titti. "Artigianato." In *Enciclopedia Italiana – VI, Appendice*. Milan: Treccani, 2000, 703; App. III, i, 139.

Castronovo, Valerio. "La storia economica." In *Storia d'Italia vol. 4 Dall'unità a oggi*, edited by Ruggiero Romano and Corrado Vivanti, 5–129. Torino: Einaudi, 1975.

Cerri, Maria Grazia, Daniela Biancolini Fea, and Laura Pittarello, eds. *Alfredo d'Andrade. Tutela e restauro*. Florence: Vallecchi, 1981.

Ciampani, Andrea, Pierre Tilly, and Vincent Viaene. "Italia e Belgio nell'Ottocento europeo. Nuovi percorsi di ricerca, Atti del convegno internazionale di Roma." *Rassegna storica del Risorgimento* III, suppl. (2002): 1–184.

Congrès international de l'enseignement 1880. Brussels: Hayez, 1880.

Constant, Jean-François. "Entre mémoire et avenir. La nation aux expositions nationales de Bruxelles et de Montréal (1880–1884)." In *Vivre en ville. Bruxelles et Montréal*

(XIX–XX siècles), edited by Serge Jaumain and Paul–André Linteau, 351–371. Brussels: P.I.E. Peter Lang, 2006.

Corneli, René and Pierre Mussely. *Anvers et l'exposition universelle de 1885*. Antwerp: Bellemans, 1886.

Corona, Giuseppe. *L'Italia Ceramica*. Rome: Tip. Eredi Botta, 1880.

Crifò, Sofia. *Raffaello Ojetti architetto nei primi cinquant'anni di Roma capitale*. Florence: Polistampa, 2004.

De Maeyer, Jan. *De Sint-Lucasscholen en de neogotiek 1862–1914*. Leuven: KADOC, 1988.

Del M.A.I.: storia del Museo artistico industriale di Roma. Rome: ICCD, 2005.

Del M.A.I.: storia del Museo artistico industriale di Roma: Collezioni d' arte antica. Inventari 1876, 1884, 1956. Rome: ICCD, 2011.

Dellapiana, Elena. *Il design della ceramica in Italia, 1850–2000*. Milan: Electa, 2010.

Dellapiana, Elena. "Il mito del medioevo" and "Camillo Boito (1836–1814)." In *Storia dell'Architettura Italiana*, edited by Amerigo Restucci, 400–421 and 622–641. Milan: Electa, 2005.

Dumoulin, Michel. "Hommes et cultures dans les relations Italo-Belges, 1861–1915." *Bulletin de l'Institut Historique Belge de Rome*, t. LII, (1982): 271–565.

Dumoulin, Michel and Herman Van der Wee. *Hommes, Cultures et Capitaux dans les relations italo–belges aux XIXe et XXe siècles*. Rome: IHBR, 1993.

Dupont, Christine. *Modèles italiens et traditions nationales. Les artistes belges en Italie (1830–1914)*. Rome: IHBR, 2006.

École des arts décoratifs. Eléments de l'enseignement. Organisation. Bruxelles: Mertens, 1879.

Erculei, Raffaele. "L'esposizione di arte sacra in Orvieto." *Arte italiana decorativa e industriale* 2 (1897): 16–20.

Erculei, Raffaele. "L'insegnamento artistico–industrial in Europa. Belgio." *Arte italiana decorativa e industriale* 5 (1894): 42–43.

Errera, Alberto. *Studi sull'istruzione primaria, industriale professionale e commerciale nel Belgio*. Rome: Tipografia Eredi Botta, 1880.

"Esposizione Internazionale di Anversa." *Bollettino di notizie commerciali*, s.II v. 1, 9 (1884): 283–285.

"Esposizione Internazionale di Anversa. Circolare diretta alle Camere di commercio del Regno." *Bollettino di notizie commerciali*, s.II v. 1, 6 (1884): 172–173.

Exposition nationale de 1880. Catalogue officiel. Brussels: Mertens, 1880.

Exposition Nationale, IV section, Industries d'Art en Belgique antérieures au XIXe siècle. Catalogue Officiel. Brussels: Vanderauwera, 1880.

Exposition universelle. Paris. 1878. Rapports du jury international. Groupe III.-Classes 17 et 18. Rapport sur les meubles à bon marché et les meubles de luxe, ouvrages du tapissier et du décorateur. Paris: Imprimerie nationale, 1880.

Fumière, Théophile. *Des expositions et de l'enseignement des arts décoratifs. Leur développement en France et leur avenir en Belgique*. Brussels: Guyot, 1882.

Fumière, Théophile. *L'exposition d'Amsterdam et la Belgique aux Pays-Bas*. Brussels: Guyot, 1883.

Fumière, Théophile. *La section italienne à l'exposition universelle d'Anvers*. Brussels: Guyot, 1885.

Fumière, Théophile. *Les arts décoratifs à l'exposition du Cinquantenaire belge*. Brussels: Guyot, 1880.

Ghisalberti, Carlo. *Storia costituzionale d'Italia 1848/1948*. Bari: Laterza, 2000.

Gola, Sabina. *Un démi–siècle de relations culturelles entre l'Italie et la Belgique (1830–1880)*. Rome: IHBR, 1999.

Knockaert, Mandy and Jan-Lodewijk Grootaers, eds. *De panoramische droom: Antwerpen en de wereldtentoonstellingen 1885, 1894, 1930*. Antwerp: Antwerpen 93, 1993.

Lamberts, Emiel and Jacques Lory. *1884: un tournant politique en Belgique*. Brussels: FUSL, 1984.

Leblanc, Claire ed. *Art et industrie: les arts décoratifs en Belgique au XIXe siècle.* Brussels: MRAH, 2004.

Leblanc, Claire. *Art nouveau & Design. les arts décoratifs de 1830 à l'expo 58.* Brussels: Racine, 2005.

Levra, Umberto, and Rosanna Roccia. *Le esposizioni torinesi. 1805–1911. Specchio del progresso e macchina del consenso.* Turin: Archivio storico del Comune di Torino, 2003.

Lombaerde, Piet. "A la recherche d'une identité historique: l'architecture néo-Renaissance à Anvers au 19ème siècle." In *Le XIXᵉ siècle et l'architecture de la Renaissance*, edited by Frédérique Lemerle, Yves Pauwels, Alice Thomine-Berrada, 165–178. Paris: Picard, 2010.

[Majorana Calatabiano, Salvatore]. "Il Museo Italiano d'arte industriale. Lettera del Ministro di Agricoltura, Industria e Commercio al Sindaco di Roma." *Annali dell'Industria e del Commercio* 2 (1879): 1–24.

Melani, Alfredo. "L'Arte industriale nuova. Un'occhiata all'estero." *Arte italiana decorativa e industriale* 10 (1901): 51–52; 65–67.

Midant, Jean Paul. "La création de l'École des arts décoratifs et l'enseignement de l'architecture à l'Académie des Beaux-arts." In *Académie de Bruxelles. Deux siècles d'architecture*, edited by Jean-Paul Midant, 309–323. Brussels: Archives d'architecture moderne, 1989.

Mihail, Benoît. "Un mouvement culturel libéral à Bruxelles dans le dernier quart du XIXᵉ siècle, la 'néo-Renaissance flamande." *Revue belge de philologie et d'histoire* 76, no. 4 (1998): 978–1020.

Musée Royal de l'Industrie. Catalogue des collections de gravures, de dessins, de modèles, etc., formées dans le but de propager les applications des arts aux diverses industries. Brussels: Dehou, 1876.

"Notizie e documenti sulle scuole commerciali e popolari in Italia e all'estero." *Annali dell'Industria e del Commercio* 10 (1879): 7–277.

Odescalchi, Baldassarre. *I musei d'arte e d'industria in Italia: considerazioni e proposte.* Rome: tip. Eredi Botta, 1880.

Pavoni, Rossana ed. *Reviving the Renaissance. The Use and Abuse of the Past in Nineteenth-Century Italian Art and Decoration.* Cambridge: University Press, 1997.

Pellegrino, Anna. "L'Italia alle esposizioni universali del XIX secolo: identità nazionale e strategie comunicative." *Diacronie. Studi di Storia Contemporanea: Le esposizioni: propaganda e costruzione identitaria* 2, no. 18 (2014): 1–21.

Pellegrino, Anna. *La città più artigiana d'Italia. Firenze 1861–1929.* Milan: Franco Angeli, 2012.

Pesando, Annalisa B. "Un inedito d'Andrade: innovatore nell'insegnamento delle arti decorative." *Bollettino S.P.A.B.A.*, LII (2004): 265–286.

Pesando, Annalisa B. *Opera vigorosa per il gusto artistico nelle nostre industrie. La Commissione centrale per l'insegnamento artistico industriale e il "sistema delle arti" (1884–1908).* Milan: Franco Angeli, 2009.

Pesando, Annalisa B. and Daniela N. Prina. "To educate taste with the hand and the mind. Design reform in post-unification Italy." *Journal of Design History* 25, no. 1 (2012): 32–54.

Pinto, Sandra. "La promozione delle arti negli Stati italiani dall'età delle riforme all'Unità." In *Storia dell'Arte Italiana*, edited by Federico Zeri, v. 6, 791–1079. Torino: Einaudi, 1982.

Prina, Daniela N. "Belgian Decorative Arts in the Later Nineteenth Century. Needs for a National Museum and Debates Surrounding Didactic Collections in Brussels." *Journal of the History of Collections* 24, no. 2 (2012): 257–274.

Prina, Daniela N. "Design in Belgium before Art Nouveau: Art, Industry, and the Reform of Artistic Education in the Second Half of the Nineteenth Century." *Journal of Design History* 23, 4 (2010): 329–350.

Prina, Daniela N. "From Centralization to Local Policies: Design Reform Dynamics in Belgium and the Creation of Antwerp's Higher Institute (1830–1914)." In *Tradition, transition, trajectories: major or minor influences?* edited by Helena Barbosa and Anna Calvera, 557–562. Aveiro: Universidad de Aveiro, 2014.

Prina, Daniela N. "L'unité des Arts avant l'Art Nouveau. La réforme de l'enseignement artistique et industriel en Belgique pendant la deuxième moitié du XIXe siècle." PhD diss., Politecnico di Torino & K.U. Leuven, 2009.

Rapports présentés à l'administration communale de Bruxelles. Brussels: Baertsoen, 1885.

"Relazione del Regio Commissario (marchese Maffei di Boglio) sui premi ottenuti dagli espositori italiani." *Bollettino di notizie commerciali,* s.II v. 11, 35 (1885): 648–664.

Sacco di Albiano, Ugo Colombo. *Un omaggio a oltre 150 anni di amicizia Italo-Belga attraverso luoghi e protagonisti della diplomazia.* Rome: Servizi Tipografici Colombo, 2014.

Segreto, Luciano. "Storia d'Italia e storia dell'industria." In *Storia d'Italia. Annali. L'industria,* edited by Franco Amatori, Duccio Bigazzi, Renato Giannetti and Luciano Segreto, 7–21. Torino: Einaudi, 1999.

Smets, Marcel. *Charles Buls et Les principes de l'art urbain.* Liège: Mardaga, 1995.

Van Tricht, Victor. *L'exposition universelle d'Anvers, revue scientifique.* Brussels: Vromant, 1885.

Viollet-le-Duc, Eugène Emmanuel. *Dictionnaire raisonné de l'architecture française du XIe au XVIe siècle.* Paris: B. Bance-A. Morel, 1854–1868.

Willis, Alfred. "Flemish Renaissance Revival in Belgian Architecture (1830–1930)." PhD diss., Columbia University, 1984.

Wouters, Wilfried. *Van tekenklas tot kunstacademie. De Sint-Lucasscholen in België 1866–1966.* Kortrijk-Heule: UGA, 2011.

Zucconi, Guido. *L'invenzione del passato. Camillo Boito e l'architettura neomedievale.* Venice: Marsilio, 1997.

Chapter 4

Aynsley, Jeremy. *Nationalism and Internationalism, Design in the 20th Century.* London: V & A Publications, 1994.

Dybdahl, Lars. *Dansk keramik 1850–1997.* Exhibition catalogue. Lyngby: Sophienholm, 1997.

Gaustad, Randi. "'Dette eiendommelige norske', Norge på den Nordiske industri-, landbrugs-og kunstudstillingen i København 1888." In *Den stora nordiska utställningen i Köbenhamn 1888,* edited by Severi Parko, 47–51. Helsinki: Nordic Forum of Decorative Arts, 1989.

Gelfer-Jørgensen, Mirjam. *Influences from Japan in Danish Art and Design 1870–2010.* Copenhagen: Arkitektens Forlag, 2013.

Grandjean, Bo. *Frederik Wilhelm Scholander och drömmen om Renässancen.* Stockholm: Nordiska Museet, 1979.

Jensen, Nicolai Falberg. "Den nordiske Industri-, Landbrugs-og kunstudstilling i København 1888-en introduction til arkivet og dets anvendelsesmuligheder." *Erhvervshistorisk Årbog* 1 (2015): 52–92.

Johansson, Gothardt. "Design in Scandinavia." In *Design in Scandinavia,* edited by Arne Remlov, 10–28. Oslo: Kristies Bogtrykkeri, 1954.

Juul, Ida. "De danske håndværker-uddannelser i krydsfeltet mellem det europæiske og det nationale." *Uddannelseshistorie* 47 (2013): 60–80.

Madsen, Karl. "Dekorationsforeningen." *Tidskift for Kunstindustri* 4 (1888): 145–154.

Meier, F.J. "Noget om dansk Keramik på Udstillingen." *Tidsskrift for Kunstindustri* 4 (1888): 72–91.

Nyrop, Camillus. *Bidrag til Dansk Håndværker-Undervsinings Historie.* Copenhagen: The Technical Society, 1893.

Nyrop, Camillus. *Den nordiske Industri-, Landbrugs-og Kunstudstilling: Officiel beretning.* Copenhagen: The Exhibition Committee, 1890.

Nyrop, Camillus. "Kunstindustrien på den nordiske udstilling." *Tidsskrift for Kunstindustri* 4 (1888): 264–285.

Raavad, Alfred. "En national stil." *Tidsskrift for Kunstindustri* 4 (1888): 25–33.

Rasmussen, Verner. "De tekniske skolers historie." *Uddannelseshistorie* 3 (1969): 7–41.

Chapter 5

Abe, Stanley K. "Inside the Wonder House: Buddhist Art and the West." *Curators of the Buddha: The Study of Buddhism under Colonialism*, edited by Donald S. Lopez, Jr., 63–106. Chicago/London: University of Chicago Press, 1995.

Beasley, W.G. *Japanese Imperialism 1894–1945*. Oxford: Clarendon Press; New York: Oxford University Press, 1987.

Burnham, Daniel. Daniel H. Burnham Collection, Ryerson and Burnham Archives, The Art Institute of Chicago, Chicago.

Choi, Don. "Shaping the Architect at the Imperial College of Engineering." Presentation for the Society of Architectural Historians 65th Annual Conference, Detroit, MI, April 18–22, 2012.

Choi, Don. "The End of the World as They Knew It: Architectural History and Modern Japan." In *Seeking the City: Visionaries on the Margins*, Proceedings of the 96th ACSA Annual Meeting. Washington: ACSA Press, 2008, 736–742.

Coaldrake, William H. *Architecture and Authority in Japan*, Nissan Institute/Routledge Japanese Studies Series. London/New York: Routledge, 1996.

Conant, Ellen P. "Japan 'Abroad'" in *Challenging Past and Present: The Metamorphosis of Nineteenth-Century Japanese Art*, edited by Ellen P. Conant, 257–263. Honolulu: University of Hawai'i Press, 2006.

Conder, Josiah. "The Condition of Architecture in Japan" Paper read before the International Conference on Architects, World's Columbian Exposition, August 2nd, 1893, reprinted in *The Japan Weekly Mail*, September 30, 1893, 392–394.

Fenollosa, Ernest F. *Epochs of Chinese and Japanese Art: An outline of History of East Asiatic Design. Vol. I, Vol. II New and revised edition, with copius notes by Professor Petrucci*. New York: Frederick A. Stokes Co., 1912. Reprinted in one volume, Berkeley, CA: Stone Bridge Press/Tokyo, Japan: Yohan Classics, IBC Publishing, 2007.

Fenollosa, Ernest F. "Contemporary Japanese Art: With Examples from the Chicago Exhibit." *Century Illustrated Magazine* 46, no. 4 (August 1893): 577.

Guth, Christine M.E. "Kokuhô: From Dynastic to Artistic Treasure." In "Mémorial Anna Seidel. Religions traditionnelles d'Asie orientale. Tome II, sous la direction de Hubert Durt." Special issue, *Cahiers d'Extrême-Asie* 9 (1996), 313–322.

Hiroshi, Watanabe. *The Architecture of Tôkyô, An Architectural History in 571 Individual Presentations*. Stuttgart/London: Axel Menges, 2001.

Ives, Halsey C. *The Dream City, A Portfolio of Views of the World's Columbian Exposition*. St. Louis, MO: N.D. Thompson Publishing Co., 1893.

Karatani, Kôjin. "Japan as Art Museum: Okakura Tenshin and Fenollosa." In *A History of Modern Japanese Aesthetics*, edited by Michael F. Marra, 43–52. Honolulu: University of Hawai'i Press, 2001.

Marra, Michael F. *Modern Japanese Aesthetics, a Reader*. Honolulu: University of Hawai'i Press, 1999.

Masamichi, Kuru. "Dai Nihon Kodai Kenchiku Enkaku" ("The Development of Ancient Japanese Architecture"). *Kenchiku Zasshi (Journal of Architecture and Building Science)* 5, no. 49 (January 1891, 5:49), 6–12.

Mason, Penelope. *History of Japanese Art*. Second edition revised by Donald Dinwiddie. Upper Saddle River, NJ: Pearson Prentice Hall, 2005.

Mishima, Masahiro. "1893 nen Shikago bankokuhaku ni okeru Hôôden no kensetsu ikisatsu nitsuite" ("On the Circumstances of the Construction of the Hôôden in the 1893 Chicago World's Fair"). *Nihon Kenchiku Gakkai ronbun Hôkokushû (Transactions of the Architectural Institute of Japan)* 429 (November, 1991): 151–163.

Mishima, Masahiro. "Hôôden no keitai to sono seiritsu yôin nit suite" ("The Factors Surrounding the Form of the Hôôden"). *Nihon Kenchiku Gakkai ronbun hôkokushû (Transactions of the Architectural Institute of Japan)* 434 (April 1992): 107–116.

Murai, Noriko. "Okakura's Way of Tea: Representing Chanoyu in Early Twentieth-Century America." *Review of Japanese Culture and Society Vol. 14, Meiji Literature and the Artwork,* 60–77. Honolulu: University of Hawai'i Press on behalf of Josai University Educational Corporation. December 2002.

Noma, Seiroku. *The Arts of Japan: Ancient and Medieval.* New York: Kodansha, 1978, rpt. 2003.

Notehelfer, F.G. "On Idealism and Realism in the Thought of Okakura Tenshin." *Journal of Japanese Studies* 16, no. 2 (Summer, 1990): 309–355.

Oakes, Julie Christ. "Contestation and the Japanese National Treasure System." PhD diss., University of Chicago, 2009.

Okakura, Kakudzo. *The Hôôden (Phoenix Hall), An Illustrated Description of the Buildings Erected by the Japanese Government at The World's Columbian Exposition Jackson Park, Chicago.* Tokyo: Ogawa, 1893.

Reynolds, Jonathan M. *Maekawa Kunio and the Emergence of Japanese Modernist Architecture.* Berkeley/LA/London: University of California Press, 2001.

Rimer, Thomas, ed. *Since Meiji, Perspectives on the Japanese Visual Arts 1868–2000.* Honolulu: University of Hawai'i Press, 2012.

Satô, Dôshin. *Modern Japanese Art and the Meiji State, the Politics of Beauty.* Translated by Nara Hiroshi. Los Angeles, CA: Getty Publications, 2011.

Shikago "Shikago no miyagebanashi I" ("Chicago Souvenirs I"). *Kenchiku Zasshi (Journal of Architecture and Building Science)* 7, no. 82 (December 1893), 283–288.

Shikago "Shikago no miyagebanashi II" ("Chicago Souvenirs II"). *Kenchiku Zasshi (Journal of Architecture and Building Science)* 7, no. 84 (December 1893), 355–363.

Smith, Walter. *The Masterpieces of the Centennial International Exhibition Vol. I, Industrial Art.* Philadelphia: Gebbie & Barrie, 1875.

Tanaka, Stefan. *Japan's Orient, Rendering Pasts into History.* Berkeley, LA/Oxford: University of California Press, 1993.

Toshiya, Kaneko. "Cultural Light, Political Shadow: Okakura Tenshin (1862–1913) and the Japanese Crisis of National Identity, 1880–1941." PhD diss., University of Pennsylvania, 2002.

Tseng, Alice Y. *The Imperial Museums of Meiji Japan: Architecture and the Art of the Nation.* Seattle/London: University of Washington Press, 2008.

Van Brunt, Henry and William A. Coles. *Architecture and Society, Selected Essays of Henry Van Brunt.* Cambridge, MA: Belknap Press, 1969.

Watanabe, Toshio. "Vernacular Expression or Western Style? Josiah Conder and the Beginning of Modern Architectural Design in Japan." In *Art and the National Dream: The Search for Vernacular Expression in Turn-of-the-Century Design,* edited by Nicola Gordon Bowe, 43–52. Dublin: Irish Academic Press, 1993.

Weston, Victoria. *Japanese Painting and National Identity: Okakura Tenshin and his Circle.* Ann Arbor, MI: Center for Japanese Studies, University of Michigan, 2004.

White, P.B. "Japanese Architecture in Chicago, Part II." *The Inland Architect and News Record* 20, no. 6 (January 1893), 60–62.

Chapter 6

Actualité. "L'Actualité: Le Projet Paschal Grousset et le Feu Intérieur." *L'Éclair* (March 28, 1895).

Actualité. "L'Actualité: Le Village Suisse à l'Exposition de 1900." *L'Eclair* (December 15, 1899).

Archéologie. "Archéologie: Pompéi morte ou vivante?" *Journal de Genève* (August 21, 1899).

Auslander, Leora. *Taste and Power: Furnishing Modern France.* Berkeley: University of California Press, 1996.

Au Village. "Au Village suisse." *Le Temps* (December 12, 1899).

Claretie, Jules. "Les Chambres." *Le Journal* (November 7, 1900).

"Les Clous de 1900: Les derniers projets examinés par la commission," *L'Eclair* (9 Aug. 1897).

d'Allemagne, Henry-René. *La Maison d'un vieux collectionneur*. Paris: Librairie Gründ, (1948).

Debord, Guy. *Society of the Spectacle*. Detroit: Co-Op, A Black & Red Translation, 1970.

de Gléon, Delort. *La Rue du Caire: L'Architecture Arabe des Khalifes d'Égypte a l'Exposition Universelle de Paris en 1889*. Paris: Librairie Plon, 1889.

de Tourette, J. "L'Exposition de 1900: Paris en 1400. – La Cour des Miracles." *La Petite Gironde, Bordeaux* (January 23, 1899).

Emery, Elizabeth. "Albert Robida, Medieval Publicist." In *Cahier Calin. Essays in Honor of William Calin*, edited by Richard Utz and Elizabeth Emery, 51–53. Kalamazoo, MI: Studies in Medievalism, 2011.

Emery, Elizabeth, and Laura Morowitz. *Consuming the Past: The Medieval Revival in Fin-De-Siècle France*. Aldershot, England and Burlington, VT: Ashgate, 2003.

Exposition rétrospective. "Exposition rétrospective et musées centennaux." *Revue Popularie des Beaux-Arts* (1898).

Friedberg, Anne. *Window Shopping: Cinema and the Postmodern*. Berkeley, Los Angeles and London: University of California Press, 1994.

Froissart, Rossella. "Les Arts décoratifs au service de la nation, 1880–1918." *Arts & Societies* (Séminaire du 17 juin 2005). www.artsetsocietes.org/f/f-froissart.html (visited February 29, 2016).

Gaillard, Eugène. *Nos arts appliqués modernes. Le mobilier au Salon d'Automne de 1910. Impressions et opinions*. Paris: E. Floury, 1910.

Giberti, Bruno, *Designing the Centennial: A History of the 1876 International Exhibition in Philadelphia*. Lexington: University Press of Kentucky, 2002.

Grimm, Thomas. "Exposition de 1900: Le Village Suisse." *Le Petit Journal* (August 7, 1899).

Hallays, André. "En Flânant." *Feuilleton du Journal des débats* (October 12, 1900).

Harris, John. *Moving Rooms: The Trade in Architectural Salvages*. New Haven and London: Yale University Press, 2007.

Houssaye, Henry. "Voyage autour du monde a l'exposition universelle." *Revue des deux mondes* (July 15, 1878).

Lara-Betancourt, Patricia. "Displaying Dreams: Model Interiors in British Department Stores, 1890–1914." In *Architectures of Display: Department Stores and Modern Retail*, edited by Anca I. Lasc, Patricia Lara-Betancourt, and Margaret Maile Petty. London and New York: Routledge, 2017.

Lasc, Anca I. "Interior Decorating in the Age of Historicism: Popular Advice Manuals and the Pattern Books of Édouard Bajot." *Journal of Design History* 26, no. 1 (February 2013): 1–24.

Lasc, Anca I. "*Le Juste Milieu*: Alexandre Sandier, Theming, and Eclecticism in French Interiors of the Nineteenth Century." *Interiors: Design, Architecture, Culture* 2, no. 3 (November 2011): 277–306.

Lasc, Anca I. "A Museum of Souvenirs: Adolphe Thiers, Collector of the Nineteenth Century." *Journal of the History of Collections* 28, no. 1 (March 2016): 57–71.

Le Corbeiller, Maurice. *Musée Centennal des classes 66, 69, 70, 71, 97. Mobilier & Décoration à l'Exposition universelle international de 1900, à Paris. Rapport de la commission d'installation*. St.-Cloud: Bélin, 1902.

Mandell, Richard D. *Paris 1900: The Great World's Fair*. Toronto, ON: University of Toronto Press, 1967.

Marguillier, Auguste. "Bibliographie des ouvrages publiés en France et à l'étranger sur les beaux-arts et la curiosité pendant le deuxième semestre de l'année 1902." *Gazette des beaux-arts* 28 (1902): 512–528.

Mourey, Gabriel. "L'Art décoratif à l'Exposition universelle." *Revue encyclopédique* 10, no. 371 (1900): 801–810.

Ogata, Amy. "The Union Centrale des Arts Décoratifs Pavilion, Art Nouveau, and the *Cabinet d'Amateur* at the Fin de Siècle." In *Salvaging the Past: Georges Hoentschel and French Decorative Arts at the Metropolitan Museum of Art 1907–2013*, edited by Daniëlle Kisluk-Grosheide, Deborah L. Krohn, and Ulrich Leben, 193–203. New York, New Haven and London: The Bard Graduate Center, The Metropolitan Museum of Art, and Yale University Press, 2013.

Palais du Costume. "Le Palais du Costume." *Le Temps* (December 17, 1897).

Paris en 1400. "Paris en 1400." *Le Petit Journal* (May 10, 1899).

Paris Exposition 1900. *Guide Pratique du visiteur de Paris et de l'exposition*. Paris: Hachette & Cie, 1900.

Pons, Bruno. *French Period Rooms, 1650–1800: Rebuilt in England, France, and the Americas*. Dijon: Faton, 1995.

Projet Félix. *Exposition Universelle de 1900: Palais du Costume – Le Costume de la femme à travers les âges. Catalogue*. Paris: Imprimeries Lermercier, 1900.

Scott, Katie and Melissa Hyde, eds. *The Rococo Echo: Art, Theory and Historiography from Cochin to Coppola*, edited by Oxford: Voltaire Foundation, 2014.

Saint-Amand, Imbert de. *La Cour de l'impératrice Joséphine*. Paris: Dentu, 1883.

Silverman, Debora. *Art Nouveau in Fin-de-Siècle France: Politics, Psychology, and Style*. Berkeley: University of California Press, 1989.

Sowerwine, Charles. *France since 1870: Culture, Society and the Making of the Republic*. New York: Palgrave Macmillan, 2009.

Thiébault-Sisson. "Promenades à l'Exposition: Dix siècles d'histoire du vêtement." *Le Temps* (September 22, 1900).

Troy, Nancy J. *Modernism and the Decorative Arts in France: Art Nouveau to Le Corbusier*. New Haven, CT: Yale University Press, 1991.

Village suisse. "Village suisse à Paris." *Journal de Genève* (December 4, 1898).

Volcan à Paris. "Un Volcan à Paris." *Le Temps* (October 17, 1897).

Chapter 7

Aav, Marianne. "The First Golden Age of Finnish Design." In *Now the Light Comes from the North: Art Nouveau in Finland*, edited by Ingeborg Becker and Sigrid Melchior, 89–94. Berlin: Bröhan Museum, 2002.

Ahrenberg, Jac. "Expositionen af finska konstverk, afsedda att utställas i Paris 1900." *Finsk tidskrift för vitterhet, vetenskap, konst och politik* 48 (1900): 68–70.

Kuutti, Tuija et al. *Akseli Gallen-Kallela: Eurooppalainen mestari*. Helsinki: Helsingin taidemuseo, 2011.

Ashby, Charlotte. "Nation Building and Design: Finnish Textiles and the Works of the Friends of Finnish Handicrafts." *Journal of Design History* 23 (2010): 351–365.

Ashby, Charlotte. "The Pohjola Building: Reconciling Contradictions in Finnish Architecture Around 1900." In *Nationalism and Architecture*, edited by Raymond Querk, Darren Deane, and Sarah Butler, 135–146. Farnham and Burlington, VT: Ashgate Publishing Press, 2012.

Baglo, Cathrine. "På ville veger? Levende utstillinger av samer i Europa og Amerika." PhD diss., Tromsø University, 2009.

Becker, Ingeborg. "The Wilderness and the City Lights – Northern Polarities, or Finland's Artists between National Romanticism and the International Scene. The Case of Akseli Gallen-Kallela." In *Now the Light Comes from the North: Art Nouveau in Finland*, edited by Ingeborg Becker and Sigrid Melchior, 14–21. Berlin: Bröhan Museum, 2002.

Facos, Michelle. *Nationalism and the Nordic Imagination: Swedish Art of the 1890s*. Berkeley: University of California Press, 1998.

Facos, Michelle, Thor J. Mednick, and Janet S. Rauscher. "National Identity in Nordic Art: Perceptions from Within and Without in 1889 and 1900." *Centropa*, 3 (September 2008), 215.

"Finland på världsutställnigen," *Hangö*, April 14, 1900.

Finnland. "'Finnland' auf der Pariser Weltausstellung," *Dekorative Kunst* 6 (1900): 457–466.

Finska "Den finska paviljongen vid världutställningen," *Hufvudstadsbladet*, June 26, 1898.

Finska "Finska paviljongen," *Borgå Nya Tidning*, May 4, 1900.

Gallen-Kallela-Sirén, Janne. "Axel Gallén and the Constructed Nation: Art and Nationalism in Young Finland, 1880–1900." PhD diss., New York University, 2001.

Goss, Glenda Dawn. *Sibelius: The Composer's Life and the Awakening of Finland*. Chicago, IL: University of Chicago Press, 2009.

Grendahl, G. *Питкаранта:Краткое описание питкарантского месторождения, рудников и заводов*. St. Petersburg: Tipo-Litografiya A. E. Vineke, 1896.

Hallas-Murula, Karin. *Soome-Eesti. Sajand arhitektuurisuhteid*. Tallinn: Eesti Arhitektuurimuuseum, 2005.

Hausen, Marika. "Finlands paviljong på världsutställningen i Paris." In *Finkst sekelskifte. En konstbok från Nationalmuseum*, edited by Eva Nordenson, 27–59. Stockholm: Nationalmuseum, 1971.

Hautola-Hirvioja, Tuija. "The Image of the Sámi in Finnish Visual Arts before the Second World War." *Acta Borealia* 2 (2006): 97–115.

Howard, Jeremy. *Art Nouveau: National and International Styles in Europe*. Manchester and New York: Manchester University Press, 1996.

Howard, Jeremy. "Latvian National Romanticism and *Art Nouveau*: Origins and Synthesis." In *Romantisms un neoromantisms Latvijas mākslā*, edited by Elita Grosmane, 128–153. Rīga: Izdevniecība AGB, 1998.

Konttinen, Riitta. *Sammon takojat: Nuoren-Suomen taiteilijat ja suomalaisuuden kuvat*. Helsinki: Otava, 2001.

Korvenmaa, Pekka. "Signals from the Periphery: National Romanticism in Finland & Central Europe." *Centropa*, 2 (January 2002).

Korvenmaa, Pekka. "Who Are We? Where Do We Come from? Where Are We Going?" In *Now the Light Comes from the North: Art Nouveau in Finland*, edited by Ingeborg Becker and Sigrid Melchior, 70–77. Berlin: Bröhan Museum, 2002.

Lukkarinen, Ville and Annika Waenerberg. *Suomi-kuvasta mielenmaisemaan. Kansallismaisemat 1800-ja 1900-luvun vaihteen maalaustaiteessa*. Helsinki: Suomalaisen Kirjallisuuden Seura, 2004.

Meddelanden "Meddelanden från Industristyrelsen i Finland," XXVI–XXX, Helsinki: 1899.

Miller Lane, Barbara. *National Romanticism and Modern Architecture in Germany and the Scandinavian Countries*. Cambridge and New York: Cambridge University Press, 2000.

Ojanperä, Riitta, ed. *The Kalevala in Images: 160 Years of Finnish Art inspired by the Kalevala*. Helsinki: Ateneum Art Museum, Finnish National Gallery, 2009.

O'Neill, Itha. *Beda Stjernschantz. Ristikkoportin takana – Bakom gallergriden*. Helsinki: Suomalaisen Kirjallisuuden Seura, 2014.

Paris Exposition Reproduced from the Official Photographs Taken under the Supervision of the French Government for Permanent Preservation in the National Archives…. New York, Akron, OH, and Chicago: The R. S. Peale Company, 1900.

Pushaw, Bart. "Art and Multiculturalism in Estonia and Latvia, circa 1900." In *A Companion to Nineteenth-Century Art. From Revolution to World War*, edited by Michelle Facos. Boston: Wiley-Blackwell, 2018.

Pushaw, Bart. "Ruotsalainen Karjala? Beda Stjernschantzin teos *Kaikkialla ääni kaikuu* (1895) – Ett svenskt Karelen? Beda Stjernschantz verk *Överallt en röst oss bjuder…* (1895)." In *Beda Stjernschantz. Ristikkoportin takana – Bakom gallergriden*, edited by Itha O'Neill, 180–196. Helsinki: Suomalaisen Kirjallisuuden Seura, 2014.

Ringbom, Sixten. *Stone, Style, and Truth: The Vogue for Natural Stone in Nordic Architecture, 1880–1910*. Helsinki: Suomen Muinaismuistoyhdistys, 1987.

Schoolfield, George C. *Helsinki of the Czars: Finland's Capital, 1809–1918*. Columbia, SC: Camden House, 1996.

Smeds, Kerstin. *Helsingfors-Paris: Finlands utveckling till nation på världsutställningarna.* Helsinki: Finska Historiska Samfundet, 1996.

Smith, John Boulton. *The Golden Age of Finnish Art: Art Nouveau and the National Spirit.* Helsinki: Otava, 1985.

Soulier, Gustave. "Le Pavilion de Finlande à l'Exposition universelle." In *Art & Décoration* 8 (1900): 1–11.

Varnedoe, Kirk, ed. *Northern Light: Realism and Symbolism in Scandinavian Art, 1880–1910.* New York: Brooklyn Museum, 1982.

Walton, William, André Saglio, and Victor Champier. *Chefs d'Œvre of the Exposition Universelle.* Philadelphia: Georgie Barrie and Son, 1900.

Wäre, Ritva. *Rakkenettu suomalaisuus: Nationalismi viime vuosisadan vaihteen arkkitehtuurissa ja sitä koskevissa kirjoituksissa.* Helsinki: Suomen Muinaismuistoyhdistys, 1991.

Chapter 8

Ács, Piroska. "The Exhibition Pavilions of Géza Maróti in Milan and Venice." In *Hungarian Ceramics from the Zsolnay Manufactory,* edited by Éva Csenkey and Ágota Steinert, 211–216. Exhibition catalogue. Bard Graduate Center for Studies in the Decorative Arts, Design, and Culture/New Haven: Yale University Press, 2002.

Ács, Piroska, ed. *"Mi vagyunk Atlantisz": Vederemo!: Mároti Géza, 1875–1941.* Budapest: Iparművészeti Múzeum, 2003.

Alofsin, Anthony. *Frank Lloyd Wright: The Lost Years, 1910–1922. A Study of Influence.* Chicago, IL: University of Chicago Press, 1993.

Alofsin, Anthony. *When Buildings Speak: Architecture as Language in the Habsburg Empire and Its Aftermath, 1867–1933.* Chicago, IL: University of Chicago Press, 2006.

Bajkay, Éva, Katalin F. Dózsa, and Marianna Hergovich, ed. *Zeit des Aufbruchs: Budapest und Wien zwischen Historismus und Avantgarde.* Exhibition catalogue. Vienna: Kunsthistorisches Museum, 2003.

Bennitt, Mark, Frank Parker Stockbridge, et al. ed. *History of the Louisiana Purchase Exposition: Comprising the history of the Louisiana Territory, the story of the Louisiana Purchase and a full account of the Great Exposition, embracing the participation of the sates and nations of the world, and other events of the St. Louis World's Fair of 1904.* St. Louis, MO: Universal Exposition Publishing Company, 1905.

Buel, James W. *Louisiana and the Fair. An Exposition of the World, Its People and their Achievements.* 10 vols. Saint Louis: World's Progress Publishing Co., 1904.

"Chinese Village," *World's Fair Bulletin* 5, no. 9 (July 1904): 88.

Clark, Robert Judson,l and Andrea P. A. Belloli. *Design in America: The Cranbrook Vision, 1925–1950.* New York: Abrams, in association with the Detroit Institute of Arts and the Metropolitan Museum of Art, 1984.

Clough, Josh. "'Vanishing' Indians" Cultural Persistence on Display at the Omaha World's Fair of 1898." *Great Plains Quarterly* 25 (Spring 2005): 67–86.

Coir, Mark. "The Cranbrook Factor." In *Eero Saarinen: Shaping the Future,* edited by Eeva-Liisa Pelkonen and Donald Albrecht, 29–44. New Haven, CT: Yale University Press, 2006.

Crowley, David. "The Uses of Peasant Design in Austria-Hungary in the Late Nineteenth and Early Twentieth Centuries." *Studies in the Decorative Arts* 2, no. 2 (Spring 1995): 2–28.

Csenkey, Éva and Ágota Steinert, eds. *Hungarian Ceramics from the Zsolnay Manufactory.* Exhibition catalogue. Bard Graduate Center for Studies in the Decorative Arts, Design, and Culture/New Haven: Yale University Press, 2002.

Czakó, Elemér. *A Magyar iparművészet szereplése a milanoi nemzetközi világkiallitáson 1906-ban.* Budapest: Országos Magyar iparművészeti museum, 1907.

Demski, Dagnosław, Ildikó Sz. Kristóf, and Kamila Baraniecka-Olszewska, ed. *Competing Eyes: Visual Encounters with Alterity in Central and Eastern Europe.* Budapest: L'Harmattan, 2013.

Facos, Michelle and Sharon Hirsch, ed. *Art, Culture and National Identity in Fin-de-Siècle Europe.* Cambridge: Cambridge University Press, 2003.

Falke, Jacob von. "Das Kunstgewerbe." In *Kunst und Kunstgewerbe auf der Wiener Weltausstellung 1873,* edited by Carl von Lutzöw, 41–180. Leipzig: Verlag von E.A. Seemann, 1875.

Filipová, Marta. "Peasants on Display: The Czechoslavic Ethnographic Exhibition of 1895." *Journal of Design History* 24, no. 1 (2011): 15–36.

"A Finnek." *Magyar Iparművészet* 11, no. 1 (1908): 1–26.

Francis, David R. *The Universal Exposition of 1904.* 2 vols. St. Louis, MO: Louisiana Purchase Exposition Company, 1913.

Fred, W. [Alfred Wechsler]. "Interieurs und Möbel auf der Pariser Weltausstellung." *Kunst und Kunsthandwerk,* 3 (1900): 331–352.

Fred, W. [Alfred Wechsler]. "The International Exhibition of Decorative Art at Turin: The Austrian Section." *The Studio* 27, no. 116 (November 1902): 130–134.

Géller, Katalin, Mária G. Merva, and Cecília Őriné Nagy, ed. *A gödöllő művésztelep 1901–1920/ The Artists' Colony of Gödöllő.* Gödöllő: Gödöllő Városi Múzeum, 2003.

Gerle, János. "Hungarian Architecture from 1900 to 1918." In *The Architecture of Historic Hungary,* edted by Dora Weibenson and József Sisa, 223–244. Cambridge, MA: MIT Press, 1998.

Gerle, János. "Maróti Géza építészet és szobrászat határán." *Magyar Építőművészet* 83, no. 6 (1983): 50–51.

Györffy, István. *A cifraszűr.* Budapest: Nap Kiadó, 1930.

Györgyi, Kálmán. "Az iparművészet a párisi kiállitáson." *Magyar Iparművészet* 3, no. 5 (September 1900): 209–304.

Györgyi, Kálmán. "Az iparművészet a Milánói kiállításon." *Magyar Iparművészet* 9, no. 4–5 (July–September 1906): 162–219; 285–296.

György, Kálmán. "A magyar hímző-háziiparról." *Magyar Iparművészet* 3, no. 5 (1926): 51–53.

Földi-Dozsa, Katalin. "How the Hungarian National Costume Evolved." In *The Imperial Style: Fashions of the Habsburg Era,* edited by Polly Cone 75–88. New York: The Metropolitan Museum of Art, 1980.

Hanson, John Wesley. *The Official History of the Fair: St. Louis, 1904. The Sights and Scenes of the Louisiana Purchase Exposition. A Complete Description of the Magnificent Palaces, Marvelous Treasures and Scenic Beauties of the Crowning Wonder of the Age.* St. Louis, MO: John Wesley Hanson, 1904.

Hlavačka, Milan, Jana Orlíková, and Peter Štembera, eds. *Alfons Mucha—Pařiž 1900: Pavilion Bosny a Hercegoviny na světové výstavěl/Alphonse Mucha—Paris 1900: The Pavilion of Bosnia and Herzegovina at the World Exhibition.* Prague: Obecní dům, 2002.

Holme, Charles, ed. *Peasant Art in Austria and Hungary.* London: The Studio, 1911.

Horti, Pál. "A St. louisi világkiállítás." *Magyar Iparművészet* 7, no. 6 (November 1904): 249–313.

Houze, Rebecca. "National Internationalism: Reactions to the Austrian and Hungarian Decorative Arts at the 1900 Paris Exposition Universelle." *Studies in the Decorative Arts* 12, no. 1 (Fall-Winter 2004–2005): 55–97.

Houze, Rebecca. *Textiles, Fashion, and Design Reform in Austria-Hungary Before the First World War: Principles of Dress.* Farnham: Ashgate, 2015.

"Hungary" and "Hungarian Section." *World's Fair Bulletin* 5, no. 10 (August 1904): 10, 36.

Hutchinson, Elizabeth. *The Indian Craze: Primitivism, Modernism, and Transculturation in American Art, 1890–1915.* Durham: Duke University Press, 2009.

Ivánfy-Balough, Sára and Imre Jakabffy. "Géza R. Maróti." *Ars Decorativa* 4 (1976): 127–147.

Kallestrup, Shona. "Romanian 'National Style' and the 1906 Bucharest Jubilee Exhibition." *Journal of Design History* 15, no. 3 (2002): 147–162.

Kaplan, Wendy, ed. *The Arts and Crafts Movement in Europe and America: Design for the Modern World*. Exhibition catalogue. Los Angeles, CA: Los Angeles County Museum of Art, 2004.

Keserű, Katalin. "The Workshops of Gödöllő: Transformations of a Morrisian Theme." *Journal of Design History* 1, no. 1 (1988): 1–23.

Kinchin, Juliet. "Hungary: Shaping a National Consciousness." In *The Arts and Crafts Movement in Europe and America: Design for the Modern World*, edited by Wendy Kaplan, 142–177. Exhibition catalogue. Los Angeles, CA: Los Angeles County Museum of Art, 2004.

Kinchin, Juliet. "Modernity and Tradition in Hungarian Furniture, 1900–1938: Three Generations." *Journal of Decorative and Propaganda Arts* 24. Special Issue. Design, Culture, Identity: The Wolfsonian Collection (2002): 64–93.

Koós, Judith. *Horti Pál élete és művészete 1865–1907*. Budapest: Akadémiai Kiadó, 1982.

Koós, Judith. "A Hungarian Pioneer of Art Nouveau: Pál Horti." *Ars Decorativa* 2 (1974): 173–189.

Kriesch, Aladár Körösfői. "Hungarian Peasant Art." In *Peasant Art in Austria and Hungary*, edited by Charles Holme, 31–46. London: The Studio, 1911.

Lackner, Mönika and Péter Granasztói. *Cifraszűr/Hirtenmantel. Vom alltäglichen Kleidungsstück zum nationalen Symbol*. Kittsee: Ethnographisches Museum Schloss Kittsee/Vienna: Österreichisches Museum für Volkskunde, 2003.

Livingstone, Karen and Linda Parry. *International Arts and Crafts*. Exhibition catalogue. London: V&A Publications, 2005.

Lyka, Karl [Károly]. "Drei ungarische Künstler: Alexander Nagy, Aladár Kriesch, Eduard Wigand." *Dekorative Kunst* 14 (April–September 1904): 642–648.

Lyka, Károly. *Official Illustrated Catalogue Fine Arts Exhibit, Hungary, St. Louis Exposition, 1904*. Budapest: V. Hornyánszky 1904.

Melani, Alfredo. "Hungarian Art at the Milan Exhibition." *The Studio* 39, no. 162 (September 1906): 300–309.

Minkus, Fritz. "Die erste internationale Ausstellung für moderne dekorative Kunst in Turin." *Kunst und Kunsthandwerk* 5 (1902): 402–450.

Mourey, Gabriel. "Round the Exhibition. Part I. The House of the 'Art Nouveau Bing'." *The Studio* 20, no. 89 (August 1900): 164–180.

Németh, Ferenc. *A torontáli szőnyeg. A szőnyegszövés negyed százada Bánátben Streitmann Antaltól Kovalszky Saroltáig* (The Torontál Carpet: A Quarter Century of Weaving in Banat from Antál Streitmann and Sarolta Kovalszky). Budapest: Forum Könyvkiadó, 1993.

Németh, Lajos. "Art, Nationalism, and the Fin de Siècle." In *A Golden Age: Art and Society in Hungary 1896–1914*, edited by Gyöngyi Éri and Zsuzsa Jobbágyi, 19–29. Budapest: Corvina, 1997.

Park, T. W. "The Addresses by Prominent Personnages and the Ceremonies of the Opening Day Witnessed by the Largest Attendance of People in the History of World's Fairs." *World's Fair Bulletin* 5, no. 8 (June 1904), 4 ff.

Radisics, Eugéne [Jenő] de. *Le Pavillon Historique de la Hongrie á l'Exposition universelle de Paris en 1900*. Paris: Librairie Centrale des Beaux-Arts, 1900.

Redondi, Pietro, ed. *Cittá effimera: Arte, technological esotismo all'esposizione internazionle di Milano del 1906*. Milan: Edizione Gabriele Mazzotta, 2015.

Skiff, Frederick J.V. and Halsey C. Ives, ed. *Official Catalogue of Exhibitors: Universal Exposition St. Louis, U.S.A., 1904. Department B. Art*. St. Louis, MO: The Official Catalogue Company, Inc., 1904.

Somogy, Zsolt. "Pál Horti's Late Works in the United States of America." *Ars Decorativa* 25 (2007): 105–122.

Somogy, Zsolt. "An Adaptable Applied Artist: Pál Horti's American Furniture." *Ars Decorativa* 26 (2008): 131–143.

Starr, George. "Art and Architecture in the Hungarian Reformed Church." In *Seeing Beyond the Word: Visual Arts and the Calvinist Tradition,* edited by Paul Corby Finney, 301–342. Grand Rapids, MI: William B. Eerdmans Publishing Company, 1999.

Stoklund, Bjarne. "How the Peasant House became a National Symbol: A Chapter in the History of Museums and Nation-Building." *Ethnologia Europaea* 29 (1999): 5–18.

Sturcz, János. "'Vederemo—It's Just Been Won!' Géza Maróti's Competition Design for the Rockefeller Center." *Ars Decorativa* 11 (1991): 67–86.

Switzer, Terri. "Hungarian Self-Representation in an International Context: The Magyar Exhibited at International Expositions and World's Fairs." In *Art, Culture and National Identity in Fin-de-Siècle Europe,* edited by Michelle Facos and Sharon Hirsch, 160–185. Cambridge: Cambridge University Press, 2003.

Szczerski, Andrej. "Central Europe." In *International Arts and Crafts,* edited by Karen Livingstone and Linda Parry, 238–251. London: V&A Publications, 2005.

Szécsén, Ferenc. "A St. Louisi Világkiállítás." *Magyar Iparművészet* 7, no. 2 (March 1904): 89–92.

Székely, Miklós. "The Resetting of the Main Historical Group From the Millennium Exhibition to the Paris Universal Exhibition of 1900." In *Ephemeral Architecture in Central-Eastern Europe in the 19th and 20th Centuries,* edited by Miklós Székely, 33–50. Paris: L'Harmattan, 2015.

Székely, Miklós. "From Figure to Pattern: The Changing Role of Folk Tradition in Hungarian Representations at Universal Exhibitions (1867–1911)." In *Competing Eyes: Visual Encounters with Alterity in Central and Eastern Europe,* edited by Dagnosław Demski, Ildikó Sz. Kristóf, and Kamila Baraniecka-Olszewska, 190–212. Budapest: L'Harmattan, 2013.

Székely, Miklós. "La critica italiana e ungherese sulle esposizioni universali in Italia fra il 1900 e il 1914." *Nuova Corvina* 21 (2009): 114–127.

Wieber, Sabine. "The Warp and the Weft: Tradition and Innovation in Skærbæk Tapestries, 1896–1903." *Journal of Design History* 28, no. 4 (May 2015): 331–347.

Chapter 9

Bennitt, Mark. *History of the Louisiana Purchase Exposition.* St. Louis, MO: Universal Exposition Publishing Company, 1905.

Bickers, Robert. "Purloined Letters: History and the Chinese Maritime Customs Service." *Modern Asian Studies* 40, no. 3 (July 2006): 691–723.

Boorman, Howard, ed. *Biographical Dictionary of Republican China.* New York: Columbia University Press, 1968.

Brunero, Donna. *Britain's Imperial Cornerstone in China: The Chinese Maritime Customs Service, 1854–1949.* New York: Routledge, 2006.

Cai Kejiao. "Jindai Zhongguo bolan ye de xianqu Chen Qi ji qi zhushu." *Jindai shi yanjiu* 1 (2001): 307–313.

Chen Qi. "Kaiban shiwu suo yanshuo ci." *Nanyang quanye hui tonggao,* no. 1, XT 1.2 (February– March 1909): 42–43.

Chen Qi. "Nanyang quanye hui guan hui zhinan bianyan." In *Guan hui zhinan,* by Nanyang quanye hui shiwu suo. N.p.: Nanyang quanye hui shiwu suo, 1910.

Chen Qi and Chen Huide. *Xin dalu Shengluyi bolan hui youji.* N.p., 1905.

Chen Shiqi. *Zhongguo jindai haiguan shi.* Beijing: Renmin chuban she, 2002.

Dirlik, Arif. "The End of Colonialism? The Colonial in the Making of Global Modernity." *Boundary* 2, 32, no. 1 (Spring 2005): 1–31.

Duanfang, *Duan Zhongmin gong zou gao.* 4 vols. Taipei: Wenhai chubanshe, 1967.

Fairbank, John King, Katherine Frost Bruner, and Elizabeth MacLeod Matheson, eds. *The I.G. in Peking: Letters of Robert Hart, Chinese Maritime Customs, 1868–1907.* 2 vols. Cambridge, MA: Belknap Press, 1975.

Feng Ziyou, ed. *Banama Taiping yang wan guo da saihui youji.* San Francisco, CA: Shaonian Zhongguo bao, 1915.

Fernsebner, Susan. "Objects, Spectacle, and a Nation on Display at the Nanyang Exposition of 1910." *Late Imperial China* 27, no. 2 (December 2006): 99–124.

Gerth, Karl. *China Made: Consumer Culture and the Creation of the Modern Nation.* Cambridge: Harvard University Asia Center, 2003.

Godley, Michael R. "China's World Fair of 1910." *Modern Asian Studies* 12 (1978): 503–522.

Godley, Michael R. "The Late Qing Courtship of the Chinese in Southeast Asia." *The Journal of Asian Studies* 34, no. 2 (February, 1975): 361–385.

Godley, Michael R. *The Mandarin-Capitalists from Nanyang: Overseas Chinese Enterprise in the Modernization of China, 1893–1911.* Cambridge: Cambridge University Press, 1981.

The Greatest of Expositions Completely Illustrated. St. Louis, MO: Official Photographic Company, 1904.

Hevia, James. *English Lessons: The Pedagogy of Imperialism in Nineteenth-Century China.* Durham, NC: Duke UP, 2003.

Hsia, C.T. *Modern Transportation and Communications in the Republic of China: Report Presented by Mr. C.T. Hsia, Special Commissioner of the Ministry of Communications of Peking, China, to the Panama-Pacific International Exposition, Palace of Transportation.* San Francisco, CA: Marlow Printing Company, 1915.

Hsu, Madeline Y. *Dreaming of Gold, Dreaming of Home: Transnationalism and Migration Between the United States and South China, 1882–1943.* Stanford, CA: Stanford UP, 2000.

Jackson, Anna. "Imagining Japan: The Victorian Perception and Acquisition of Japanese Culture." *Journal of Design History* 5, no. 4 (1992): 245–256.

Jiangsu shengzhang gongshu shiye ke. *Jiangsu ban li Banama saihui baogao shu.* Shanghai: Shangwu yin shu guan, 1917.

Li Xuangong. *Banama Taipingyang wan guo bolan hui yaolan.* Shanghai: Shangwu yin shu guan, 1914.

Ma, Eve Armentrout. *Revolutionaries, Monarchists, and Chinatowns: Chinese Politics in the Americas and the 1911 Revolution.* Honolulu: University of Hawai'i Press, 1990.

MacMechen, Thomas R. "The True and Complete Story of the Pike and Its Attractions." *World's Fair Bulletin* 5, no. 6 (April 1904): 4–34.

Mitchell, Timothy. *Colonising Egypt.* Berkeley: University of California Press, 1991.

Nanyang quanye hui jinian ce. Shanghai: Jicheng tushu gongsi, 1910.

Nanyang quanye hui shiwu suo. *Guan hui zhinan.* Nanyang quanye hui shiwu suo, 1910.

Ngai, Mae. *The Lucky Ones: One Family and the Extraordinary Invention of Chinese America.* Boston, MA: Houghton Mifflin Harcourt, 2010.

Nong-Shang bu. *Zhonghua you Mei shiye tuan baogao.* Shanghai: Shangwu yin shu guan, 1916.

Starr, George. "Truth Unveiled: The Panama Pacific International Exhibition and Its Interpreters." In *The Anthropology of World's Fairs: San Francisco's Panama Pacific International Exposition of 1915,* edited by Burton Benedict, 134–175. Berkeley: University of California Press, 1983.

Strong, Josiah. *Expansion under New World Conditions.* New York: Baker and Taylor, 1900.

Thum, Rian. *The Sacred Routes of Uyghur History.* Cambridge, MA: Harvard University Press, 2014.

University of Nanking Magazine. *Guide to Nanking and the Nanyang Exposition.* May 1910.

Wang Shuyan. *Nanyang quanhe hui za yong.* N.p., 1910.

Watanabe, Toshio. "The Western Image of Japanese Art in the Late Edo Period." *Modern Asian Studies* 18, no. 4 (1984): 667–684.

Wyatt, M. Digby and Eliza Paul Kirkbride Gurney. *The Industrial Arts of the Nineteenth Century: A Series of Illustrations of the Choicest Specimens Produced by Every Nation, at The Greatest Exhibition of Works of Industry, 1851.* 2 vols. London: Day and Son, 1851–1853.

Xian Wei. "Banama yunhe zhi chengji yu Zhongguo qiantu zhi guanxi." *Chung Sai Yat Po* (19 May 2015; 20 May 1915).

Xiang Ruikun. "Nanjing yu Nanyang quanyehui." *Nanyang quanye hui tonggao*, no. 2, XT 1.5 (June 1909): 103.

Yu, Renqiu. *To Save China, to Save Ourselves.* Philadelphia, PA: Temple University Press, 1992.

Zarrow, Peter. *After Empire: The Conceptual Transformation of the Chinese State, 1885–1924.* Stanford, CA: Stanford UP, 2012.

Zhang Jian. Preface to *Zhonghua you Mei shiye tuan baogao,* edited by Nong-Shang bu. Shanghai: Shangwu yinshu guan, 1916.

Zhili xie zan hui. *Nanyang quanye hui Zhili chupin lei zuan hebian.* 4 vols. N.p., 1910.

Chapter 10

Allamand, Ana Francisca. *Pedro Lira: El maestro fundador.* Santiago: Origo Ediciones, 2008.

Allamand, Ana Francisca. *Alberto Valenzuela Llanos: Visión entrañable del mundo rural.* Santiago: Origo Ediciones, 2008.

"American Art in South America." *Art and Progress* 1, no. 7 (May 1910): 207.

Belanger, Noelle and M. Elizabeth Boone. "'Art' Smith, Flying at Night, and the 1915 San Francisco World's Fair." *Panorama: Journal of the Association of Historians of American Art* 1 (Winter 2015): http://journalpanorama.org.

Bindis, Ricardo. *Pintura chilena: Doscientos años.* Santiago: Origo Ediciones, 2006.

Boone, M. Elizabeth. "'Civil Dissension, Bad Government, and Religious Intolerance:' Spanish Display at the Philadelphia Centennial and in Gilded Age Private Collections." In *Collecting Spanish Art: Spain's Golden Age and America's Gilded Age*, edited by Inge Reist and José Luis Colomer, 42–63. New York and Madrid: The Frick Collection in association with Centro de Estudios Europa Hispánica, 2012.

Boone, M. Elizabeth. "Marginalizing Spain at the Chicago Columbian Exposition of 1893." *Nineteenth-Century Studies* 25 (2011): 199–220.

Boone, M. Elizabeth. "'A Renewal of the Fraternal Relations that Shared Blood and History Demand': Latin American Painting, Spanish Exhibitions, and Public Display at the 1910 Independence Celebrations in Argentina, Chile, and Mexico." *Revue d'art canadien/Canadian Art Review (RACAR)* 38, no. 2 (Fall 2013): 90–108.

Boone, M. Elizabeth. *"Una cualidad lírica de un encanto duradero: La pintura norteamericana y chilena en el Centenario de Chile en 1910."* Santiago de Chile: Museo Nacional de Bellas Artes, 2014.

Braddock, Alan C., and Christoph Irmscher, eds. *A Keener Perception: Ecocritical Studies in American Art History.* Tuscaloosa: University of Alabama Press, 2009.

"Carlos Wiedner." In *Memoria Chilena.* Santiago: Biblioteca Nacional de Chile. http://www.memoriachilena.cl/602/w3-article-94663.html, accessed July 18, 2017.

Castro-Klarén, Sara and Charles Chasteen, eds. *Beyond Imagined Communities: Reading and Writing the Nation in Nineteenth-Century Latin America.* Baltimore, MD: Johns Hopkins University Press, 2003.

Clark, Charles Teaze. "In Pursuit of Higher Truth: The Landscape Paintings of Charles Morris Young." *Antiques* 168 (November 2005): 163–169.

Coates, Benjamin A. "The Pan-American Lobbyist: William Eleroy Curtis and U.S. Empire, 1884–1899." *Diplomatic History* 38, no. 1 (2014): 22–48.

Cortés Aliaga, Gloria. "El paisaje habitado: de geografía humana a objeto de deseo." In *Puro Chile: Paisaje y territorio*, 198–203. Santiago: Centro Cultural de La Moneda, 2014.

Daireaux, Godofredo. "La Exposición I. de Arte: Impresiones." *Athinae: Revista Argentina de Bellas Artes*, año III, 2 (July 1910): 11.

"La exposición artística norteamericana." *La Tribuna Popular* [Montevideo], 6 February 1911, 4.

Exposición Internacional de Arte del Centenario, Buenos Aires 1910; Catálogo Ilustrado. Buenos Aires: Est. Gráfico M. Rodríguez Giles, 1910.

Exposición Internacional de Bellas Artes, Santiago de Chile; Catálogo Oficial Ilustrado. Santiago de Chile: Imprenta Barcelona, 1910.

Exposición de pinturas y bronces por artistas norte-americanos; Catálogo. Montevideo: Talleres gráficos A. Barreiro y Ramos, 1911.

Fox, Claire F. *Making Art Panamerican: Cultural Policy and the Cold War.* Minneapolis: University of Minnesota Press, 2013.

Frascina, Francis, ed. *Pollock and After: The Critical Debate*, 2nd ed. London: Routledge, 2000.

Gienow-Hecht, Jessica C.E. and Mark C. Donfried, eds. *Searching for a Cultural Diplomacy.* New York: Berghahn Books, 2010.

Greet, Michele. *Beyond National Identity: Pictorial Indigenism as a Modernist Strategy in Andean Art, 1920–1960.* University Park: Pennsylvania State University Press, 2009.

Isaza Londoño, Juan Luís, et al. "Centenarios de la independencia." *Apuntes* [Pontificia Universidad Javeriana, Bogotá] 19, no. 2 (July–December 2008): 172–305.

"Julio Bozo (Moustache) (1879–1942)." *Memoria Chilena.* Santiago: Biblioteca Nacional de Chile. http://www.memoriachilena.cl/602/w3-article-4927.html, accessed July 18, 2017.

Larkin, Susan G. *American Impressionism: The Beauty of Work.* Greenwich, CT: Bruce Museum of Arts and Science, 2005.

Leonard, Thomas M., ed. *United States-Latin American Relations: Establishing a Relationship, 1850–1903.* Tuscaloosa: University of Alabama Press, 1999.

Lira, Pedro. "La Exposición Internacional de Bellas Artes." *Selecta* 2, no. 7 (October 1910): 271–273.

Majluf, Natalia. "'Ce n-est pas le Pérou', or, the Failure of Authenticity: Marginal Cosmopolitans at the Paris Universal Exhibition of 1855." *Critical Inquiry* 23, no. 4 (1997): 868–893.

McPherson, Alan, ed. *Anti-Americanism in Latin America and the Caribbean.* New York: Berghahn Books, 2006.

Moreno Luzón, Javier and Rodrigo Gutiérrez Viñuales, eds. *Memorias de la independencia: España, Argentina y México en el primer centenario (1908–1910–1912).* Madrid: Acción Cultural Española, 2012.

Pérez Vejo, Tomás, et al. "Los Centenarios en Hispanoamérica: La historia como representación." *Historia Mexicana* 60, no. 1 (July–September 2010): 7–717.

Rodó, José Enrique. *Ariel.* Translated by F.J. Stimson. Boston, MA: Houghton Mifflin Company, 1922.

Sadlier, Darlene J. *Americans All: Good Neighbor Cultural Diplomacy in World War II.* Austin: University of Texas Press, 2012.

Saunders, Frances Stonor. *The Cultural Cold War: The CIA and World of Arts and Letters.* New York: The New Press, 1999.

Subercaseaux, Benjamín Vicuña. "Don Pedro Lira." *Revista Selecta* 2 (May 1909): n.p.

Taylor, Alex J. "Unstable Motives: Propaganda, Politics, and the Late Work of Alexander Calder." *American Art* 26, no. 1 (spring 2012): 24–47.

Trachtenberg, Alan. *The Incorporation of America.* New York: Hill and Wang, 1982.

Trask, John E.D. *The United States Section; International Fine Arts Expositions at Buenos Aires and at Santiago.* Philadelphia, PA: Office of the Commissioner General, 1910.

Zurier, Rebecca. *Picturing the City: Urban Vision and the Ashcan School.* Berkeley: University of California Press, 2006.

Index